**PARK
LEARNING CENTRE**

The Park, Cheltenham
Gloucestershire GL50 2QF
Telephone: 01242 532721

WEEK LOAN

Professional Content Management Systems

Handling Digital Media Assets

Dr Andreas Mauthe

Multimedia Communications Lab – KOM, Germany

Dr Peter Thomas

blue order GmbH, Germany

John Wiley & Sons, Ltd

Copyright © 2004   John Wiley & Sons Ltd, The Atrium, Southern Gate, Chichester, West Sussex PO19 8SQ, England

Telephone (+44) 1243 779777

Email (for orders and customer service enquiries): cs-books@wiley.co.uk
Visit our Home Page on www.wileyeurope.com or www.wiley.com

Other Wiley Editorial Offices

John Wiley & Sons Inc., 111 River Street, Hoboken, NJ 07030, USA

Jossey-Bass, 989 Market Street, San Francisco, CA 94103-1741, USA

Wiley-VCH Verlag GmbH, Boschstr. 12, D-69469 Weinheim, Germany

John Wiley & Sons Australia Ltd, 33 Park Road, Milton, Queensland 4064, Australia

John Wiley & Sons (Asia) Pte Ltd, 2 Clementi Loop #02-01, Jin Xing Distripark, Singapore 129809

John Wiley & Sons Canada Ltd, 22 Worcester Road, Etobicoke, Ontario, Canada M9W 1L1

Wiley also publishes its books in a variety of electronic formats. Some content that appears in print may not be available in electronic books.

British Library Cataloguing in Publication Data

A catalogue record for this book is available from the British Library

ISBN 0-470-85542-8

Typeset in 10/12pt Times by Laserwords Private Limited, Chennai, India
Printed and bound in Great Britain by Antony Rowe Ltd, Chippenham, Wiltshire
This book is printed on acid-free paper responsibly manufactured from sustainable forestry in which at least two trees are planted for each one used for paper production.

Contents

Preface

Content management is becoming a cornerstone for many operations in the media industry. Several processes in media creation and distribution are already supported by systems that manage content under specific circumstances. However, since such systems are developed and deployed for all kinds of media and use cases there is a plethora of diverse solutions, all labeled Content Management Systems (CMS). They range from fairly simple PC-based file management applications to sophisticated systems handling all kinds of media data and integrating a wide range of devices. One of the most demanding areas in content management is the handling of high quality audiovisual content in a production and broadcast environment. This book concentrates on all aspects related to content and content management systems in professional media production, exploitation, and delivery (particularly broadcasting). It takes into account experiences gathered by working in this industry sector for a number of years but also considers the research and standardization efforts on which these systems are based. Thus, it brings together expertise from different domains, in particular the broadcast and IT domain, the user and system perspective, and professional products, research efforts, and standardization activities in the area of content management.

This book does not deal with the particular aspects of content management for Web applications. Specific programs are being implemented to accommodate the distinct requirements in this application domain. However, Web pages as one media type that has to be handled by a professional content management system are considered within this book. Further, document management is also not of prime concern to this book. Documents are considered just as another content type that has to be managed in the system. The content objects the book deals with are audiovisual, visual, audio, text, and structured media objects that are produced and have to be managed in today's content-rich organizations. The main focus of this book is on media and broadcast operations. However, since more and more organizations (for instance, libraries, educational institutions, large corporate organizations, etc.) have to handle high-quality content, the discussed concepts, principles, and systems are also relevant to them.

The aspects considered in this book include a workflow analysis that also provides insights into who deals with content. Further, digitization and encoding principles and relevant encoding standards (in particular MPEG- and DV-based formats) are introduced. Metadata principles and related standards (such as MPEG-7 and Dublin Core, SMPTE Metadata Dictionary but also XML, MOS, and SOAP), and file formats are also discussed. The central part of the book is the content management system architecture and infrastructure. Further, system integration and applications are discussed. The book introduces research and practical issues alongside each other.

HOW THIS BOOK CAME INTO BEING

Both authors have been working in the area of multimedia and content management for a number of years. While working together on the development of content management systems and within projects, a lot of knowledge was accumulated about all the issues related to the handling and management of content in a professional production and broadcast environment. Initially there was a steep learning curve during which the various aspects in this area had to be understood. These include the technical issues such as relevant encoding formats, standards, equipment features, system aspects, etc. But almost more important was to gain an understanding of the workflows and user requirements. This process was facilitated in a number of research projects (funded by the European Commission and German regional and national government grants) in which the different interest groups came together to explore all the issues involved in content management. The authors realized that although there is a lot of expertise on specific aspects of content management in the various distinct domains, it has not been brought together and put into context. This is particularly important since quite different areas are coming together and a common understanding has to be established by all involved parties. These are the technicians from the broadcast and IT domain who develop, implement, run, and maintain these systems. There are also the users (such as catalogers, journalists, editors, producers, etc.) and managers who are responsible for the operations within a media-rich organization. Therefore, we decided to put the knowledge gathered in recent years on paper to provide a comprehensive overview of all the issues and possible solutions for the professional management of multimedia content.

Thus, this book aims at bringing all the relevant areas and aspects together. It is a textbook as well as a compendium on professional content management systems. It explains the basic principles and introduces the system specifics.

WHO IS THIS BOOK FOR?

This book is targeting professionals as well as students who are involved or interested in the handling and management of multimedia content. Aspects in this context include media encoding and processing, metadata and file formats, data and storage management, distributed multimedia systems, system integration, and applications. The target audience in particular is IT and broadcast engineers, archivists and catalogers, media and content managers, and everybody directly involved in the handling of content or the development and maintenance of content management systems.

Moreover, the book gives an insight to decision makers in the media industry and in content-rich organizations about the issues involved in content management in the digital age. These are, for instance, media professionals and broadcast executives. All the necessary concepts are explained in the book and it enables the reader to get a good overview of the aspects and potential pitfalls involved in executing content management projects.

ACKNOWLEDGMENTS

First of all we have to thank all those who have worked with us in various projects and made us perceptive to all the aspects of multimedia content management. These are, for instance, the project partners from the EU-funded projects EUROMEDIA, OPAL, and DIVAN, and from projects that were part of the German regionally funded research program FMS. Further, there are all the partners from the industry with whom we developed ideas and worked together in various projects. There are also professionals working in the area of standardization at EBU and SMPTE, with whom we had many fruitful discussions and who helped us to gain deeper insights. A special thank you goes to all our colleagues with whom we have worked over the years and who have had to serve as guinea pigs for our ideas and who developed many of the system aspects together with us. To name all the persons that have contributed over the years to build the background for what is presented in this book would result in quite a long list, so we decided not to. We know, however, that all these brilliant people know that they are the ones we are expressing our thanks to.

Last but not least we have to thank all those who have supported and encouraged us over the years, especially our friends and families, who sometimes had to put up with always-busy, absentminded persons preoccupied with their work and projects.

Andreas Mauthe
Peter Thomas

1

Introduction

Information creation, presentation, and exchange, but also the collection, organization and storage of information carriers, is an old craft. Libraries and archives existed in ancient Egypt, Babylon, China, and all advanced civilizations. It can be regarded as part of cultural development to handle media and preserve information. Thus, content has been managed for millennia. What makes the problem different in today's information society is the amount of information that has to be handled, the speed at which it is produced, and also the kinds of media and formats that are used. At the turn of the nineteenth century continuous media (i.e. film and audio) were added to the then traditional discrete media formats. In the last two decades of the twentieth century new digital multimedia formats (such as digital video and audio but also Web pages and hyperdocuments) emerged. All these different media formats have to be managed within organizations that produce, handle, store, and deliver content in its various forms.

Another trend that influences the way content has to be managed today is the fact that its creation is no longer a linear process. Content is assembled from various pieces representing different media and information types. Raw footage, rushes, existing archive material, and additional recordings are used to create audiovisual media objects. The production is accompanied by data sheets, scripts, production plans, processing instructions, and other documents that are generated during the content creation process. This, together with the much decreased turnover time of information, requires that all parts are accessible and available to all involved parties throughout the production process. Thus, systems that manage content are not end-of-chain storage vaults but are becoming central parts of the content creation and delivery process.

What could be observed in recent years is that the media industry is changing as well. With the advent of the World Wide Web the separation between traditional print publishers, broadcasters, and entertainment businesses has become blurred. Most organizations dealing with media are hosting their own Web site. Broadcasters are not only transmitting over the established channels but are also making use of digital broadcast technologies. Further, music records and videos are nowadays not only distributed via traditional retail outlets but are also available electronically. In fact, content (respectively information) is, apart from finance products and software, the only good that can be handled and delivered electronically.

Professional Content Management Systems: Handling Digital Media Assets   A. Mauthe, P. Thomas
© 2004 John Wiley & Sons, Ltd   ISBN: 0-470-85542-8

Moreover, a growing number of non-media organizations are faced with the problem of how to manage their content. Large enterprises have multimedia promotion material and corporate archives. Educational institutions are using audiovisual lecture material that has to be managed. Museums, tourist attractions, towns, and cities have to cope with a growing volume of content that needs to be preserved. Also, telecommunication companies are developing from pure network infrastructure providers to service companies. They too see the media and entertainment market as an area that provides potential for growth. In order to move in to this space, they have to manage the media and information they want to offer. Thus, content management is not just an issue for the traditional media industry but for a whole variety of institutions and organizations.

However, the ubiquitous, all-encompassing, generic content management system (CMS) has not emerged so far. A number of products carry the name and even more claim to provide general or at least basic support. However, they have all been developed in the context of a specific industry and therefore concentrate on special aspects and features. Until now no platform product has been developed that could be used to manage the content of a small institution as well as to support the media production process and archiving in a content-rich organization. It is even doubtful if such a system could be built since the requirements from the various use cases are quite different.

This book deals with all aspects related to the management of content (or media assets) in the context of professional media production, handling, and delivery. The main focus is on the broadcast industry but the requirements of other content-rich organizations are also considered. The entire workflow, from the introduction of the media into the system, over the production stages, to the finale archiving, is covered. Further, the main media formats and encoding principles are introduced, and the way content is represented in the system is discussed. The latter includes metadata standards and frameworks. The core of the book deals with the architecture and infrastructure of a CMS. Since such a system is part of an existing environment and possibly established operations, integration is a very important aspect that has to be considered. Applications are the most visible part of the system and therefore it is deemed appropriate to introduce them in depth. An outlook and discussion on future trends concludes the book.

In more detail, the book is organized as follows:

- The remainder of this chapter defines the term content and introduces the problem space. The various areas where content management plays a major role are discussed.
- Chapter 2 shows the different groups and individuals who deal with content. It also discusses a number of workflows that are relevant in the context of content handling and administration. This includes workflows from non-media production areas such as content management in e-commerce, education and training, and marketing and sales.
- Chapter 3 deals with relevant media formats that have to be administered in CMS. This includes the introduction of the basic encoding and compression principles. Relevant video and audio formats and their structure are discussed. The main focus is on standardized, non-proprietary formats, which are often the basis for media production and broadcasting products. However, images and structured document formats are also outlined in this chapter. It is concluded by a section introducing the principles of automatic media processing, which are the basis for video, audio, and speech analysis tools.

- Chapter 4 focuses on the representation of content and metadata aspects. In order to administer but also work with media, it has to be represented and described considering all relevant viewpoints. The different ways to represent content in the system are discussed in the first part of this chapter. Subsequently, relevant metadata description schemes and standards are introduced. A number of initiatives in this area exist, and it is important to know their principles and to be able to distinguish their objectives and the resulting structures. This chapter also includes a discussion of the relevant metadata transmission and exchange standards, such as XML and SOAP.

- Chapter 5 concentrates on file formats that have been specified considering the special needs of content production and management. File formats encompass both the actual encoded media and descriptive information about the content. The main focus is on formats that have been proposed for professional systems. However, other multimedia file formats are also covered.

- Chapter 6 proposes a content management system architecture. This is a structured framework containing all elements that are required to manage content in a professional environment. The architecture comprises a number of core elements, services, and application components that are needed to handle content and interact with the system. The different components are discussed in detail. This architecture is not a blueprint but a reference framework that comprises the major components relevant in the context of content management.

- Chapter 7 explains the content management system infrastructure and its various components. Whereas the architecture concentrates on the software modules that compose a CMS, the infrastructure represents the physical system elements that constitute the CMS and host the different software modules, but also the actual content (i.e. media in the form of essence and the descriptive metadata). It is discussed in detail how the different infrastructural elements support the management and production process. Further, operational considerations comprising migration issues, cost, and scaling strategies are also part of this chapter.

- Chapter 8 deals with system integration aspects. A CMS is part of a larger structure in a content-rich organization. In established operations it has to consider legacy content but also existing components need to be integrated or at least interfaced to. Further, the CMS of the future may be at the center of the operation. However, there will always be a number of other systems that are also used in content production and distribution. Such systems, for instance, are studio automation systems, nonlinear editing systems, newsroom systems, etc. How they can be integrated or interfaced to is explained in this chapter.

- Chapter 9 gives an overview of the application components that are relevant in the context of a professional CMS. It introduces the concept of component-based application design since CMS applications need to be highly flexible and configurable. A number of modules that can be used to build actual applications are outlined. However, example application configurations are used to explain how the different demands that are placed on the systems by the users and workflows are reflected in the application and its user interfaces.

- Chapter 10 looks at future trends in content management. Relevant initiatives and standardization efforts are outlined. Further, the experiences gained in the area of professional content management are summarized. A discussion on the future developments that are expected to take place in this area concludes the book.

1.1 WHAT IS CONTENT?

Conventionally, the term *content* is used to refer to any kind of audiovisual, visual, sound, or textual information. A specific media type representing content in this context may have a determined presentation lifetime (e.g. a video or audio broadcast). However, in the system context content is characterized by its permanent presence and availability, i.e. content can be accessed on request or is available at certain times within the system. Content can be produced, altered, transmitted, consumed, and traded in parts or in its entirety.

This general characterization of content is very broad and does not give any qualitative description, nor does it specify the different elements of content. However, this reflects common usage of the term at present. Content is used with different connotations (depending on the context) and in various ways, sometimes describing disparate concepts.

In order to overcome this problem, the Society of Motion Picture and Television Engineers (SMPTE) and the European Broadcasting Union (EBU) have set up a taskforce that addresses the problems related to content and content management. This taskforce defined the term *content* in the context of the media industry and identified its elements.

According to the taskforce's definition (SMPTE/EBU, 1998) content consists of:

- *essence* and
- *metadata*.

Essence in this context is the raw programme material itself, represented by pictures, sound, text, video, etc. The essence carries the actual message or information. Essence is also often referred to as media. However, the term media is also applied in connection with physical carriers such as videotapes, CDs, etc. Since essence refers to the general concept and is independent of the physical carrier it has to be clearly distinguished. Therefore it was decided to use this relatively uncommon term, essence, to refer to encoded information that directly represents the actual message. Throughout this book essence is used in the way it has been defined by the EBU and SMPTE taskforce.

The second content element is metadata. It is used to describe the actual essence and its different manifestations. Metadata can be classified into:

- *content-related metadata*, giving a description of the actual content or subject matter;
- *material-related metadata*, describing available formats, encoding parameters, and recording specific information;
- *location-related metadata*, describing location and number of copies, condition of carriers, etc.

Content-related metadata comprises formal data (such as title, subtitle, duration, cast, etc.), indexing information (such as keywords, image content description, classification, etc.) but also rights-related data (such as rights owner, acquired rights, etc.).

Depending on the application context and area, the two constituents of content (i.e. essence and metadata) are of varying relevance. The actual essence is consumed and operated upon. Metadata is required to describe, find, and retrieve content. Therefore it has a crucial role in all applications where content is selected rather than just being presented (i.e. in the content production process, content sales, and personalized content delivery).

A system that manages both essence and metadata is called a *content management system* (CMS). Major tasks of managing essence are the storage, administration, and delivery of high-volume, high-bandwidth, partially time sensitive digital data. In a professional environment this also includes the integration of (or interfacing to) specialist production and broadcast systems. Content management is not restricted to a specific organizational unit such as the archive. It spreads across all departments of a content-rich organization.

The management of metadata is mainly concerned with the description, storage, and location of content-related data in information systems and databases. Besides traditional manual annotation, this includes enhanced metadata descriptions and index information generated by automatic processes such as video analysis and speech recognition tools.

1.1.1 CONTENT AND INTELLECTUAL PROPERTY RIGHTS

Intellectual property rights (IPR) and *digital rights management* (DRM) are related subject areas, although together they are considered to form a self-contained problem sphere. Rights management has two parts, namely the description and documentation of rights associated with a content object (called rights management), and rights protection.

Content for which the rights situation is not known is almost unusable in a commercial context. The use of content without observing legal or contractual rules and IPR might cause major problems. Without knowing the associated IPR a content object does not necessarily represent any value to an organization since it cannot be exploited. Only if the rights are cleared can a content object be presented, broadcast or traded. Thus, a content object becomes an *asset* when its rights are known. It can only be commercially exploited if there is certainty over its rights status.

Some even declare a content object without proper rights a liability since it has to be stored, managed, and preserved without having the possibility of using it in a public environment. However, there might still be a reason for keeping content without owning the appropriate IPR. There are, for instance, institutions that are charged with preserving the cultural heritage of a nation. Thus, they have to manage all content objects belonging to this class. This does not, however, mean that they have any right to exploit the content publicly. There are also other, more pragmatic, reasons to keep content without owning the appropriate rights or knowing about them. For instance, specific content might become interesting in a certain situation. Some consider it early enough to clarify the rights to use it shortly before an actual broadcast. However, in order to have easy and quick access to it they keep a copy of the content object in their own system even if the original contract has been terminated.

According to these considerations there is a clear distinction between content and asset. This book mainly deals with content and its management. Content objects may be assets as well and the same management procedures will apply regardless of the rights situation. However, in order to ensure the authentic, well-defined use of terminology, we use the term content in the above presented definition throughout the book.

1.1.2 CMS VERSUS DAM-S AND MAM-S

There are a number of disparate systems that are called CMS. Apart from media production support systems there are, for instance, also a number of systems handling Web pages that are labelled CMS. On the other hand, people talk about digital asset management systems (DAM-S) and media asset management systems (MAM-S) but effectively refer to the management of audiovisual content, i.e. essence components and metadata in the above defined way. This is often done to highlight the fact that the objects these systems are dealing with are of high value, or are at least encoded in high-quality, high-volume formats. The prefixes 'digital' and 'media' in this context are chosen to distinguish these objects from financial assets, which are also administered in asset management systems (AMS).

For this book we deliberately chose the title *professional content management systems* since the management of rights is not an inherent part (but a side aspect) of the systems described here. Professional in this context refers to the fact that these systems are platforms for the professional handling of content in an enterprise or organization-wide context. This involves a number of systems and devices, a plethora of formats, and the support of all major workflows in content production, handling, and delivery. We also decided not to use the prefix digital in connection with CMS since the systems discussed in this book also cover content objects where part of the essence might be stored in analog formats on the shelf.

However, the subtitle *'handling digital media assets'* has been chosen to ensure that the potential reader immediately becomes aware of the main subject area. Although digital media assets are not managed (i.e. the IPR are not managed) they are still handled by the CMS and the content part is managed there. Further, the full potential of these systems can only be exploited when using digital multimedia formats while moving towards a fully digital, tapeless content production and management environment.

Thus, the focus of this book is on content and content management. This includes all applications and tools required to manage a piece of essence together with its related metadata in a system context. The main objective of such a system is to provide a consistent and comprehensive view on objects managed in the system throughout the entire life cycle. This also includes, for instance, maintaining the relationship between essence and associated metadata and the relationship between different versions of the same content object. Additionally such a system has to provide an interface for the exchange of information and objects (e.g. essence) with associated systems such as automation or rights management systems.

1.2 APPLICATION DOMAINS

Although content has been around for a long time the problem of managing and finding it has been intensified by the ever increasing amount of content being produced and the persistent increase of output channels. With emerging digital formats and digital production tools, fully digital media production without the use of physical carriers (such as videotapes) becomes possible. This also enables new ways of collaboration, communication, and commercial exploitation of content. Content management plays a key role in this process since it provides the platform for handling digital media. The number of applications requiring the management of content is steadily growing. In order to identify the requirements this places on the underlying system architecture, it is important to study the different application areas to get an idea about the various use cases.

In this section some examples of areas where content has to be managed are given to illustrate the related issues in a bit more detail and to show where a CMS can provide useful support for content handling. This list is by no means complete but gives a good overview of the different areas and requirements where content management is relevant.

1.2.1 CONTENT MANAGEMENT IN TELEVISION, RADIO, AND MEDIA PRODUCTION

Content management applications in broadcasting can at present be mainly found in archives and the production of actualities such as news, sport, and current affairs. In the archive they support the cataloging of content objects and facilitate their retrieval. The cataloging tasks are supported by the use of analysis and automatic indexing tools that provide automatically retrieved information such as keyframes and keywords. These tools are also able to segment video in different shots, allowing faster documentation utilizing the predetermined segment structure, keywords, and browse copies of the content object. Thus, a less detailed verbal description is required. The keyframes and browse copies also give a new quality to the documentation that goes beyond the pure textual description since they give an additional audiovisual impression of the content.

In order to find content, generic search tools have to enhance the traditional native database queries. This includes fulltext search capabilities. Since a number of databases exist they have to be integrated by the CMS. Integration in this context refers to support of federated search including result consolidation. Advanced search tools allowing audio-visual searches (e.g. image similarity retrieval, search by humming, etc.) are also part of some advanced CMS. To exploit the multimedia capabilities of such a system, optimally the result list(s) should also contain an audiovisual representation (e.g. keyframes, skims, preview, and pre-listen copies) of the material whenever appropriate. Continuous media has to be represented frame and timecode accurately to allow the retrieval of parts of the original content object only. Online, near-online, and offline essence has to be managed altogether by a CMS. Special application tools are required to allow expert users (such as archivists) and non-specialist users access to the content.

In production the CMS has to be tightly integrated with broadcast control and production tools such as studio automation systems, newsroom systems, and nonlinear editing systems (NLE). The application interfaces have to be integrated seamlessly into the desktop environment of the journalists and editors working with the system. A change of applications is not acceptable in this environment. Content items are managed within the CMS; however, they have to be linked to objects in other systems such as news agency feeds and program items in the newsroom system. This refers not only to essence but also to metadata, and implies that cross-system searches have to be possible. Frame and timecode accurate content selection is required for the production of new items. This is usually done using preview copies that have to be exact low-bandwidth clones of the production item. According to this selection the production material is transferred and the actual material is produced on an NLE. This happens outside the control of the CMS. Since content is shared between the different systems, references and frequent transfer of control over a specific content item can occur. Integration with special studio servers, NLE, and other broadcast equipment via interfaces, message or file exchange is also necessary. There are a number of databases that have to be either integrated or synchronized. Storage management has to control a heterogeneous storage environment with different online and near-online systems (including specialist broadcast storage systems).

The requirements of audio and radio production and archiving are similar to video and television, although other devices have to be integrated and the applications to integrate are different.

A relatively new area is Web content production and archiving. The data rates and storage requirements in this context are much smaller but the applications that require integration are different. A particular problem is the way hyperlinks to other Web pages are treated. Especially for material that is archived these links might become outdated. Thus the archiving depth has to be established, i.e. it has to be decided which referenced pages are included in the archive and where links are just disabled. However, apart from this content management in this environment is much less demanding.

1.2.2 CONTENT MANAGEMENT IN 'NON-MEDIA' ORGANIZATIONS

Organizations that are not in the business of media production and delivery also have to deal with an ever-increasing amount of content. For example, corporate archives or marketing departments of large corporate organizations have content that needs to be managed. Other organizations are museums or educational institutions such as schools, universities, and higher education colleges that have content that needs to be stored, managed, and retrieved by a number of individuals who have different roles and user rights. Although the requirements in this context are less stringent than in a media production environment, they still require integration with a potentially large set of third-party applications and systems (such as enterprise resource planning (ERP) systems, process planning systems, telemedia learning platforms, etc.).

One application area where huge amounts of audiovisual content are produced on an hourly basis is that of security services using CCTV observations in buildings or public places. Since the massive amount of data produced in this context cannot be viewed by individuals, support from electronic or automatic tools that help to identify and classify content is required. Relevant processes are, for instance, video analysis, face and object recognition, and person identification.

Essentially, the requirements of these application areas are similar to broadcast and media production. However, the workflows and third-party systems that require integration are considerably different. This implies that core functions have to be encapsulated and offered via standard interfaces that are flexible enough to serve a large variety of other components and third-party systems. Core is the integration into other applications, with other data management systems and with systems handling media/essence.

1.2.3 CONTENT MANAGEMENT IN ADVANCED MEDIA SERVICES

Content management is also required in new media delivery applications and services and is slowly also being used across organizational boundaries (e.g. to support content exchanged and collaboration).

Examples of such applications are, for instance, content exchange and collaborative content production support. Many of the requirements (such as the integration of different components and third-party systems, cross-system searches, and data management) are similar to internal content management processes. In contrast to 'traditional' content management applications, however, many of these systems are distributed over a large geographical area. This is especially true for the storage of essence that might reside at the sites of each participating partner. In a secure environment copies can be p___ different locations, though IPR aspects still have to be considered.

Another application class is e-commerce applications selling content. Important in this context is the integration of the content management functionality with the sales and back-office infrastructure (i.e. an ordering system, billing system, and ERP system). Further, the electronic delivery of content requires high-bandwidth links to all customers. At present there is no universally available infrastructure to support this kind of delivery in a cost effective way.

The possibility of outsourcing content management to large content hosting centers connected via high-speed networks to the content owners and distributors is also currently being explored. This approach promises economies of scale and potential synergies between different content areas. So far this concept has not been realized since CMS in general are of large scale (hence economies of scales are not applicable) and the available communications infrastructure does not support high-bandwidth, high-volume operations. In this context system requirements (in addition to the general CMS requirements) are mainly concerned with security and content protection. If content management is outsourced it has to be ensured that no unauthorized third party can access the content.

Other areas where content management is a substantial part of the system is in personalized content delivery and tele-teaching. In the former case the metadata of the content has to carry additional information that allows the application to select and choose according to personal preferences. Integration with data-mining tools is required in this context. Tele-teaching also requires the management of content and the integration with presentation tools. A special feature of these applications is the high degree of interactivity between the participating users. Content is sometimes created and changed on the fly. Also, it might be stored at different locations throughout the system. Thus, distributed storage and content management are required.

1.3 CONTENT MANAGEMENT: THE PROBLEM ILLUSTRATED

The above-discussed examples give an impression about the requirements and demands placed on a CMS. The application is the most visible part of the system and the workflows that need to be supported determine a large part of the system's functionality. However, other important aspects that need to be considered are the requirements coming from the characteristics of the different essence formats that are managed in the system. Further, the metadata and the way it is represented in the system are also major issues. Besides, there are a number of external systems and devices that have to be integrated or interfaced to in order to draw the maximum benefit from a CMS. All this creates the set of requirements that have to be satisfied.

Such a wide range of heterogeneous demands can probably not be satisfied by a monolithic architecture. Ideally a CMS is designed as a completely distributed management solution framework for the handling and administration of content. The actual installation of a system within a content-rich organization builds the hub for the management and distribution of essence and metadata. Thus, it needs to provide easy access to information and essence, considering the special needs and rights of the parties and user groups involved in the day-to-day operations.

In order to understand the problem space and define the scope of the book, the following sections introduce the requirements placed on a CMS in more detail. This discussion is structured according to the areas that the requirements come from.

The focus of the book is on enterprise-wide systems used by content-rich organizations for the production, handling, and/or distribution of content. These are large-scale systems that provide the platform for all content-related operations. Thus, the specific demands of workgroup and smaller applications are not explicitly considered. However, in many cases the approach and the developed solutions presented in this book are suitable for this application domain as well.

1.3.1 MEDIA AND ESSENCE HANDLING REQUIREMENTS

The main purpose of a CMS is to provide a platform for the management of multimedia objects (or essence) such as digital video, audio, images, graphics, Web pages, etc. Within a content-rich organization (such as a broadcaster) a large amount of essence but also a plethora of different media types and formats can be encountered. Not all of them are digital, i.e. the system has to support the fully digital operation but must also be able to manage material 'on the shelf.'

However, the demands placed on the system by the large number of high-quality, high-bandwidth formats are the real challenge. For instance, a medium size video archive of 100 000 hours of material will contain objects in various formats ranging from broadcast quality formats at around 4 Mb/s (e.g. MPEG-2MP@ML based formats) over standard 25 Mb/s production formats (e.g. DVCPRO) to higher quality formats or even uncompressed video at 270 Mb/s (encoded according to ITU R BT 601-5). With an assumed average bandwidth of 25 Mb/s the storage requirement of such an archive is already well above 1 petabyte. The item length can also vary considerably, ranging from 2 minute news clips to 90 minute feature films. In general this data has to be handled in real-time, sometimes faster than real-time communication is desirable. In organizations that reuse content in production, a CMS typically also has to provide functionalities that allow the selection of sub-clips from archived content and delivery of these sub-clips in production quality to a selected target destination via partial restore. This implies that the CMS must be aware of all file and encoding formats the content is archived in. Due to the still diverse technology landscape used in media production, many transfer and delivery procedures also require file and/or encoding format conversions. This has to be complemented by interfacing standard IT infrastructures and networks with standard broadcast infrastructures and networks, and may even include wide area network (WAN) delivery when the organization has production centers in different remote locations.

Alongside these high-quality video formats a content-rich organization also has to manage other data such as audio, images, graphics or even text. The requirements of audio range from a few Kb/s to 1.5 Mb/s (in the case of 44.1 kHz PCM encoded audio) or more. Increasingly often low-bandwidth browse and preview formats are kept in conjunction with high-quality formats. Further, the presentation of content objects is enhanced by keyframes and other multimedia formats.

Thus, a CMS should ideally be able to store, handle, and administer all different kinds of media types and formats regardless of their encoding scheme or any other specific features (i.e. it should be format-agnostic). IT-based systems are particularly well suited for this since they handle files essentially as bits and bytes. However, a CMS has to offer advanced support for media processing and delivery. Video and audio has to be streamed as well as delivered as a file. Automatic analysis processes are used to extract additional information about a media object. This can only be achieved if the system knows the structure and features of specific media and encoding formats. Thus, a CMS should be format-agnostic as far as the storage of essence is concerned but nevertheless provide optimal support for specific formats.

1.3.2 REQUIREMENTS FOR METADATA CREATION AND HANDLING

A CMS has to facilitate the handling and usage of content in production, broadcasting (or delivery), and archiving. One of the most important tasks in this context is to maximize the (re)use of existing material in order to minimize production costs. This is accomplished by providing easy access via intuitive user interfaces, audiovisual feedback to searches, and other important elements of the human–machine interaction. However, the user interface and application is only one aspect: the quality of search results strongly depends on the quality of the metadata with which the material has been described. Thus, the quality of the cataloging process plays a crucial role.

It is important that a CMS provides tools that automatically extract a substantial metadata set from the content itself by computer-based content analysis. Additionally, it is vital that the CMS still supports the creation of the most important metadata, i.e. the descriptions entered by experienced, highly qualified catalogers and archivists. These professionals often have intimate domain knowledge about the content and therefore ensure the required quality of the descriptive information.

Traditionally, almost all of the metadata is created and entered by the archive or cataloging department at the time when the material enters the archive. This implies that in many cases information that was created at earlier stages (such as production details or newsroom system information) is either lost or has to be manually copied into the system in order to be present as archive metadata. This puts additional workload on the cataloging staff and leaves them less time for the creative work of in-depth indexing. Thus, a CMS should collect and keep metadata during the full life cycle of a content object ranging from pre-production over ingest, retrieval, editorial work, production, post-production, airplay, to archiving and reuse. This implies that a CMS should:

- generate as much metadata as possible automatically in order to enrich the documentation without increasing the human workload;
- get involved throughout the full media usage cycle by providing applications supporting the various steps in the cycle;
- preserve the integrity of metadata while the content moves along the cycle and enable the cataloging staff to easily evaluate and modify all metadata as required.

1.3.3 USER REQUIREMENTS EXAMPLE: TELEVISION
PRODUCTION APPLICATIONS

Professional CMS are being deployed in many different industries and domains. However, the area where they are currently most prevalent is the broadcasting and media production industry. In the following the user requirements of this sector are examined to give an impressions about the needs of the users working with the system. As will be shown in detail later in this book, other domains have sometimes different but very often similar requirements.

1.3.3.1 Integration into Production

The change from VTR-based to server-based production systems demands automatic systems for storage and delivery of essence. This in turn requires an infrastructure for the management of content. The parts that have to be considered in this context are the actual essence and elements that are vital for an efficient exploitation of the essence (including associated technical and descriptive metadata) as well as the cataloging process. The manual generation and gathering of content information is extremely expensive and time consuming. Therefore, it must be possible to integrate the CMS into the editorial and production processes in order to avoid the loss of useful information along the process chain.

1.3.3.2 Essence Management

It is not sufficient to treat a video stream as a bulk of bytes without any particular syntax or semantics. Some basic functionality to manipulate the signal may be required including transcoding to different formats. Further, the available metadata included in the stream must be made accessible to archive users. Thus, the CMS must be able to manage new digital video formats applying compression to the signal, and to extract and interpret the associated encoded metadata (e.g. video parameters, compression scheme, etc.).

1.3.3.3 Browsing

In order to save bandwidth and storage capacity as well as to reduce the load on the production server low-resolution formats that can be viewed and manipulated at the desktop are required. Two different levels of browsing can be distinguished, *edit decision list (EDL) browsing* for rough-cut editing and *content browsing* for content viewing and clip selection.

EDL browsing has the following properties:

- frame rate identical to that of the original high-resolution material;
- frame-accurate presentation of timecodes;
- frame-accurate navigation;
- quality level suitable for generating rough cuts;
- the coding method must provide a picture quality equivalent to or higher than VHS or MPEG-1;
- the coding scheme must be standardized as well as supported by the industry;
- must run on a standard PC;
- timecodes have to be part of the encoding of the browsing format.

Content browsing, in contrast, has less stringent requirements concerning frame accuracy and timecode but is still a valuable tool for evaluating and selecting material. Content browsing is characterized by a very low bit rate preview quality. However, basic trick mode functionality (such as real-time playback, fast forward, fast rewind) has to be available. As this essence version will in some cases be made available to the general public,

a protection system from unauthorized commercialization (i.e. a rights protect
is desirable. Further, the coding scheme should be compatible to state-of-the
technology. For instance it should support streaming and file transfer. Moreover, it should
be scalable to enable optimum service for heterogeneous transport channels and be able
to adapt to the bit rate changes of best-effort transfer channels. This includes automatic
quality of service (QoS) management by the streaming server.

1.3.3.4 Information Management

Information management is mainly dealing with the handling of metadata and the incorporation of index libraries. The handling of metadata covers the entire material life
cycle from the point of creation to the point of archiving. A unique material identifier (UMID) identifies the essence throughout its lifetime, while more abstract unique
identifiers represent the content object. Metadata can be generated automatically during
ingest by automatic analysis processes. In parallel, any existing metadata associated with
the content is extracted. Automatic tools should support the whole process of metadata
generation. The use of standard formats is crucial in this context.

The CMS should be able to accommodate various metadata description schemes. This
includes existing data models represented by legacy databases but also various standard
encoding schemes. Even within one system it is often necessary to support multiple
databases and information systems that use different data models.

1.3.3.5 User Interface Requirements

User interfaces are crucial for the success of a content management project. Since the
CMS is a general platform for the handling of content, there are a number of users with
various skills and backgrounds who have to interact with the system. Thus, the applications
have to support expert and non-expert users alike. Separate user interfaces is one way to
accommodate the different requirements of various user groups. The interfaces have to
be ergonomic and adapted to the requirements of the tasks of specific users.

The search interfaces have to consider best IT practice. Different search concepts have
to be supported in order to serve distinct user groups. Fulltext search, for instance, is
one way of supporting non-skilled users. In contrast, archivists, catalogers, and media
managers very often would like to target their search directly using native interfaces and
attribute search. Whatever search technology is implemented in a CMS it is important
that the system remains extensible and open to the integration of new search technology.

1.3.4 OPERATIONAL REQUIREMENTS

Since the CMS is the hub for all content-related operations it is crucial that the operational
system requirements are met. These are the demands placed on the system by the standard
operations. The role a CMS plays within a content-rich organization can be compared to
an operating system (OS), i.e. besides the management of information it is also a platform
and host for other services and components. The following operational requirements have
to be considered in this context:

- Provide means of storing, organizing, finding, and retrieving assets in the same way a
  file system does. This implies that the CMS has to allow the storing, moving, renaming,

and deleting of files. Further, it has to be possible to organize files in directory trees, find them with suitable system commands, and make them accessible for applications.

- Provide system-wide central services such as user management, name services, resource reservation, and others. This is comparable to an OS providing administrative and system management capabilities.
- Provide means of running services that may access the assets archived in the CMS and that may manipulate them or derive information. This extends the capabilities of the CMS in the same way that daemons or background services running on top of an OS extend the functionality of the OS platform.
- Provide means of running applications on top of the system that allow access to information and material stored in the system, supporting all relevant steps in the workflow. This is comparable to the way applications can be built and run on top of an OS. Thus, granting access to the system's capabilities.

A CMS with these capabilities provides the background support for all content-related administration and management tasks. It is therefore not only a component deployed within a specific organizational unit but ideally provides a service for all parts of a content-rich organization.

Since CMS in organizations such as broadcasters are mission critical a high system availability has to be ensured. Provisions have to be made to avoid downtimes and cope with possible failures in a way that does not affect the operation.

1.3.5 SYSTEM REQUIREMENTS

In most cases an enterprise-wide CMS will be introduced in an existing organizational environment that has been dealing with content for some time. Thus, it is not a 'green-field' installation but has to take existing infrastructures and operations as well as legacy systems into account. Therefore, a CMS should provide well-defined interfaces to allow easy integration of existing technology and solutions into the overall framework. This is necessary since the introduction of a CMS has to be gradual, considering the existing operation. In order to achieve this, a step-by-step migration from an existing system to a new system is the best approach. One of the major tasks in this context is the integration of important legacy systems such as existing databases or information systems, production systems, newsroom solutions, or certain middleware products.

In addition, a CMS has to be open to integrate components from other vendors providing specialized equipment and devices for specific tasks, which form a vital part of an integrated content management infrastructure. Examples are advanced logging, indexing or search engines. It must be easy to integrate such technologies into the overall solution.

In order to support large, geographically distributed enterprises (such as national broadcasters) a CMS also needs to support organizations that are working or collaborating at remote locations. An example is local or regional studios, or affiliate stations. In addition, the internal structure of an organization may also require support for distributed operations. There are, for instance, different organizational units (e.g. editorial offices) that keep their own content. Distribution may also be a way to deal with systems scalability issues.

Thus, the major system requirements are:

- *Openness*: the solution has to provide well-defined interfaces in order to facilitate the integration of legacy and third-party systems.

- *Modularity*: the solution has to be flexible and component-based, providing a clear specification of the functionality that needs to be supported by each of its components.
- *Distribution*: the solution has to support distribution to allow the integration of systems and system components at different physical locations and for better scalability.

How these requirements can be met depends not only on the technical design but also on the characteristics of the content elements or information structure. For instance, in the case of metadata, openness refers to the exchange of information between independent systems regardless of their internal data structures and content representation. This implies that a standardized common metadata format has to exist that can be used for the exchange. Openness of software not only refers to open interfaces but also means that the products must be extendable and backward compatible. Further, they should support migration management strategies.

1.3.6 ARCHITECTURAL SYSTEM REQUIREMENTS

The above-discussed requirements have implications for the system architecture. Considering the plethora of the involved components a highly modular system architecture (i.e. an architecture that is extensible by adding modules to serve specific requirements) will ensure the protection of investments by allowing the replacement of individual modules when technology improves. Such an architecture is also scalable enough to meet the ever-growing demands regarding the functionality but also the size of the system.

The lifetime of the various CMS components can vary considerably. Thus, an ordered migration of the entire archive system may be impractical or extremely expensive. An open system composed of modules with well-defined functionalities and interfaces is therefore the only viable solution to protect investments. It has to be ensured that seamless import and export of content is possible to allow system upgrades. Additionally, the interoperability of the different system components has to be ensured by the use of standardized interfaces. CMS-related interoperability issues have to be addressed between different vendors' solutions, different archive modules, services and client, archive and production, and new and legacy archive systems.

1.3.7 REQUIREMENTS SUMMARY

The above consideration give an overview of the problem area and the multitude of requirements that are placed on a professional CMS that is deployed in an enterprise-wide context in media-rich organizations. It has to be able to manage a whole range of digital media formats and provide easy access to the encoded data. In established organizations a CMS will also have to manage traditional videotapes and data cassettes in parallel. In this case it also has to be possible to easily migrate from video formats to file formats. The main objective is to support the production process better, make it more efficient, and provide a platform to move towards a tapeless environment based on server technology and digital formats. While this process is ongoing, the CMS must also provide the interface between the IT world and the broadcast world. The deployment of servers requires automated storage and retrieval support.

A crucial aspect is the management of descriptive data (i.e. metadata) since this provides the means of finding content in the system and exploiting it. One of the main objectives of a CMS is to allow better reuse and commercial exploitation. This also implies that the application and user interfaces are adapted to the specific requirements of the professionals working with the system. Although a CMS is a platform, and its task and functionality lie in the management of content (i.e. it should run in the background) the user interfaces are still the most visible part and therefore crucial to the success of a system.

A CMS can only provide a positive return on investment when it solves problems effectively and in a commercially viable way. This means that a CMS must support important business processes and workflows in the enterprise, and it must be flexible in adapting to new workflows. The introduction of a CMS into an organization changes business processes and workflows, and the CMS should support the users in migrating from old to new workflows. Many content-related business processes revolve around planning, drafting, commissioning, acquisition, annotation, query, retrieval, collection, transfer, edit, approval, and delivery. The CMS should get involved in these processes and support them by retaining metadata added to an object along a business process, and by providing easy access to existing content objects wherever applicable.

Finally, a CMS alone does not solve all problems, but it does so in conjunction with other applications and systems. Examples of such applications and systems are databases and information systems, production systems, newsroom systems, automation systems, recording and playout systems, enterprise resource planning systems, and rights management systems. Thus, a CMS must provide means to interface with such systems, allowing cross-platform search, retrieval, collection, and transfer, including metadata and essence exchange, and provide messaging and event handling capabilities that enable workflows across system boundaries. The overall solution that enables enterprises to perform their business processes efficiently is a concert of systems, in which the CMS plays an important part by interconnecting otherwise non-connected systems and enabling seamless cross-system workflows.

2

Content-Related Workflows

A content management system (CMS) is primarily a repository for multimedia content objects. However, its value is in the way it facilitates the exploitation of the content from a commercial and/or cultural perspective. In order to do this the CMS has to provide application modules and interfaces that optimally support the development, creation, and exploitation of content. Various workflows have to be supported depending on the organizations dealing with content and the context in which it is handled and used.

Content-rich organizations (in particular broadcasters) are at present the main users of content and hence of CMS. It is important to consider the various requirements the different user groups and workflow processes place on the system architecture of a CMS. This is crucial to realize the full potential of a CMS and to facilitate its introduction into the organization without any major friction. Hence it is important for CMS developers as well as for the content-rich organizations themselves to analyze the processes underlying their business.

Each industry and institution has specific structures and workflows concerning the handling of content. They are different for each organization but basic work steps that are similar for most of them can be identified. Apart from broadcasters and media production houses, other important areas where CMS are required to support the business and work processes are the sale of content via e-commerce systems, teaching and training systems, and systems that support marketing and sales activities. The examples given in this chapter should help to identify important issues within an organization considering the introduction of a CMS.

2.1 WHO IS DEALING WITH CONTENT?

Within media-rich organizations (such as broadcasters, production and post-production houses, film studios, publishers, libraries, and multimedia archives) almost everybody is dealing directly or indirectly with content. There are those who are producing and handling content but also a number of professionals who are concerned with the administrative and legal issues related to content. Also, other organizations have more and more to deal with a significant amount of content. For instance marketing and public relations companies have to manage audiovisual and textual material. Further, large industrial, consumer, and

Professional Content Management Systems: Handling Digital Media Assets   A. Mauthe, P. Thomas
© 2004 John Wiley & Sons, Ltd   ISBN: 0-470-85542-8

service companies have archives and multimedia training and marketing material they have to mange within the organization. Content and its management in this context is not at the core of the operation. However, a significant number of people are handling content, and an efficient management of content can result in advantages in areas such as marketing and staff development.

Within an organization various professional groups are handling and managing content. They have different views on the content and require access to content and related information according to their roles, rights, and interests. Specific parts of the content are relevant to the different user groups. The access and presentation of this information has to consider the specific requirements of a particular group and sometimes also of an individual user. The groups dealing with content comprise production staff, directors and creative directors, journalists, editors, craft editors, post-production editors, catalogers and archivists, but also administrative and management personal who are only concerned with metadata and related information. In the following the different user groups and their role in handling content are discussed in relation to their involvement in the production, presentation, and management tasks in content production and management.

2.1.1 CONTENT ACCESS IN PRODUCTION

In the pre-production and production stages of the content creation process two main user groups can be distinguished, namely the 'artists' (such as producers and directors), and the 'craftspeople' (such as cameramen, sound engineers, production assistants, craft editors, etc). The former are concerned with conceptual and creative work, including research work, production planning, and scripting. In this process an initial set of metadata about a content object is created. Ideally this information should already be created within the system context of a CMS. This information should be kept and associated with the content object throughout its lifetime. Further information produced during production (such as production notes) might also be kept with the content. Initially the access of this user group to the content is purely textual, in the form of contributions such as scripts (i.e. free text) and planning documents (including time schedules, work plans, and budgets). In the production phase the directors and editors are also reviewing the work in process. Hence they have to be able to access the raw material efficiently and to make annotations (which is again metadata that might be kept within a CMS). The reviewing process during the production might be carried out by using low-resolution proxies. In this case it has to be ensured that the browse copy is a true representation of the original material. Access to the content has to be easy and intuitive. Since this group is used to working with broadcast production tools the application interfaces should adopt a similar style and provide similar functionality.

The production staff (such as cameramen and sound engineers) generate new content. Traditionally it is mainly essence (i.e. video or audio material) that is being produced during this process. This is also mainly done outside the context of a CMS. However, ideally the material should already be annotated with production-related metadata during this stage. This data could be transferred to the CMS during the ingest of the raw material. In order to keep the workload of this task minimal, as much information as possible should be retrieved automatically. For instance, the production media could be obtained from the equipment settings; location and time parameters could be derived using GPS modules.

At a later stage in the production craft editors access the material for editing and effect work. Therefore the essence is accessed and manipulated using special equipment. This process happens outside the control of the CMS. Thus, the material and metadata created during this work step have to be (re)imported into the CMS after the work has been concluded or occasionally as a work-in-progress version. The metadata in this context comprises edit notes but also edit decision lists (EDL). In order to provide optimal support the CMS has to be integrated with the native production tools. Consolidation points and interfaces between the different tools have to be defined. The access of this user group has to be direct, with a specific focus on production-relevant information and material. Ideally the CMS application interfaces should be integrated with native production tools to allow seamless operation within a given environment.

2.1.2 CONTENT ACCESS IN DOCUMENTATION AND CATALOGING

A central role within content management is the documentation and cataloging of content. This is the traditional task of the archive. A number of trained feed assistants, catalogers, and archivists deal with the in-depth documentation of content but also with the preservation of essence. From the point where content first enters a CMS it requires a certain amount of description to be able to search, find, and retrieve it. An initial documentation is very often carried out by so called feed assistants, who annotate material (such as agency feeds, rushes, raw material, etc.) in real-time when it enters the system. This user group requires quick and easy access to the content, i.e. any initial metadata associated with the incoming material and a browse copy (that is, a frame- and timecode-exact copy of the original material) have to be instantly available. Since the annotation has to happen in real-time the application interfaces have to be specifically tailored to the need of these users.

The task of in-depth cataloging through archivists also has to be supported by a CMS. They require direct access to the organization's documentation database. Further, a detailed free-text description of the content is also carried out in this context. For continuous media (such as video and audio) the description also uses timecodes to specifically refer to certain segments of the content. The work of the archivists and catalogers may be supported by tools such as controlled word lists or thesauri. These are often organization-specific and therefore have to be specifically implemented or adapted for different organizations. Other advanced tools such as video analysis and keyword spotting are also emerging as auxiliary instruments that help to support the archiving process. They provide additional information for the documentation of material. However, this information has to be verified since these automatic tools currently do not provide 100% accuracy.

The handling of the essence in this context is changing. So far essence is stored (and closely associated) with its physical carrier, i.e. a video- or audiotape. The task of the archive in this context is to store these carriers in a controlled environment and to preserve content by copying material from deteriorating carriers to new ones. Further, tape formats and equipment have a certain lifetime and older formats might have to be replaced because the equipment itself, spare parts, and also trained operators are not available any longer. Hence the copy process often also involves the migration to new formats represented by the new carrier of the essence. These copy operations are costly and bind a lot of capacity within a multimedia archive. In emerging CMS using mostly digital formats the encoding format and the physical carrier of the essence are decoupled. Copies of the essence can be stored online on disks or on tapes. Tapes in this context do not have to be video- or audiotapes but can be data tapes that are stored in automatic tape libraries. The actual storage and retrieval process of the essence can be entirely automatic. This also includes automatic copy processes that might be required for preservation reasons or format change. As a consequence the essence does not have to be handled directly any more but is automatically available on request. Multiple copies can exist in the system, online on servers and near-online in tape libraries. Especially online copies can be created and deleted whenever required. This changes the perception of a content object. At present very often a content object is represented by its physical carrier (e.g. the tape it is stored on). In future it might not be possible to determine the location of essence and its carrier unambiguously since it might be stored in multiple copies and various formats at different locations in the system. Thus, the notion of a content object becomes more abstract. This also has an impact on the work of archivists. More and more the focus will shift towards the logical representation (i.e. documentation) of a content object, away from the physical handling of the media.

2.1.3 SEARCH AND RETRIEVAL OF CONTENT

The value of content is in its exploitation (i.e. presentation and usage). Thus, it has to be possible to easily search and retrieve content within a CMS. Traditionally this is the task of the archive, which offers services such as mediated research. This is due to the specific documentation tools (such as databases), and special documentation procedures and languages, requiring expert knowledge. With an ever-growing amount of content and also a growing number of users who have to access it, the search procedure has to become more intuitive and easier to handle. The users who search for (representations of) content within a CMS are journalists, editors, researchers, marketing experts, and (in open systems) even the general public. Therefore the most important aspects of search and retrieval are uncomplicated and user-friendly application interfaces, a well-structured presentation of the search results, and the consideration of user profile access rights. The search tool has to allow unstructured (i.e. fulltext-based) search as well as professional databases search. Further, with emerging media processing tools it might also be possible to use search by examples (such as image similarity retrieval or query by humming) in the future. Thus, the system also has to provide interfaces for these new search operations.

The representation of content in the result list is mainly textual. However, more and more this information is also enriched with low-resolution copies of the original essence or (especially in the case or video) keyframes, skims, and other audiovisual abstracts. Ideally the interface for search and presentation of the result should be configurable to be able to adapt to the specific requirements of user groups and individual users. The selection process can also include rough-cut, which allows the users to mark parts of the content for further processing.

The users searching for content can have different rights for viewing and retrieving it. In a professional environment this can include the automatic transfer of an essence copy (or selected parts) to production or broadcast equipment. Other users might only be allowed to view low-quality copies. In a Web environment these copies might be streamed in order to prevent the receiving users from storing a copy of the content.

2.1.4 CONTENT-RELATED ADMINISTRATIVE AND MANAGEMENT TASKS

Within a media-rich organization there are a considerable number of professional groups just dealing with administrative tasks related to content. These are the financial planning and accounting departments, legal departments, and production and program planning. These users need to have access to all relevant data about the status, costs or legal issues related to a content object. Very often they are already using specific applications and software tools (such as enterprise resource planning systems, digital rights management systems, program planning systems, etc.). Their view on the content is also limited to certain aspects documented in the metadata part of a content object. They hardly ever require access to descriptive information about the actual content, the essence itself or any audiovisual proxies accompanying the metadata.

However, certain parts of the information these users are generating and using should be present in a CMS as well. For example, it might be important to have an indication about the rights situation in connection with a specific content object or to have access to a financial summary about a project. On the other hand the users concerned with the administrative aspect of the content might also require information from the CMS. For instance, information about the production or reusage of content is of interest to these user groups. Thus, information between different systems has to be exchanged. Alternatively, (restricted) access to specific parts of the systems storing particular information about content objects should be provided. In this case an integrated approach at the application level is desirable, i.e. the user should only have to deal with one application, which also presents information and views from third-party systems.

2.1.5 THE ROLE OF A CONTENT MANAGER

CMS are currently used to assist the various user groups in dealing with content. Thus, they are perceived as an auxiliary tool to support existing workflows and facilitate the work of the users concerned with content. However, CMS also change the way content is handled, and the emerging systems and (hardware and software) components require professional service and maintenance.

Since a CMS comprises database and (mass-) storage technologies, PC and server systems, communication networks, and broadcast equipment a technically versatile person with an understanding of the workflows and processes involved in the handling of content is required. Further, there needs to be somebody with a focus on the documentation tasks and the actual management of the content within the CMS. Many of the content-management-related tasks associated with the handling and administration of content are closely related to the work of traditional archivists and catalogers, or broadcast engineers. However, since the handling of content is becoming a central part of the operation there needs to be a dedicated *content manager* who is responsible for the system and the content stored within it. The focus of such a content manager has to be twofold, i.e. s/he has to have a technical and engineering focus as well as a content and workflow focus. The former refers to the technical maintenance of the system and infrastructure whereas the latter is concerned with the actual support of the processes and users handling content.

At present this kind of job does not exist. Different departments and individuals are responsible for the operation of a CMS within an organization. However, it is envisaged that with a growing penetration of such systems a clear line of responsibility for the system and its content has to be established. The content manager will have a key role in this context since s/he is responsible for the operation of a system that provides a service across departmental boundaries. In contrast to common infrastructure and IT tasks this also comprises the actual handling of content (i.e. the non-technical part of content management).

Depending on the size of the system and the organization, the role of a content manager might be divided with the actual content manager or cataloging editor focusing on more content-related aspects (such as essence and media handling, system resourcing and provisioning, etc.) and the CMS engineer concerned with system maintenance aspects (such as hardware, software, and network maintenance). How these new roles can be integrated into existing organizational structures is also dependent on the envisaged change of workflows through the introduction of a CMS.

2.2 WORKFLOWS IN BROADCAST

The content creation and distribution processes in the broadcast industry are changing. This is the result of the introduction of new output channels, new digital production methods, and a general demand for an increasing amount of quality content that has to be created faster than ever before. Further, technological developments nowadays also enable a much more integrated workflow than previously possible.

The content creation process typically starts with the planning of a program combined with the transmission schedule and production planning. At this stage some initial metadata is already created. This set of metadata will continuously be enriched in the course of the actual program production process and remains associated with a content object throughout its lifetime.

Subsequently acquired material from electronic news gathering (ENG) teams, external feeds, and in-house productions is ingested[1] into the system and associated with new or already existing content objects. While ingesting, an initial annotation of the content is conventionally generated. This can either be done manually by so-called *feed assistants* or automatically with the help of specific extraction, indexing, and annotation tools. This information (ideally produced in real-time) supports basic querying and allows content search and retrieval even while the ingest is still ongoing. Hence a selection of suitable, most up-to-date material parallel to its recording (even during an ongoing event) is possible.

Editors access the CMS primarily by querying for metadata. The required information is searched using fulltext or structured search and query operations. The search produces a result list, which can contain newly ingested material as well as existing content from the archive.

The *cataloging department* is responsible for the complete and detailed description of content selected for long-term archiving. Catalogers create or augment the formal and context-related content description and refine existing metadata if required. It is also within their responsibility to verify the quality of annotations extracted by automatic tools.

The *rights and licence department* adds specific metadata concerning the copyright and intellectual property rights (IPR) status of the content. At present the complexity of

---

[1] *Ingest* refers to the process of introducing the material into the CMS by means of recording or encoding of the raw content.

legal issues related to content makes the autonomous administration of assets within an asset management system (i.e. a CMS that also handles IPR) difficult. A more pragmatic approach for deriving the IPR status of content is to optimize the information exchange between the CMS and the rights department to identify the rights holders and clarify the legal status.

In *post-production* the created content is conformed into a program ready for transmission. This results in a new (or final) version of a content item that has to be (re)introduced into the CMS. The conformed program material can later be reused as entire programme or by using selected sequences for new productions. Metadata created during this stage (e.g. EDL) should also be transferred to the CMS.

Further, information that should be maintained in a CMS is the information gathered during the transmission (e.g. as-run logs). Other entities such as those responsible for program exchange and marketing of program products will also create additional metadata that should be made available in a CMS.

The following sections are concerned with workflow models that describe the processes occurring during the creation and delivery of broadcast content. Further, the ingest and logging workflow and the retrieval workflow are examined in more detail. These two workflows are the basic building blocks for many processes within the content production and exploitation process in the broadcast industry.

2.2.1 SEQUENTIAL VERSUS CONTENT-CENTRIC WORKFLOW MODEL

Traditionally the content creation process follows a push model as depicted in Figure 2.1. The content creation starts with a plan and draft stage, where the original idea is developed and a project is set up. After approval, the details of the project are developed further and the actual production process is planned. In each of the production (capture and analysis) and post-production (synthesis, composition, and packaging) steps the content is processed and after the work has been concluded is passed on to the next step. This transfer between the work steps mostly happens on a physical carrier (e.g. tape). The further steps up to archiving are concerned with the delivery of the content. Again, the main medium for transferring it between the steps is on a physical carrier. This also implies that the content as it finally arrives for archiving mainly consists of the actual essence and only very few metadata elements. Within the archive the material is documented using the available metadata and the information that can be retrieved by examining the content. Finally, the content is put into storage. The drawback of this approach is that it is an entirely sequential workflow with no possibilities to access the content or information about the production process outside a specific work step. Further, since the different workflow steps are traditionally carried out using isolated systems for each step, and since these systems are not designed to share metadata, a lot of metadata about the creation process is lost along the chain. As a consequence, when the content object is finally archived this metadata needs to be manually recovered and re-keyed into a database.

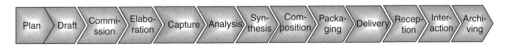

Figure 2.1   Linear push workflow model

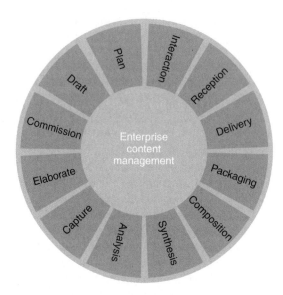

Figure 2.2   Content-centric pull workflow model

Facilitated by upcoming CMS with much higher integration, there is a move away from the sequential push model towards a content-centric pull model as shown in Figure 2.2. In this model the content is at the core of the operation represented by the CMS. All the information created during the production process is stored in the CMS. The CMS becomes the central hub for all content-related information that is created and used throughout the content creation process. This information is constantly updated with every work step. The content is not pushed to the next workflow step after processing has been concluded. Only the information about the termination of a specific workflow process is passed on to the following processes. The progress of the work can be observed by all participants who have access to the content. This allows faster production, and even working on a content item in parallel is possible. Further, since all information during the creation process is stored in the CMS there is a much richer set of metadata for a content object. Easier and faster access to content also facilitates the reuse of existing content.

Within the EBU and SMPTE working groups, the content-centric workflow model has been analyzed in more detail, specifically considering the data flow that occurs in this context. It distinguishes between the flow of metadata and the flow of essence. Figure 2.3 shows the model with the CMS in the center surrounded by the different work steps occurring within the production and delivery of broadcast content.

In the initial production steps metadata is created and exchanged between the different entities. No essence needs to be accessed directly during interaction and commissioning. In the elaboration step existing content can be reviewed and might be selected for further production. New essence is ingested into the system during capture and used during production, post-production, and delivery. Essence leaves the system at reception. During the different work steps the essence may be altered and further metadata is gathered during each step. Essence and metadata in the CMS are updated during each work step in which they have been altered.

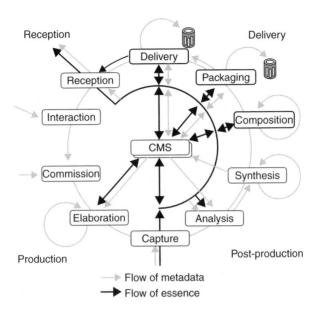

Reception · Delivery

Production · Post-production

Flow of metadata
Flow of essence

Figure 2.3 Essence and metadata flow in the content-centric process model

2.2.2 BASIC BROADCAST WORKFLOWS

The workflows within a broadcast CMS can be based on two basic workflow processes, namely the ingest and logging workflow, and the search, query, and production workflow.

2.2.2.1 Ingest and Logging Workflow

Content is introduced into the CMS during acquisition in a process called ingest (when material is directly recorded into system) or import (when pre-recorded material is placed in the CMS). Content is acquired from different media and carriers, e.g. tape, cable, satellite. Essence is digitized by a suitable encoder and recorded on a server and/or videotape. From this point on it is available in the system and can be played back and transported to other locations through the studio network. In parallel to the recording of the high-resolution material a low-resolution preview stream is created and recorded onto a server. The actual recording process can be fully automated. During the ingest process the incoming essence can be automatically analyzed to create indexing information and audiovisual auxiliary information such as keyframes, skims, etc. Figure 2.4 shows a generic model of the ingest and logging workflow. Depending on the context where the ingest takes place, there might be additional steps or slight alterations. However, the basic workflow steps are similar for the majority of ingest processes.

The ingest depicted in Figure 2.4 starts with the digitization and automatic ingest of the material. High-resolution and low-resolution preview and browse copies are created at the same time. The material is recorded on different devices and storage media under the control of the CMS.

The encoding and recording format of the material depends on the digitization and recording equipment used. Ideally the material should be accessible while the digitization

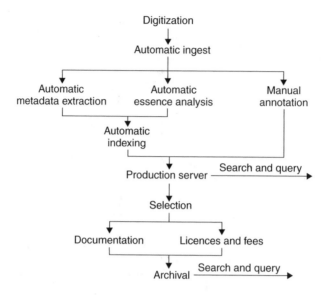

Figure 2.4 Workflow ingest and logging

and recording is still ongoing to allow immediate access as soon as the content first arrives. This enables the manual annotation of the content by feed assistants during recording. This process is called logging and is necessary to provide some initial information about the content in order to make it searchable.

During logging, automatic metadata extraction, and automatic essence analysis of the incoming content the following information is extracted and stored in the CMS:

- automatically extracted metadata accompanying the ingested signal that includes encoding parameters, videotext, and other data carried within the vertical blanking interval (VBI);
- information provided by the automatic essence analysis such as edit points, keyframes, speakers, faces, and keywords;
- textual description entered manually during logging. The logging process follows semantic rules specified by the organization within which the logging process takes place;
- additional information provided by intelligent tools using the previously automatically extracted information enhanced by semantic description standards.

The manual logging is of particular importance in this context (especially since automatic analysis tools such as speech recognition are still not providing the accuracy required in a professional media production environment). It creates the first manually generated information in the production process and is instrumental for a successful search and further detailed documentation. Logging therefore requires a high job qualification that combines in-depth knowledge of documentation with an intimate knowledge of the requirements of the editors who have to evaluate the relevance of the incoming material for their production.

The incoming essence and the created metadata can be made available to journalists and editors immediately even while the initial logging process is progressing. Thus it is instantly accessible and usable in the production process.

After some time archivists select the material that is deemed relevant and suitable for long-term preservation. Criteria for this selection are not only the historical value and relevance of the material for future production but also the kind of usage rights that have been acquired with the content. The content selected in this process is documented in more detail, the IPR status is recorded and the objects are marked as permanently archived. Permanently archived material is not removed from the system any more and is available for future use in broadcast and production.

2.2.2.2 Search, Query, and Production Workflow

The second fundamental workflow process is search and query for content, its selection, and preparation in the production context. Not considering the content creation phase where ideas are developed, selected, and finally transformed into a detailed description for a program, the actual content (re)use phase starts with the search and query request for content that can be used for production. This process can be carried out by editors or journalists; in more complex research cases experts from the archive might be involved.

The search and query produces a hit list that is reviewed by the user. Each hit provides information about the content item made up of metadata, keyframes, and preview video and/or audio. This information supports the selection of the material to be used in the content production. Figure 2.5 shows the different workflow steps in the search, selection, and production process.

Pre-listen and preview copies of the items found in a hit list facilitate the actual review and selection process. Selection of relevant segments rather than selecting production material on an item basis is possible using the preview or pre-listen copies and rough-cut. Using rough-cut as a selection tool results in a list of segments and relevant program elements represented by a rough-cut list. In order to make the material selection as accurate as possible the rough-cut should be frame-accurate and should refer to timecodes that can be used during the subsequent craft edit. The rough-cut list is used to automatically transfer the selected elements from the CMS to the nonlinear editing systems (NLE). Here the final program edit and conform is carried out. These processes take place outside of the control of the CMS since the changes happen on a third-party system that is not necessarily tightly integrated.

The final conform refers to the formal acceptance process. This is carried out by the responsible editor based on technical, legal, and contextual criteria. After successful completion of these work steps the new program item is then transferred to the playout server (in the case of direct program playout) or other storage media (if the program is used at a later stage). The final step of the production process is the archiving of the program.

Depending on the infrastructure and the system context, the physical movement of essence is either performed by the CMS autonomously (if suitable control software is implemented) or is controlled by a responsible editor, cataloger, or content manager.

A crucial aspect in the context of this workflow is the conformation process in the editing stage. Conforming content here refers to the physical creation of a program item from the different assembled parts. At first sight it seems advantageous to confine archival to rushes or raw material. New program contributions then only have to reference them

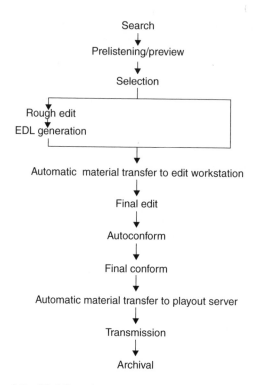

Search
↓
Prelistening/preview
↓
Selection
↓
Rough edit
↓
EDL generation
↓
Automatic material transfer to edit workstation
↓
Final edit
↓
Autoconform
↓
Final conform
↓
Automatic material transfer to playout server
↓
Transmission
↓
Archival

Figure 2.5   Workflow from search and query to final conform

via an EDL without the need for creating new items. This would, for instance, allow minimizing the volume of material stored in the archive since only first-generation material
is used. However, in practice this assumption can have a number of significant disadvantages that make it unfeasible to implement such a virtual conformation process. Among
the main arguments against this method are:

- Metadata will allow the tracking of the original material anyway.
- There is an increased risk of failure during transmission if the transmission system
  needs to assemble the program online from fragments based on an EDL.
- The administration of all links required to represent a program item is extremely complex. Loss of information on the links will lead to a total loss of all programs. This
  is becoming even more complex with second- and third-generation program items that
  are created from already edited material.
- If programs are fragmented in such a way, the programme fragments may be distributed
  over a large number of servers and tapes in the mass storage system. This means that
  when retrieving a programme from the archive, it may be necessary to access a large
  number of tapes in order to stage all required programme fragments to a playout server
  in order to recreate the complete programme.

Thus, the benefit of potentially minimizing archive volume and the guaranteed access to
original material is compromised by reduced operational security and bad system response.

Thus, it is strongly recommended that within the production chain content is transferred to the next step in the production chain as physically conformed entities only.

2.2.3 IMPACT OF CONTENT MANAGEMENT ON BROADCAST WORKFLOWS

The introduction of a CMS not only supports existing workflows but also provides the potential for changing and adapting workflows. It is not expected that the introduction of a CMS into a broadcast organization will lead to fundamental change in the established work process immediately. However, significant changes within certain work steps will occur right from the beginning. For example, the introduction of real-time logging as well as innovative methods in documentation of content already significantly influences editorial and cataloging work.

The need to create a closed loop for metadata generation along the production chain combined with the possibility to transfer content automatically from system to system will change the individual staff members' work and will eventually lead to the definition of new job profiles.

The process of logging (i.e. the manual annotation of ingested material, which is often carried out by feed assistants today) can be taken on by personnel combining the qualification of catalogers with profound knowledge of the detailed requirements of editors. This change shifts the responsibility from editorial staff to staff in the cataloging and archive department. On the other hand it is expected that the requirement for research skills will shift from the archive to the editors, although the archive will still have to offer assistance in the case of complex queries.

The automatic extraction of metadata from the production process will relieve catalogers from routine operations such as manual re-entry of existing metadata. Thus catalogers can be involved earlier in the content production process, for instance at the stage of logging. They can also concentrate on more qualitative and challenging tasks such as conceptual descriptions.

2.3 WORKFLOWS IN E-COMMERCE SYSTEMS

Digital content is one of the few goods that can be entirely traded electronically, i.e. the offer, order, *and* delivery can happen completely in the digital domain. Two basic kinds of content-related e-commerce interactions have to be distinguished, namely business-to-business and business-to-consumer processes. The former case includes all business processes between companies and organizations that trade content. In the latter case content is offered to the wider public over a public communication network such as the Internet or cable television. The main difference in this context is the way the content is made available and the requirements placed on the underlying system infrastructure in terms of storage, bandwidth, and access. This section concentrates on the business process and internal workflow of an operation providing content. The focus is on the business-to-business processes. The business-to-consumer workflows are either similar to the broadcast workflows (i.e. in the case of scheduled content delivery) or close to business-to-business e-commerce transactions (i.e. in the case of video-on-demand). Especially in the latter case the demands placed on the underlying system infrastructure may

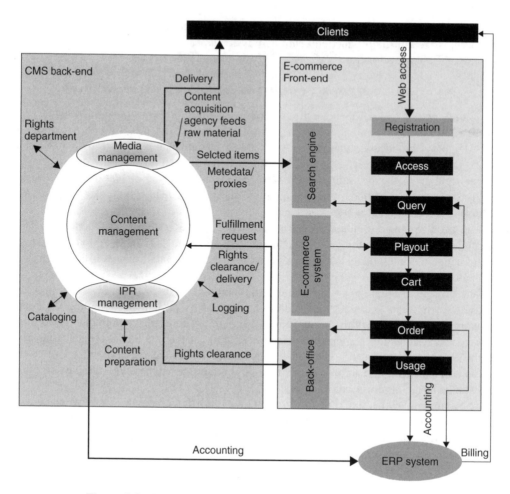

Figure 2.6   Content management and e-commerce system structure

be higher than in the business-to-business case but the basic processes and workflows are the same.

As shown in Figure 2.6, the structure of an e-commerce system offering content can be divided into two major components, i.e. the CMS back-end and the e-commerce front-end. The CMS back-end is concerned with all operations relevant to managing the content that is offered by the e-commerce system. It is also concerned with the fulfillment of the orders, i.e. the delivery of the content after the contractual issues have been cleared. The e-commerce front-end contains the storefront with the content objects on offer (represented by proxies), a search function, the e-commerce module (responsible for dealing with the customer requests), and the back-office (coordinating the ordering process and the actual fulfillment). This part of the system is also linked to an enterprise resource planning (ERP) system that is in charge of billing and accounting.

The separation of the two subsystems is motivated by security and business process reasons. The content is the business object, i.e. the actual asset to be traded. It has to be protected from intruders and any possible misuses. Hence the system should not be accessible from the outside. In contrast, the e-commerce front-end has to be open to the public and attract as many potential customers as possible. Therefore it is a system that is accessible via public networks and might also attract attempts to defraud the system.

In terms of business processes it is usually not the case that all the content belonging to a content-rich organization (e.g. a broadcaster with a sales department) should or can be offered at the same time. For instance, the organization may not have the necessary IPR for all content items in the CMS to sell to outside partners. Some of the material might also be kept back because it is to be used in another context (e.g. a broadcast on the organization's own channels) and some time should elapse before it should be released for sale. Thus, there needs to be an active selection process during which issues such as rights are already checked. Further, not all the metadata required for the management of content is relevant for content sales. Only the subset that is necessary for the sale should be accessible to the customer. Content sales also require the active presentation of objects to the clients. It is not just a system in which the client actively searches for content. Hence, during the selection of content for sale presentation issues should already be considered.

For systems only concerned with the sale of content it might be an option to implement one system with two modules (i.e. a CMS and an e-commerce module). However, advantages and disadvantages of such a solution should be carefully considered. In any case it has to be ensured that the actual content objects (i.e. high-resolution essence and all internal metadata) are protected.

2.3.1 WORKFLOW IN THE CMS BACK-END

There are two main workflows within the CMS back-end, i.e. the ingest, documentation, and management of the material in the system, and the answering of requests from the e-commerce front-end.

2.3.1.1 Ingest, Documentation, and Content Management

The workflow in the CMS back-end of an e-commerce system is similar to that of a broadcasting operation. Material enters the system via different channels (e.g. on tape, via line-feed, satellite, etc.). The incoming material is recorded, low-resolution proxies are created, it is annotated with some basic metadata, and the ERP system is informed about the new content. The initial documentation of the content also comprises the registration of the acquired rights that form the basis for the sales process given that content can only be marketed if the respective rights have been acquired as well.

The initial annotation (i.e. logging) and the in-depth documentation can be separate processes since (as in the case of broadcasting) certain material might be most interesting when it first arrives (e.g. actual information about politics, business, stock markets, etc.) and hence should be offered as soon as possible to the customers. Depending on the kind of offer, the content might be processed and prepared before it is offered. This is a standard editing process that also provides summaries, etc. This is closely related to the journalistic and craftwork carried out within a broadcaster.

Cataloging refers to the in-depth documentation within the system. In the context of an e-commerce system it is important that the documentation is carried out in a manner reflecting the skills and knowledge of the target customer group. Searching and retrieving content should be straightforward for the potential customer. For instance, if medical content is provided for doctors and medical experts the documentation should use the appropriate technical terms, whereas if the general public is the target group for the same content the documentation should use the terminology understood by an average medical layperson. The rights documentation has to record contract details and ensure that copyrights and IPR are honored. This is usually done within a separate rights management system that is linked to the CMS.

The actual management of the high-resolution material (i.e. essence) is performed in conjunction with a media management module. This is a component that controls the media transfer between systems and also is responsible for the electronic fulfillment of orders. The management of the essence requires control of the storage subsystem and technical quality control of the material. This includes migration of copies to other storage media if required. The content management process is at the center of the operation. Its basic characteristics are equivalent to those encountered within broadcast.

2.3.1.2 Fulfillment and Delivery

One of the main tasks of the CMS back-end of an e-commerce system is the fulfillment of requests. Such a request is generated in the e-commerce front-end and passed to the CMS electronically. It contains the content ID and delivery information. Before the delivery can take place the rights situation is verified. This information is also passed on to the e-commerce front-end for accounting and billing. If all rights are cleared for a specific customer request the CMS can go ahead with the delivery of the content.

There are different ways to deliver content depending on the kind of content, system capabilities (of the CMS, communication network, and at the customer side), and customer preferences. For instance, electronic delivery of audio content via existing data networks such as the Internet is possible. High-resolution video content requires broadband or satellite connections. In this context the media management component will control the delivery of the content. Content might also be delivered via traditional means such as mail or courier services. In this case the material has to be copied onto tape, packed, and posted. This requires human intervention.

The delivery is recorded and the information is passed on to the ERP system.

2.3.2 WORKFLOW IN THE E-COMMERCE FRONT-END

Two different processes have to be distinguished in the e-commerce front-end of a system trading content, i.e. the selection and presentation process of the content objects for sale, and the actual sales process. The basics of these processes are similar to those occurring in any other e-commerce system. Most interesting in the context of CMS is the interaction that happens between the e-commerce front-end and the CMS back-end.

2.3.2.1 Selection and Presentation of Content

The selection of content is an active process that is carried out by staff responsible for the sales activities. The selection can happen by actively searching the CMS back-end for content that should be offered at a certain time in a specific context (e.g. material about personalities or events becomes more relevant before anniversaries). If the e-commerce system offers content dealing with current events and ongoing affairs, up-to-date content has to be offered as soon as it comes in. Hence there should be a trigger that content has arrived in the system and is ready to be offered to the customers. This can happen in parallel to the content preparation in the back-end.

During the selection copies of (a subset) the descriptive metadata and low-resolution copies of the essence are placed in the e-commerce front-end. A client can actively search in the e-commerce system for content. However, since this is the storefront of the 'shop' the client should be actively offered products and should be guided in the sales process as well. Thus, this part of the system needs to be designed specifically taking into account the kind of offered content, the organization it represents, and the target customer group.

2.3.2.2 The Sales Process

Before a user can access an e-commerce system there is usually a registration process in which the customer details are taken and stored in the system. After this a customer can log-in and search the system for specific content or look at the items offered. This includes not only the presentation of descriptive metadata but also the possibility to watch and/or listen to browse copies of the content. Depending on the policy of the provider these can be keyframes, skims, short samples of the content object, or the full item only in low quality. It should be possible to refine or change the search in case the desired item could not be found. Entire items as well as segments might be purchased. Thus there should be a possibility to select and mark segments of an item a customer would like to purchase. This can be supported by rough-cut capabilities similar to those that support the content selection during the production process. Items (or segments of items) that are going to be purchased are placed in a shopping cart. This links the details of the customer with those of the content object to be purchased. Depending on the content and the kind of customer, it might be necessary to give details about the intended use. For instance in a business-to-business context it has to be assured that all the IPR for a certain use are available and part of the contract.

The content items in a shopping cart are ordered by actively concluding the purchase process from the customer side. The order triggers a message from the e-commerce back-office to the CMS back-end. This message contains the details of the customer and items to be purchased. The rights situation is again verified and (if the rights are cleared) the delivery process is initiated. There is a message back to the e-commerce back-office confirming the sale. This information together with the intended usage is passed on to the ERP system that deals with the accounting and billing process. The usage information is required for the accounting since this is a pricing factor in the professional use of content. Billing is then a task of the ERP system.

2.4 CONTENT-RELATED WORKFLOWS IN CORPORATE AND OTHER ORGANIZATIONS

Large corporate, educational, and government organizations increasingly have to deal with audiovisual content. There are three main areas where content is playing a prominent role, i.e. within archives, for training and staff development, and within marketing and sales. Corporate and other audiovisual archives are closely related to any other type of multimedia archive, and the workflow and processes are similar to those of the broadcast archives as described in Section 2.2. The areas with direct business impact and where additional aspects have to be considered are training, education and staff development, and the support of marketing and sales. The former is important for educational, governmental, and corporate institutions whereas the use of CMS within marketing and sales is mainly relevant for commercial organizations.

2.4.1 CONTENT MANAGEMENT IN TRAINING AND E-LEARNING

Electronic media such as video, audio, and interactive programs are more and more being used for educational and training purposes. Their deployment reaches from the use of video and audio within courses and lectures to tele-media courses and lectures. The structure of these systems is similar to the one introduced for e-commerce systems with a CMS back-end and a special e-learning front-end. In contrast to the e-commerce systems, however, this front-end module does not store any data itself but accesses the CMS back-end for the content retrieval.

Within the CMS back-end the course material is handled, stored, and made available for use. A course is assembled from audio, video, text, slides, Web pages, etc. There can be also programs where direct user interaction determines the presentation of the content. The way the course material is presented in this context is determined by the computer program that guides the student through the course but does not necessarily prescribe a linear presentation. However, there are a finite number of content objects and possible combinations that are relevant for a specific course module. Hence, for each course it is possible to prepare all relevant modules for playout as soon as it is accessed by a student.

For tele-media and e-learning systems it is important that the content objects are atomic modules, i.e. basic components of a course. The course itself can be either represented by a script similar to an EDL (according to which the presentation is controlled) or by a computer program that governs the presentation and also allows interaction. For the creation and assembly of the course material there are special programs. Hence a course is imported into the CMS together with all the relevant content objects or with references to existing objects in the system. These references are part of the course metadata.

There are two cases during the creation process where access to content stored within the CMS is required, namely when a new course is assembled from existing material and when a course is annotated during its presentation. The latter case is equivalent to an update or versioning of existing material. This happens outside the control of the CMS and has to be regarded as an export with subsequent re-import. In order to support the creation of new courses from existing content, integration of CMS applications modules with the course creation platform is required. The integration has to allow either the inclusion of references to other modules or copying of the content that is to be included in the new course.

The presentation can be either through scheduled broadcasts/multicasts or via programs that retrieve the different course modules on request. In the latter case the user interacts with a tele-media learning platform. This program accesses the CMS and requests a specific content module for presentation. In this context it is important that the CMS enables proactive preparation of all the course modules that might be required for playout once the first course module has been requested.

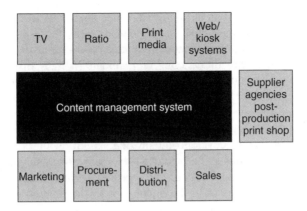

Figure 2.7   CMS in the context of corporate marketing and sales

2.4.2 CONTENT-RELATED WORKFLOWS IN MARKETING AND SALES

The importance but also the cost of audiovisual content used as promotional material for marketing and sales is growing. More and more corporate organizations are starting to manage this content themselves instead of leaving it with the agencies and production houses that originally created it. Apart from keeping the material under control and allowing easier reuse, this helps to coordinate the marketing and sales activities over different organizational units and output channels. Further, it allows quicker reaction to (unforeseen) events. Figure 2.7 shows the different organizational units and processes concerned with the creation and exploitation of content within the marketing and sales context of a corporate organization.

Corporate organizations have different organizational structures depending on the line of business they are in, their business objectives, and organizational history. However, a set of tasks within marketing and sales activities can be identified that are common to the majority of large corporate enterprises. In particular, these occurs in the marketing department, procurement, distribution, and sales. Apart from the internal departments, there are marketing agencies, (post-)production houses and print shops that provide content and need access to the systems. The different output channels for the material are television, radio, print media, and, more recently, the Web and kiosk systems. The relevant workflows that have to be considered in this context are the planning, creation, and production stage within marketing, and the distribution of the content to the different output channels.

2.4.2.1 Content Creation for Marketing and Sales

Within marketing and sales content has the sole purpose of promoting a product and supporting the sales process. The content creation process happens in parallel to the product developed in the context of a marketing campaign. During the life cycle of a product the marketing and sales material is developed further and new campaigns are developed for the promotion of the product. The marketing department is in charge of developing the ideas and initial concepts for the campaign. At this stage external agencies might already be involved. This initial phase can be compared with the planning and drafting that take place in the production of a television program. During the development of new ideas old material might be analyzed and considered. Ideally this material is available in the CMS with metadata and preview proxies for a quick review. Notes, scripts, and protocols about a new campaign should be stored in the CMS. Part of this material should be made available to other departments (namely distribution and sales) to inform them early on about a new campaign. The procurement department has to be involved at this point to get details about any third-party components that are part of the product.

The next stage in the campaign development is the gathering of material such as taping of commercials, photo shoot, development of slogans, etc. This is usually carried out by agencies that also work closely with production and post-production houses and print shops. The CMS is the central content hub and repository. The different parts and versions developed at the various stages from the initial idea to the final strategy are stored and exchanged via the CMS. Thus the external partners have to have access to the relevant projects and content objects. This places high demands on the underlying system and communication infrastructure. In the system context it is important that the different user groups only have access to those parts of a project and content object they are entitled to see. The different user rights restrict read and write access. In order to optimally support the creation process, the communication infrastructure has to support the exchange of high-quality audiovisual material. In case this is not possible certain preview operations might be based on preview copies only. High-resolution material might be brought into the system conventionally, i.e. on tape.

At this stage of the development process there are various iterations from the first idea over concepts and alternative proposal to the final version of the advertisement material. The CMS should support this work process by providing versioning and elaborate support for acceptance procedures. Further, a process view of the projects is crucial in this context. Integration of the CMS with a workflow engine and project-planning module also aids the work process.

Various country and language versions might be created in parallel. These might be sub-projects linked to the main campaign. Within these sub-projects national and regional marketing departments might contribute their own material.

In the final phase of the development of the marketing campaign the distribution and sales departments should be involved more closely since they are directly affected by it and have to execute certain aspects of it. They need access to the latest version of the marketing material to develop their own strategies. Further, the different national and regional distribution and sales organizations have to be able to review and order the marketing material they require.

2.4.2.2 Content Distribution

The different channels in which the marketing material is used have to be provided with the final version of the material. Television and radio commercials have to be transferred to the broadcasters at which airtime has been booked. This can happen via tape, satellite transmission or file transfer. For the print media the adverts have to be sent to magazines and newspapers. Further, if flyers and information material are printed they have to be distributed to the sales and distributions organization for further use. All these processes are fed from the CMS but there is an active transfer and no direct playout.

For marketing material used on the Web and in kiosk systems this is different. They display the material on request in an interactive manner. Usually they are built on top of an organization's own repository similar to a CMS. However, direct integration of these systems with the enterprise content management system that is also used during the development stage of marketing campaigns is not advisable. The reason for this is again security and process considerations as in the case of e-commerce systems. These systems are open to the public whereas the enterprise CMS should be protected from any illegal intrusion. Further, marketing material should be released actively and only material that is approved should be accessible in the Web and kiosk systems.

3

Essence

Essence is the physical representation of content in different forms and formats. It can be produced, altered, stored, exchanged, transmitted or broadcast. The various use cases for essence place a multitude of requirements onto essence formats used for different purposes. One universal format that could represent all different media types (such as video, audio, text, images, graphics, etc.) and satisfy all the different requirements does not exist. However, ideally a CMS should be media- and format-agnostic, i.e. it should be able to handle and manage all kinds of media types and essence formats. However, as a CMS is not only a file repository it has to understand the file structures, media syntax, and even semantics to a certain extent. The function and services beyond pure file retrieval that a CMS provides are, for instance, media indexing, automatic information retrieval, and streaming. Even for partial file retrieval the system requires some understanding of the media type and file format. Thus, for the development and operation of a CMS it is crucial to know about the relevant essence formats, their characteristics, and features.

Further, in order to assess different media types and essence formats, and to decide which is most suitable for a specific purpose, it is necessary to look at the requirements and constraints within a certain application context. The media type in which specific information is represented is usually predetermined. This is, however, not the case for the format(s) a content object is encoded in. A number of encoding schemes and standards have been defined for the encoding and (digital) representation of the various media types. Media-rich organizations have to select carefully the format(s) most suitable for their operation and objectives. A basic understanding of the fundamental principles, functionalities, and features of the encoding schemes behind them is therefore essential.

Within a CMS the role of the different essence formats is twofold. On the one hand they represent the objects that have to be handled and managed. The characteristics of the essence determines if and how certain functionality can be realized. On the other hand essence is used to represent content within the CMS. At present different formats are used for production and presentation purposes, and for the pure representation of content in a CMS. The specific requirements of each task have to be considered carefully to ensure that the required functionality can be optimally provided.

Professional Content Management Systems: Handling Digital Media Assets   A. Mauthe, P. Thomas
© 2004 John Wiley & Sons, Ltd   ISBN: 0-470-85542-8

Apart from the pure management of essence within a CMS, it can be automatically processed to retrieve additional information about the content and to create new representations of a content object. Video analysis programs, for instance, allow detection of cuts and edits and creation of a condensed version of the content object in form of skims or storyboards. Transcoding processes use one format to create a copy in another format that is more suitable for a specific purpose.

This chapter discusses essence-related issues relevant in the context of content management. It concentrates on the basic principles and concepts, and the most important standards. This includes also a discussion on essence processing where the principles of the major techniques are explained.

3.1 THE DIFFERENT FORMS OF ESSENCE

Essence in the broadcast context is defined as the raw program material, i.e. the actual program items. A more generic definition of essence describes it as the part(s) of a content object that represents the actual information or message (i.e. the content in its literal meaning) encoded in various formats. In contrast, metadata is data about data, and hence describes a content object (see Chapter 4). Hence, essences are all forms of media that directly express and convey a message, original idea, intention or impression.

Depending on the media, essence can have different forms. There are audiovisual, visual, audio, and textual essence elements. Further, essence can be combined to form structured formats or documents. Within a CMS the management and administration aspects (i.e. format-related issues) are of main concern. The interpretation of the content is left to the users who deal with the essence.

A CMS manages all kinds of essences related to a content object. Different formats are suitable for different purposes. High-resolution (also called high-res) essence formats in general require higher bandwidth and more storage capacity. Thus, most CMS use low-resolution (also called low-res) proxies for browse and selection purposes. Not all formats are suitable in this context. It is important that they provide an accurate representation of the original content.

3.1.1 BASIC ESSENCE ELEMENTS

Basic essence elements are the fundamental components. Content items are represented by a specific basic essence (or media) type such as audio, video, images, graphic, text, etc. Different media types (for instance audio and moving images) might be combined. However, this combination establishes a fixed relationship. It is not expressed by relative links that would indicate options, but instead is a one-to-one correlation.

Essence elements can be classified according to their timing and representation characteristics. This classification is analogous to the definitions used for multimedia systems (Steinmetz and Nahrstedt, 1995). It is derived from the way in which media is perceived by humans. In general it is distinguished between *discrete* (i.e. time-independent) and *continuous* (i.e. time-dependent) media or essence types. A continuous media stream is characterized by consecutive, time-dependent information units. The time dependency in this context refers to the presentation of the media (or distinct parts of it) to the human user. In this case time attributes semantics to the media. In order to provide information intelligible to the users the presentation of continuous media has to take place within certain time limits (Mauthe *et al.*, 1992). For example, in PAL-encoded video there have to be 25 video frames per second where each frame has to be presented in the right order within $\frac{1}{25}$ second of the previous frame otherwise the information represented by a frame is not valid and the system can be considered as failing. However, unlike in real-time systems where failing a deadline can cause major damage (for instance in anti-lock braking systems), within continuous media processing a certain failure rate can be tolerated and might even pass unnoticed. Thus, a failure rate can be defined up to which the essence and its processing results are still considered acceptable and within tolerable limits. The time-dependency of continuous media essence is represented by time-lines and the time-codes they are associated with. The basic continuous media essence elements are video, audio, and animated graphics.

In contrast, discrete media has no inherent timing requirements and the represented content is valid regardless of its presentation time. Examples of discrete media are text, images, and graphics.

3.1.2 STRUCTURED ESSENCE FORMATS

Apart from basic essence elements a CMS has to manage structured essence formats such as Web pages, XML documents, and multimedia files. In structured essence formats different basic essence elements or other structured essence elements are combined using references and links. A relationship is established between the different components of the structured essence formats that governs or determines the information presentation. For instance the links within Web pages point to other essences that are relevant within the context of a specific page. These links then also represent the relationship between the basic components.

Relationships within structured essence formats are not only established by explicit references but can also be expressed by timing constraints associated with the different basic elements. For instance, absolute or relative timecodes can be assigned to continuous but also discrete basic essence elements. In the latter case this determines at which point the information represented by a discrete media module within structured essence formats should be presented to the user. This places timing restrictions onto the discrete media module that are similar to those of continuous media elements. However, these restrictions are only valid within the context of a specific content object represented by the structured essence format and are not inherent as is the case for continuous media.

The challenge in the management of structured essence formats is to keep the information they represent valid and consistent. Since the different basic modules related to content objects represented by structured essence formats may be stored independently and referenced from a number of objects they can be relocated and changed. This can happen without changing the references and links in the structured essence object. Thus, inconsistencies can occur. Within CMS dealing with structured essence formats this has to be taken into account. In order to guarantee content object integrity a CMS has to explicitly manage external links of structured essence objects and prevent operations that might lead to inconsistent content objects.

3.1.3 HIGH-RESOLUTION AND BROWSE FORMATS

In order to evaluate the suitability of essence formats for production, presentation, and storage it is crucial to consider application use cases alongside the technical properties of essence formats. Application use cases provide additional selection criteria when choosing certain essence formats for specific purposes and specific systems. The technical properties place restrictions on the use and handling of the essence. This again can have repercussions for the possible workflows. For example, high-resolution, high-quality video usually implies high bit rates. Since with current state of the art communication and storage technologies the direct access of these formats by all users is not feasible, alternative formats (i.e. browse formats) have to be used that provide an accurate representation of the original essence material. The choice of the high-resolution format(s) is determined by quality, production, broadcasting, and archiving considerations. In contrast, the requirements for the low-resolution format come from the workflow that has to be supported and the CMS infrastructure.

In order to retrieve and select content, fast and unambiguous visual preselection and inspection of the available material is important. In a typical workflow dealing with audiovisual material a first selection of video content is based on the display of keyframes. Subsequently, the detailed choice and selection of specific content is made by browsing (viewing and listening) the selected audiovisual segments using low-resolution copies. In order to use browse copies in a professional context they have to support certain functionalities such as standard trick modes (i.e. fast-forward, reverse speed, frame-by-frame shuttle, etc.). Further, since instant access to browse material is desirable the browsing signals should be highly compressed to save network bandwidth and storage. However, heterogeneity is even encountered at this level. For example for in-house users it might be possible to access 1.5 Mb/s browse video whereas a user connected via a dial-up modem might only have a data rate of 64 Kb/s. Therefore scalable bit rate encoding that provides adaptable quality levels according to system capabilities is desirable.

Within CMS it is common practice to use low-resolution browse copies alongside the high-quality material. Two different approaches can be identified with regard to the specification, selection, and use of appropriate browsing formats within a CMS:

- The definition and use of a generic standard format or set of formats with interoperability on coding level. This guarantees a well-specified quality and functionality but limits the options to follow technological progress.
- The development of application components that can seamlessly process and display multiple formats. This may limit the quality or functionality in certain cases but makes the best use of technological progress.

Both approaches are valid and have to be taken into account at the development and implementation of a CMS. In general a CMS should be able to support multiple formats to take advantage of new formats while maintaining interoperability. EBU and SMPTE have addressed this issue by sending out a *Request for Technology* for a Browsing Interchange Format (BIF). The goal of this RFT is to state the requirements that a browsing format has to fulfill in order to be considered suitable as a low-resolution browse format.

3.1.3.1 Basic BIF Requirements

The BIF format is intended to be used by media professionals for browsing and selection. It therefore has to provide the following basic feature set:

- video (frame- and timecode accurate, color);
- two-channel audio;
- associated content-related metadata (including information on access rights and IPR management and protection);
- material-related metadata (timecode, unique ID, etc.).

The BIF should support both streaming and file transfer. Further, it should provide the technical functionalities to facilitate at least two general browsing modes, namely *edit decision list (EDL) browsing* and *very low-bandwidth browsing*.

An EDL browser is a tool that allows somebody to view content and choose appropriate sections for further production. This format has to allow the selection of material and the creation of an EDL. It should also allow the seamless reviewing of preselected content based on an EDL at browsing quality level. This selection can, for instance, be a draft for a new program item and not a random collection. To support the creation of an EDL the incorporation of time stamps (or timecodes) in the audiovisual stream and therefore within the BIF is required. The representation of each single frame is therefore mandatory to enable the creation of an accurate EDL. Frame synchronous audio is a further requirement because edit decisions are often based on audio events. Generally for preview purposes a video signal with one quarter of SDTV resolution within a suitably sized window on a workstation terminal is considered sufficient. A higher error rate than the one defined for SDTV can be tolerated.

The low-bandwidth browse format should allow content browsing from a remote location via relatively low bandwidth wide area network (WAN) connections (such as the Internet) or dial-up connections. In this case the data rate should be scalable. A data rate of approximately 64 Kb/s could be sufficient for this application area. To achieve this, compromises with respect to resolution and compression artifacts have to be accepted. A representation of each frame is preferred but is not mandatory. Similarly, frame synchronous audio is preferred but not mandatory. For this type of video browsing, the dataflow does not need to be continuous; relatively high error rates can be tolerated.

3.2 ENCODING AND COMPRESSION BASICS

In order to appreciate the possibilities but also requirements that different essence formats place on a CMS it is important to understand their characteristics and how they work in principle. There are certain fundamental encoding and compression concepts and techniques that all essence types have in common. Most encoding and compressions schemes (even proprietary formats) use combinations of these basic procedures. Hence these basic principles and techniques should be understood before discussing any specific encoding and compression formats.

3.2.1 ENCODING: FROM THE ANALOG TO THE DIGITAL DOMAIN

The recording of essence and its handling in computer-based CMS usually includes a transformation from the analog into the digital domain. This process is called encoding. Whereas in the analog world the information is represented by continuous signals, in the digital domain it is turned into discrete numerical values represented in the binary number space. In this transformation process information is lost since each discrete representation

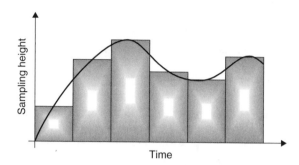

Figure 3.1 Waveform and sampled waveform

of a continuous value can only be an approximation. Figure 3.1 illustrates how an analog value (represented by a wave form) is represented by a number of numerical values. The information loss that occurs in this process is unavoidable and has to be accepted.

The quality of digital formats depends on the granularity or duration of the interval that one discrete value represents, and of the number of bits used to represent an individual value. The former is called *sampling*; the latter process refers to the so-called *quantization*. The *sampling rate* is the rate at which an analog value (e.g. a continuous wave form) is sampled. The higher the sampling rate and the more bits are used in the quantization to represent a discrete value, the better is the approximation of the original value and hence the quality of the digitized material. International standards such as ITU-R BT 601-5 (International Telecommunication Union, 1995) for video specify the transformation process from the analog into the digital space.

3.2.2 COMPRESSION: REDUCING THE BIT RATE

Digitization results in relatively high data rates, i.e. a high amount of data. Applying ITU recommendation ITU-R BT 601-5 to standard 4:3 television video signals, the resulting bit rate is 270 Mbit/s, which leads to 121.5 GBytes of data for one hour of video. In order to reduce the bit rate, and hence the bandwidth and storage requirements, compression is applied. Compression reduces the number of bits by exploiting the redundancies in the bit stream, and certain properties in the media or the receptors (i.e. the human senses) that makes it hard to notice if part of the information is missing. *Lossless* compression and *lossy* compression can be distinguished. With lossless compression the decompressed data stream is identical with the original data stream. In contrast, with lossy compressions the decompressed stream is not identical with the original stream since certain information is either removed or approximated.

Compression techniques can be classified into *entropy coding, source coding*, and *hybrid coding* schemes (Effelsberg and Steinmetz, 1998). Entropy coding is a lossless compression technique and does not make any use of media- or stream-specific characteristics. In entropy encoding, the data is considered as a sequence of bits only. As an example for entropy encoding schemes, Figure 3.2 shows how run-length encoding works. Run-length encoding replaces sequences of identical encoding patterns by replacing the sequences by a mark-up, followed by the pattern (once), and the number of its occurrences. Since a run-length encoded pattern itself has a minimum length, replacing the pattern is only done

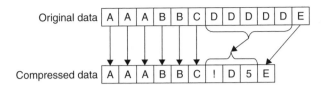

Figure 3.2   Example of run-length encoding

for repeated pattern sequences that exceed this length. Figure 3.2 proposes an exclamation mark ('!') as the mark-up for an encoded pattern sequence. Obviously, in this case only patterns that are longer than three letters benefit from run-length encoding. Other entropy coding techniques are vector quantization, pattern substitution, Huffman coding, and arithmetic coding (Effelsberg and Steinmetz, 1998).

Source coding (in contrast to entropy coding) takes advantage of certain properties of the human senses (i.e. eyes and ears). Source encoding also accepts a decrease in quality when compared to the original signal. This compression is therefore lossy, but can achieve much higher compression ratios. Examples of source coding are, for instance, removing parts from an audio signal that the human ear cannot hear within the overall context of a sound, or subsampling color signals compared to the luminance signal (which exploits the fact that the human eye is much more sensitive to differences in brightness than in color). Further examples are interpolation and transform encoding (where the data is transformed into another mathematical domain more suitable for compression).

Most compression standards use hybrid coding techniques, i.e. a combination of entropy and source coding. Examples are JPEG, H.261, and MPEG-1, MPEG-2, and MPEG-4. The ways the different compression techniques are used and combined depend on the standards and the intended effect.

For continuous media the compression schemes sometimes also exploit spatial redundancies and temporal relationships occurring in adjacent information units within the continuous data stream. In this case not an entire object is encoded but references are made to objects representing the same information unit that come in frames before or after (i.e. only the differences and changes occurring on this object between the consecutive information units are encoded). For video, for example, this means that for some frames not the full frame, but only the difference between this frame and one or two other frames is encoded. Hence, in order to fully decode such a frame all the frames that have been involved in computing the frame difference have to be available and already decoded. This technique is used in the MPEG standards (see Section 3.3.2).

3.3 VIDEO

Video is an essential part of a CMS. Different (digital) formats have to be handled, stored, and administered in the system. The storage and communication requirements placed on the underlying system architecture and infrastructure are very high and demanding. Video also has to be exchanged frequently with other, third-party systems. Thus, interoperability plays a key role in this context. Further, low-resolution video is used as browse material and represents the content available in the system. These formats have to be chosen carefully to provide the best possible quality within given technical and financial limitations. Therefore it is crucial to understand the basic principles of video encoding and compression and the major video formats.

In this section the principles on which the major video formats are based are introduced. The most relevant formats within video production are either based on MPEG or DV encoding. A number of video formats deployed in a CMS are based on these two standards.

3.3.1 VIDEO ENCODING: BASICS AND PRINCIPLES

Video images are represented by a number of *pixels* (picture elements), which are the smallest unit within a picture. The *aspect ratio* in this context gives the ratio of picture width to height and conventionally is 4:3. Therefore the vertical resolution equals the number of pixel separately presented in the picture height, and the number of pixels in the picture width is equal to the horizontal resolution multiplied by the aspect ratio (Steinmetz, 2000). In the case of NTSC there are for instance 525 lines and hence 700 vertical rows.

Color vision is determined by a combination of three signals representing the relative intensity of red, green, and blue (RGB) light. These signals are usually represented separately and in their combination define the color impression. However, during the transmission of the signal a different representation in the form of one luminance and two chrominance signals (YUV) is used.[1] YUV encoding does not separate colors, but the brightness (i.e. the luminance factor Y) is separated from the color information (U and V). This is due to the human perception of colors, which is more sensitive to alterations in the brightness than any chrominance information. Therefore the luminance information is more important and can be coded using higher bandwidth. This difference in component bandwidths of the coding is often expressed in the ratio between the luminance and the two chrominance components (Steinmetz and Nahrstedt, 1995).

The fact that the human eye cannot distinguish single frames if they are presented faster than 15 frames a second is exploited to convey the impression of moving images. The European PAL standard uses a replication rate of 25 Hz (i.e. 25 frames/second) whereas the American NTSC standard uses 29.97 Hz (29.97 frames/second). In order to achieve a flicker-free perception of moving images a refresh rate of 50 Hz is necessary. Lower refresh frequencies can be used if measures are taken to counter the flicker effect. For instance a television picture can be divided into two half pictures, which consist of interleaved scanning lines. Half pictures are then transmitted after one another using the interleaving method. The half pictures are scanned at the double rate, i.e. $2 \times 25\,\text{Hz} = 50\,\text{Hz}$ or $2 \times 29.97\,\text{Hz} = 59.94\,\text{Hz}$.

In order to process video by computers or transmit it over computer networks the images have to be transferred from the analog into the digital domain (see Section 3.2.1). The basic steps are sampling, quantization, and coding. For video the gray (or color) levels are sampled into an $M \times N$ array of points in this process. Subsequently the continuous values are mapped onto discrete values representing a quantization interval (e.g. divided into 256 intervals).

The video coding deals with the encoding of the different color components of an image. For the coding there are two basic alternatives, namely sampling and coding

---

[1] Note, in NTSC the YIQ representation is used, which is similar to the YUV signal.

of the entire analog video signal (i.e. *composite coding*) or the separate coding of the luminance and chrominance components (i.e. *component coding*). In the former case all signal components are jointly transformed into the digital domain whereas in the latter case the luminance and chrominance signals are digitized separately. In order to transfer them jointly they can be multiplexed for transmission.

Since the luminance and chrominance components can be sampled independently a higher sampling rate (e.g. 13.5 MHz) can be used for the more important luminance signal whereas the two chrominance signals can be sampled with half the luminance sampling rate (i.e. 6.75 MHz) only. This kind of sampling, where the luminance sampling rate is twice that of the two chrominance values, is known as 4:2:2 sampling.

In the case where an 8-bit quantization is used with 864 sampling values per line for the luminance and 432 for each chrominance component, the cumulated rate without blanking interval is 216 Mb/s. In order to reduce the initial bit rate of the digitized video some formats use different sampling frequencies and rates per line. This of course already has an impact on the quality of the digitized video even before any compression has been applied. Table 3.1 gives an overview of some of the major formats and the sampling rates and frequencies they use.

3.3.2 MPEG-BASED FORMATS

The Moving Picture Experts Group (MPEG) within the ISO/IEC JTC1/SC29/WG11 has been developing standards related to video encoding since 1988. The relevant video standards comprise MPEG-1 (MPEG, 1996) for the coding of video of up to 1.5 Mb/s, MPEG-2 (MPEG, 2000) and MPEG-4 (MPEG, 2002a). The different standards have been

Table 3.1   Component coding standards

Format	Signal	Sampling frequency (MHz)	Sampling/line	No. lines	Data rate (Mb/s)	Cumulated rate (Mb/s)
4:4:4 ITU 601	R	13.5	864	625	108	324
	G	13.5	864	625	108	
	B	13.5	864	625	108	
4:2:2 ITU 601	Y	13.5	864	625	108	216
	Cr	6.75	432	625	54	
	Cb	6.75	432	625	54	
4:2:2	Y	13.5	720	576	83	166
	Cr	6.75	360	576	41.5	
	Cb	6.75	360	576	41.5	
4:2:0 [a]	Y	13.5	720	576	83	124.5
	Cr	6.75	360	576	41.5	
	Cb					
4:2:0 SIF	Y	6.75	360	288	20.7	31.1
	Cr	3.375	180	288	10.4	
	Cb					

[a] In 4:2:0 formats, the values for Cr and Cb alternate in subsequent frames (progressive scan) or fields (interlaced).

developed at different points in time and also address different application areas. However, they share many of the basic principles and belong to one family of formats.

3.3.2.1 MPEG-1

The original motivation for the MPEG-1 standard was to define a format suitable for digital storage technologies (such as CD). The standard comprises three major parts, namely MPEG video, MPEG audio, (see Section 3.4.2), and MPEG systems (defining how to multiplex and synchronize audio and video). The average bandwidth defined for an MPEG-1 encoded audiovisual data stream is 1.1 Mb/s for the encoded video part and 128 Kb/s for audio (stereo audio is supported). Since the original target application area of MPEG-1 was stored media, the chosen compression scheme is more suitable for asymmetric compression processes. In this case the compression is more processing intensive than decompression.

In contrast to most other compressions schemes the MPEG-1 standard does not specify the encoder but defines the syntax and semantics of the MPEG-1 video and audio bit stream. Thus, it specifies how an MPEG-1 stream has to look. MPEG-1 compliant decoders have to be able to decode such a bit stream.

Sampling and Quantization in MPEG-1

The starting point of the compression in MPEG-1 is a YUV image with the so-called macroblock as the basic component. A macroblock is divided into an array of 16×16 luminance samples (decomposed into four 8×8 blocks) and two 8×8 arrays for the Cb and Cr chrominance samples. These 8×8 blocks are then transferred from the two-dimensional image domain into the frequency domain using discrete cosine transformation (DCT) (Effelsberg and Steinmetz, 1998). The result of this process is a set of 64 DCT coefficients for each block of the image. The coefficient in the upper-left-hand corner of this set represents the gray and color values and is called the DC coefficient; the other values are called AC coefficients.

Subsequently quantization is applied to the DCT coefficients. In this process real numbers are mapped onto integer values. Different quantization steps and values reflecting the relevance of DC and AC coefficients are used. There is one quantizer step size for the DC coefficients of 8, and 31 even value quantizer step sizes that are multiples of two, with values ranging from 2 to 62 for AC. The quantization values range from -255 to 255. One common quantization table for the luminance and chrominance components is used. More information on MPEG-1 quantization can be found in Hung (1993).

After quantization MPEG-1 applies entropy encoding to achieve an initial reduction of the bit rate. MPEG-1 specifies run-length level variable coding.

MPEG-1 Frame Types

In order to achieve a high compression ratio MPEG-1 not only uses information within a single frame for compression but also references encoded information from previous and subsequent frames. This method is called inter-frame encoding. Two conflicting requirements have to be considered in this context. On the one hand, the highest possible compression should be achieved. This is the case when large blocks of information are encoded only once and referenced in other frames. On the other hand, it should be possible to randomly access a video stream. A stream can only be accessed randomly at frames that do not reference information from other streams.

In MPEG-1 four different frame types have been defined, which can be distinguished according to the degree upon which they rely on information encoded in other frames. The combination of these frame types within an MPEG-1 stream influences the compression ratio and the intervals in which an MPEG-1 stream can be accessed randomly. The different MPEG-1 frame types are I, P, B, and D frames:

- *I frames* (intracoded pictures) are pictures encoded without reference to any other frames in the video sequence. I frame compression is similar to the JPEG (Joint Photographics Expert Group) compression scheme (Wallace, 1991).
- *P frames* (predictive coded pictures) require information from previous I or P frames. In order to decode a P frame the preceding I or P frames have to be decoded first. The reference areas within a frame are based on macroblocks. Motion-compensated prediction is used to 'predict' areas (i.e. macroblocks) in the current frame from previous reference frames. Motion vectors indicate the direction of 'movement' of certain macroblocks from their original position in the reference frame to their current position in the P frame.
- *B frames* (bidirectional predictive coded pictures) require the information from preceding *and* subsequent I and/or P frames. A B frame represents the difference image to its reference frames. It can never be used as a reference for other images. Motion-compensated interpolation is used to find matching macroblocks in one frame in the past and one frame in the future (either I or P frames).
- *D frames* (DC coded pictures) are special images for fast-forward and fast-backward. A D frame is intracoded (like I frames) but only DC parameters are encoded. D frames are not used as reference frames.

The MEPG-1 inter-frame coding exploits the fact that large areas in a picture do not change from one frame to another. Hence information already encoded in macroblocks of previous and subsequent images only has to be referenced and does not need to be encoded again. The highest compression can be achieved in B frames. P frames also have a much higher compression rate than the intra-frame encoded I frames. Figure 3.3 depicts the relationship between the different frames.

The number and occurrence of the different frame types within a video stream is determined by the group of pictures (GoP). A GoP defines the number and order of P and B frames between two I frames. In order to achieve a high compression rate the majority of pictures should be B frames. However, random access at every single frame requires an

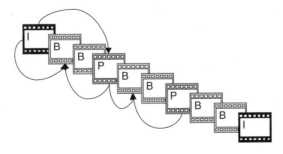

Figure 3.3   MPEG-1 I, P, and B frame group of pictures

I-Frame-only stream. Thus there is a trade-off between compression ratio and the ability to randomly access a video stream. The number of I frames also has implications for the error resilience of a video stream. If an I frame is corrupted or missing in a video stream all subsequent P and B frames cannot be decoded properly. In practice a group of pictures with a sequence of IBBPBBPBBI... has proven suitable since it gives random access at every 9th frame (i.e. every 330 milliseconds) (Steinmetz, 2000).

Two different Standard Interchange Formats (SIF) (namely for PAL and NTSC) have been defined. The constrain parameter set (CPS) specified in this context is a minimum set of parameters that have to be supported in order to be MPEG-1 standard compliant.

3.3.2.2 MPEG-2

MPEG-1 was defined for a specific purpose, namely as a format suitable for digital storage technologies. Since it specifies a maximal data rate of 1.5 Mb/s a significant improvement of the given quality with existing techniques was not possible. Thus, MPEG-1 is unsuitable for applications that have high quality requirements. In order to satisfy the demands of high quality video in a media production and television context MPEG-2 has been defined (MPEG, 2000; Mitchell *et al.*, 1996). The standardization of MPEG-2 was a joint activity between ISO/IEC, ITU-TS, ITU-RS, EBU, and SMPTE.

The aim of the MPEG-2 standard is to provide efficient encoding of audiovisual information at a wide range of resolutions and bit rates. It specifies a data rate of up to 100 Mb/s considering the requirements for higher image resolutions according to ITU-R 601. The MPEG-2 standard even takes HDTV into account. Further, it provides features that make it suitable for interactive multimedia services such as random access for interactive television, trick modes (e.g. fast forward and backward, slow motion, etc.), multi-track audio, etc. Another requirement considered during the MPEG-2 standardization process was the transmission of video over lossy channels.

MPEG-2 shares the basic encoding principles with the MPEG-1 standard. The source for the compression is a digital video stream (i.e. a previously digitized stream). As in the case of MPEG-1, MPEG-2 also only specifies the video bit stream syntax and decoding semantics and not the encoding process. There are four different picture types (I, P, B, and D frames) equivalent to those defined by MPEG-1. During compression the images are decomposed into macroblocks of a block size of 8×8. The blocks are transformed using DCT transformation, quantized, and 'zig-zag' scanned, followed by motion estimation and compensation. The entropy encoding step uses variable run-length encoding. Hence all the basic techniques developed in the MPEG-1 standard are also used within MPEG-2.

In contrast to MPEG-1, however, MPEG-2 also supports the compression of interlaced video. This is achieved by MPEG-2's two different picture modes, namely *field pictures* (where fields are encoded independently) and *frame pictures* (where each interlaced field pair is interleaved into a frame, divided into macroblocks, and encoded). In this context MPEG-2 also offers an alternate scan mode for the case where the zig-zag scan would produce sub-optimal results. These cases occur when adjacent scan lines come from different fields, which can result in a reduced correlation in scenes with motion.

MEPG-2 Layered Coding

In order to provide better scalability MPEG-2 allows the encoding of video information in different *layers*. The different layers build on top of each other. Each layer improves the quality provided by the preceding layers starting with the so-called base layer. The layered coding modes defined by MPEG-2 are:

- *Spatial scalability mode*: this mode allows video to be encoded with a range of different horizontal and vertical resolutions. All layers together build up to the full resolution. Supported resolutions are, for instance, 352×288 pixels, 360×240 pixels, 704×576 pixels (equivalent to ITU-601), and the HDTV resolution with 1.250 lines at 16×9 at the luminance level. The chrominance components in this context are sampled at half the luminance values (i.e. 4:2:2 sampling). This feature allows transmitting video of different quality simultaneously, e.g. standard TV and HDTV.
- *Temporal scalability mode*: in this mode the base layer contains sequences at a lower frame rate. The enhancement layers then scale the video up to the full frame rate. In this context the division of the different frame types between the layers is important, i.e. the distribution of I, P, and B frames within a video sequence is not only relevant for a GoP but also for the different layers.
- *Data partitioning mode*: in this mode data is encoded according to the priority of the data. The high-priority streams include, for instance, low-frequency DC coefficients and motion vector headers. This allows the delivery of different qualities to different users. Further, it supports progressive image build-up.
- *Signal-to-noise ratio scalability (SNR) mode*: this mode allows encoding of pictures in a basic quality version. The enhancement layers carry the information required for the full quality image.

MPEG-2 Profiles and Levels

In order to support professional applications alongside consumer and other formats the standard defines a number of extensible *profiles*. These profiles support different chrominance sampling modes, i.e. (4:2:0), (4:2:2), and (4:4:4). A profile specifies a set of coding features that supports a (sub)set of the full MPEG-2 syntax. Successive profiles comprise the preceding profiles. Apart from profiles the standard also defines *levels* that specify a (sub)set of spatial and temporal resolutions supporting a large set of image formats. Table 3.2 gives an overview of the different profiles and levels defined by the standard.

Only a subset of the possible profile and level combinations is defined. They are intended to support particular classes of applications. For instance the MPEG-2 main profile was defined to support video transmission in the range of 2–80 Mb/s. Within television production often a special profile of MPEG-2 called MPEG-2 4:2:2 profile is used. For news gathering and production, a GoP of IB and 18 Mb/s is deemed appropriate. For feature production an I-frame-only GoP at 50 Mb/s is considered suitable.

MPEG-2 Delivery

Apart from the definition of video and audio encoding the MPEG-2 standard also specifies (in the MPEG-2 Systems specification) how the different components can be combined. Two data stream formats are defined, namely the MPEG-2 *transport stream* (designed to carry multiple programs simultaneously) and the MPEG-2 *program stream* (designed to support system processing in software). The program stream is optimized for multimedia applications and also considers MPEG-1 compatibility. The transport stream especially considers applications where data loss may be likely. The transport stream was specified having a range of applications from video telephony to digital television in mind. Supported networks range from fiber over satellite, cable, ISDN to ATM networks.

Table 3.2 MPEG-2 profiles and levels

	Simple profile (No B frames, 4:2:0, not scalable)	Main profile (B frames, 4:2:0, not scalable)	SNR scalable profile (B frames, 4:2:0, SNR scalable)	Spatial scalable profile (B frames, 4:2:0, SNR scalable)	High profile (B frames, 4:2:0 or 4:2:2, spatial or SNR scalable)
High level $1920 \times 1152 \times 60$[a]	Not defined	$\leq$80 Mb/s	Not defined	Not defined	$\leq$100 Mb/s
High-1440 level $1440 \times 1152 \times 60$	Not defined	$\leq$60 Mb/s	Not defined	$\leq$60 Mb/s	$\leq$80 Mb/s
Main level $720 \times 572 \times 30$	$\leq$15 Mb/s	$\leq$15 Mb/s	$\leq$15 Mb/s	Not defined	$\leq$20 Mb/s
Low level $352 \times 288 \times 30$	Not defined	$\leq$4 Mb/s	$\leq$4 Mb/s	Not defined	Not defined

[a] Upper bound of sampling density in pixels/line $\times$ line/frame $\times$ frames/s.

A transport stream is made up of a sequence of fixed-sized packets that are identified by a *packet identifier* in the header of the packet. A number of program streams can be multiplexed into an MPEG-2 transport stream (e.g. video and audio streams). Packet identifiers distinguish packets belonging to a certain stream. In contrast to time division multiplexing this method allows the insertion of packets of any sub-stream into the transport stream at any time.

Similar to file formats (see Chapter 5), a transport stream might correspond to a single multimedia stream with a number of component streams for video and audio tracks (representing for instance a TV program). In MPEG-2 this is called a single program transport stream. A multiple program transport stream can combine single transport streams and also contains other data such as control information, program-specific information, and metadata. Figure 3.4 shows the structure of an MPEG-2 multiple program transport stream.

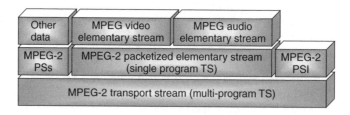

Figure 3.4 Structure of MPEG-2 multiple program transport stream

3.3.2.3 MPEG-4

MPEG-4, officially called Coding of Audio-Visual Objects (MPEG, 2002a; Pereira and Ebrahimi, 2002), has a far broader scope than either MPEG-1 or MPEG-2 (which are mainly concerned with encoding and compression issues). Initially the main objective of MPEG-4 was to specify a standard that can achieve a significantly better compression ratio compared to those achieved with conventional encoding techniques. This was later changed to encompass the needs of emerging multimedia applications and devices, ranging from mobile telephones over interactive multimedia applications to media production and broadcasting. The predicted convergence of media and technology areas (i.e. communications, computing, and the television and entertainment sector) was one of the main motivations for the standard. Functionality areas for the standard important for the envisaged audiovisual applications were identified considering:

1. Support for *content-based interactivity*, including multimedia access tools, content-based manipulation and bit stream editing, hybrid, natural, and synthetic data coding, and improved temporal random access.
2. *Optimized compression* through improved coding efficiency and the coding of multiple concurrent data streams.
3. Support *for universal access* in different environments (ranging from high-speed and dedicated networks to low-bandwidth, error-prone wireless communication) but also considering content objects and the resulting content-based scalability.

MPEG-4 specifies a tool set that is based on the above functionality and a set of identified requirements. This set comprises:

- requirements for systems;
- requirements for natural video objects;
- requirements for synthetic video objects;
- requirements for natural audio objects;
- requirements for synthetic audio objects;
- requirements for a delivery multimedia integration format (DMIF);
- requirements for MPEG-J (the definition of a JAVA application engine);
- requirements for multi-user environment;
- requirements for an animation framework;
- requirements for intellectual property management and protection (IPMP).

This list shows the actual extent and scope of the standard, where the encoding of audiovisual objects is just one aspect. However, efficient compression has been a central requirement for the standard. Standard Parts II and III deal with video and audio coding and related aspects. Further, Part X of the standard is concerned with advanced video coding issues. Table 3.3 gives an overview of all the major parts of the MPEG-4 standard.

MPEG-4 supports speech encoding from 2 Kb/s to 24 Kb/s and video encoding with bit rates from 5 Kb/s to 1 Gb/s. The video formats can be progressive or interlaced and the resolution can vary from QCIF to 4K × 4K studio resolution. MPEG-4 is a very comprehensive standard that supports content-based coding functionalities such as the separate decoding of video objects/shape encoding, face and body animation, and 2D and 3D mesh coding. Examples of synthetic objects include text and graphic overlays, animated faces, and objects. The standard envisages that different natural and synthetic objects can be integrated within one scene.

Table 3.3   MPEG-4 standard parts

Part	Description
I Systems	Specifies scene description, multiplexing, synchronization, buffer management, and management and protection of intellectual property rights
II Visual	Specifies the coded representation of natural and synthetic video objects
III Audio	Specifies the coded representation of natural and synthetic audio objects
IV Conformance Testing	Specifies conformance conditions for bit streams and devices for the test of MPEG-4 implementations
V Reference Software	Includes software and sample implementations of most Parts of MPEG-4 that can be used for implementing compliant products
VI Delivery Multimedia Integration Framework (DMIF)	Specifies a session protocol for the management of multimedia streaming over generic delivery technologies
VII Optimized Visual Reference Software	Includes software and sample implementations for visual tools such as fast motion estimation, fast global motion estimation, and fast and robust sprite generation [a]
VIII Carriage of MPEG-4 Contents over IP Networks	Specifies the mapping of MPEG-4 content into several IP-based protocols
IX Reference Hardware Description	Intended for portable synthesizable or simulatable very high-speed integrated circuit hardware description language (VHDL) descriptions of MPEG-4 tools
X Advanced Video Coding	Intended for the specification of video syntax and coding tools within the Joint Video Team activity based on the H.264 video coding specification

[a] Note, Part VII is a Technical Report only and not a standard specification.

MPEG-4 Object-Oriented Coding

In the object-oriented video coding approach audiovisual scenes are composed of basic units of visual, sound, and audiovisual content objects, called AVO (audiovisual objects). These objects can be of arbitrary shape and have a spatial and temporal extent. The coded data representing AVOs are carried in separate elementary streams. The AVOs are described by object descriptors (OD) and include information about their (fixed) spatio-temporal location and scale. The AVO's local coordinate system serves as a handle for manipulating the AVO with reference to this information.

A scene is a composition of various AVOs. The information about how to compose such a scene is kept in the *scene descriptor information*. It defines the spatial and temporal position and relationship between AVOs, their dynamic behavior, and the kind of interactivity possible in the context of such a scene. The scene descriptor contains pointers to the object descriptors of the different AVOs that are part of the scene. These pointers represent the link between the scene and the basic objects. As a formal scene description language MPEG-4 defines the *Binary Format for Scenes* (BIFS). BIFS provides a comprehensive set of functionality for 2D and 3D compositions, and text

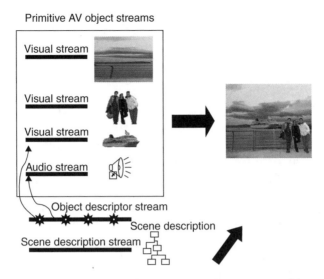

Figure 3.5   Example of an object-based scene composition

and graphic primitives. Further, interactivity based on events triggered by users is also supported by BIFS. Figure 3.5 shows an example of an object-oriented MPEG-4 scene composition.

As in MPEG-1 and MPEG-2, MPEG-4 does not specify the encoder but only the bit stream it produces and the decoding procedure. The video coding scheme is based on motion-compensated hybrid DCT encoding. The motion decoding and compensation tool, and the texture decoding tool result in a standard block-based decoder comparable to MPEG-1 and MEPG-2. Thus, the standard is compatible with MPEG-1 and MPEG-2. However, compared to these the video coding algorithms have been improved. For instance, increased compression efficiency and error resilience are two of the major improvements in MPEG-4 (Pereira and Ebrahimi, 2002).

By adding the shape decoding tool, the standard is turned into an object-based video coding solution, i.e. the coded content does not have to be rectangular but can be of arbitrary shape. An object-coded MPEG-4 bit stream can be described in a hierarchy. The top level of this hierarchy is the visual sequence composed of the visual objects. A visual object is defined as a sequence of video object planes corresponding to an image sequence defined by luminance and chrominance values and its shape.

MPEG-4 Non-Video Parts

In order to better support emerging services within telecommunication, computer animation, and the multimedia industry, the MPEG-4 standard considers requirements such as support for alpha-blending and the encoding of multiple levels of details of a scene. This is specifically considered in MPEG-4 visual texture coding (VTC). It provides the efficient compression of textures (considering a wide range of qualities), and supports random access to (some or all) objects of a scene. Therefore it specifies various coding modes with different levels of complexity. Wavelet transformation and zero-tree coding are used to achieve a more efficient compression (Davis and Chawla, 1999). Wavelet transformation replaces DCT. It has very good energy compaction properties and exploits structures within an image for better compression.

MPEG-4 Profiles and Levels

MPEG-4 also specifies profiles and levels to reduce the complexity of supporting all possible tools for all possible application scenarios within one decoder. For each profile the set of tools that have to be supported is defined. The level sets complexity bounds, e.g. for required memory, number of objects, bit rate, etc. Therefore the supported profile and level have to be specified for each MPEG-4 compliant decoder. This is also necessary to ensure interoperability between systems. A profile and level combination sets a well-defined conformance point.

MPEG-4 distinguishes between bit stream and decoder conformance. A stream is considered bit stream compliant when it only contains the syntactic elements specified for a certain profile and when the parameters stay within the boundaries set by a specific profile/level combination. A decoder is compliant with a particular profile@level when it is able to interpret all allowed values of all allowed syntactic elements (static compliance) and has all necessary resources for the required decoding operations according to the decoding semantic for the syntax supported by this profile@level at the required place (dynamic compliance). In the context of object-oriented coding the profile@level does not define the maximum complexity per individual object, it rather specifies the bounds on the complexity of all objects in a scene. It is important to note that the number of objects in a scene and their complexity have to be taken into account when determining the complexity required in a decoder for a specific video.

MPEG-4 Object Types

In addition to profiles and levels, MPEG-4 also specifies object types. They define the kind of tools required to code an object in the scene and specify restrictions on how object types can be combined. Additionally, object types specify the syntax and semantics of an object. Table 3.4 gives an overview of the object types defined for rectangular video and arbitrarily shaped video.

MEPG-4 Video Profile Level

The profiles determine which object types are allowed in a scene, i.e. they basically specify a list of allowed objects. Table 3.5 gives a list of video profiles for rectangular and arbitrarily shaped video. It also states the number of levels specified for a certain profile.

The different profiles specify different bit rates reflecting the quality and bandwidth requirements of the intended use. The Simple profile, for example, specifies a bit rate of 64 Kb/s at level 0 and 384 Kb/s at level 3. The range of the Advanced Simple type is much broader, ranging from 128–768 Kb/s suitable for Internet applications to 3–8 Mb/s for more high quality applications. The Main profile allows at its highest level 38 Mb/s intended for high-quality television services. Finally, for production purposes bit rates from 180 Mb/s up to 1800 Mb/s are specified in the Simple Studio profile and up to 900 Mb/s in the Core Studio profile.

At present hardly any products provide support for MPEG-4 encoded video. Therefore, it is not possible to assess the impact MPEG-4 might have in the future in the media and content domain. However, within a CMS it should be possible to handle MPEG-4 video.

Table 3.4 MPEG-4 video object types

Coding type	Object type	Definition
Rectangular video	*Simple*	Defines an error-resilient, rectangular natural video object of arbitrary height/width ratio. Coding tools are based on intra (I) and predicted (P) VOP (i.e. frames)
	Advanced Real-Time Simple (ARTS)	Superset of simple object type for real-time coding. Defines a back channel for monitoring information such as throughput, resolution, and error rate
	Advanced Simple	Adds tools to enhance compression efficiency such as $1/4$ pel motion estimation, global motion estimation, and B frames
	Fine Granularity Scalable (FGS)	Enhances the advanced simple object type with temporal and fine-granular SNR scalability
	Simple Scalable	Extension to simple object type that provides temporal and spatial scalability
Arbitrarily shaped video	*Core*	Superset of the simple object type adding binary shape coding and B VOP. Further, it supports temporal scalability using extra P VOP
	Core Scalable	Superset of the core object type that adds rectangular temporal and spatial scalability and object-based spatial scalability
	Main	Adds coding of progressive material and gray-scale shape, sprites, and interlaced content
	Advanced Coding Efficiency (ACE)	Similar to the main object type excluding sprites. It adds coding efficiency tools such as $1/4$ pel motion compensation and global motion compensation
	N-Bit	Equivalent to the core object type extended by a variable pixel depth from 4 to 12 bits for luminance and chrominance planes
	Simple Studio	I VOP only object for high quality video. Supports arbitrary shape and multiple alpha planes. MPEG-2 like syntax
	Core Studio	Equivalent to simple studio object type enhanced by P VOP

3.3.3 DV-BASED FORMATS

Digital Video (DV) was originally conceived as a digital tape recording format for consumer products. It has later been adopted for professional applications. At present there are two major standards, namely the original Consumer DV standard based on International Electrotechnical Commission (2001, 2002) and the Professional DV standard as published by SMPTE (Society of Motion Picture and Television Engineers, 1998a, 1999) specifying the characteristics and functionality of DV.

The DV specification is not only concerned with the encoding of video but also addresses issues such as tape cassette (e.g. Society of Motion Picture and Television

Table 3.5 MPEG-4 video profile level

Coding type	Video profile	Description
Rectangular video	*Simple*	Intended for very low complexity and bit rate video. Accepts only objects from type Simple. Number of levels: 4
	Advanced Simple	Intended for Internet streaming applications. Accepts object types Simple and Advance Simple. Scalable to TV size picture and quality. Number of levels: 6
	Fine Granularity Scalability (FGS)	Adds FGS layers (i.e. FGS object types) to the Simple and Advanced Simple profile. Number of levels: 6
	Simple Scalable	Adds scalability to the simple profile. Accepts object types Simple and Simple Scalable. Number of levels: 2
	Advanced Real-Time Simple (ARTS)	Adds real-time features such as back channel and adaptive encoding to the Advanced Simple profile. Accepts object types Simple and ARTS. Number of levels: 4
Arbitrarily Shaped Video	*Core*	Intended for medium quality interactive services. Accepts object types Simple and Core. Number of levels: 2
	Core Scalable	Is a superset of the Simple and Simple Scalable profile. Enhances the Core profile with scalability. Accepts object types Simple, Simple Scalable, Core, and Core Scalable. Number of levels: 3
	Main	Intended for TV broadcast services. Supports progressive and interlaced material. Accepts video object types Simple, Core, and Main. Number of levels: 3
	Advanced Coding Efficiency (ACE)	Provides enhanced coding efficiency, e.g. based on global motion compensation $1/4$ pel motion compensation. Accepts object types Simple, Core, and ACE. Number of levels: 4
	N-Bit	Intended for surveillance and medical applications. Accepts video object types Simple, Core, and N-Bit. Number of levels: 1
	Simple Studio	Intended for content production and studio applications. Accepts object type Simple Studio. Number of levels: 4
	Core Studio	Enhances the Simple Studio profile. Allows object types Simple Studio and Core Studio. Number of levels: 4

Engineers, 1998b), error correction on magnetic tapes, etc. The focus in the context of this section is on the actual encoding functionality specified by the Professional DV standard.

3.3.3.1 DV Format: Encoding Basics

As in the case of MPEG, the DV compression is also based on the discrete cosine transformation (DCT). However, unlike MPEG it only uses intra-frame compression. The compression ratio within DV is about 5:1 depending on the motion within a frame. This is due to the inter-field compression (i.e. a higher compression ratio can be achieved when a image has many similar areas and not much motion).[2] The result is a variable bit rate. Since DV-encoded video streams require a constant bit rate an adaptive intra-frame spatial compression is used (i.e. as the amount of motion in a scene increases the spatial compression increases as well). The DV encoding is composed of a three-level hierarchical structure. First, a picture frame is divided into a rectangle (or clipped rectangle) shaped block. This block is further divided into 278×8 DCT macroblocks.

DV uses in the consumer case a 4:2:0 format for component signals with a 13.5 MHz sampling rate and 8-bit encoding. For professional applications this was adapted to 4:1:1 quality.

Considering the example of an NTSC video, a frame of compressed video data consists of 10 tracks and 138 data blocks. Each block contains 76 bytes of video data and a 1-byte header. With 29.97 frames/s this corresponds to a video rate of 25.146 Mb/s and hence is equivalent to the mentioned compression ratio of 5:1, compared to the ITU-R BT 601 non-standard 4:1:1 encoded video with 125.5 Mb/s.

The capability to correct single bit errors is provided by error-correcting rank (inner) and file (outer) parity codes accompanying the video data. For large blocks of data more sophisticated error correction schemes such as Reed–Solomon can be used.

Apart from the video sector the encoded stream also contains auxiliary video data (VAUX). This includes recording data and time, shutter speed, lens aperture, color balance, and other camera information. It is seen as an advantage that additional metadata can be kept with the video coding information.

3.3.3.2 DV Professional Compression According to SMPTE

The SMPTE Standards 306M (Society of Motion Picture and Television Engineers, 1998a) and 314M (Society of Motion Picture and Television Engineers, 1999) define the content, format, and recording method of video and associated audio and auxiliary data for DV-encoded video and associated audio. Whereas SMPTE 306M only deals with video at 25 Mb/s, SMPTE 314M is concerned with both 4:1:1 video at 25 Mb/s and 4:2:2 video at 50 Mb/s.

The standard defines one video channel and *two* independent audio channels capable of independent editing for DV at 25 Mb/s, and one video channel and *four* audio channels for DV at 50 Mb/s. Further, the standards consider NTSC (525 lines and 480 active lines, 29.97 frames/s) and PAL (625 lines and 576 active lines, 25 frames/s) as television signals.

---

[2] Note, this is one of the major differences to MPEG compression, which also uses inter-frame compression.

SMPTE 306M takes into account the recording of essence on 6.35 mm tapes in cassettes as specified in (Society of Motion Picture and Television Engineers, 1998b). The recording is specified for tapes with helical track where each record track contains an Initial Track Information (ITI) sector (containing start sync and track information), an audio sector, and a sub-code sector (containing time and control code data plus any optional data). The recording of a frame is specified with 10 tracks for an NTSC and 12 for a PAL system.

The audio sector consists of the audio preamble, the audio sync block, and the audio postamble (comprising, apart from the actual audio data, audio auxiliary data and error correction and detection information). The actual audio data is part of the audio sync block. Similar to audio, the video sector has a video preamble, video sync block, and a video postamble. The video sync block comprises 149 data sync blocks.

For audio encoding the standards specify data packets containing 72 bytes of audio data. Each of the two (or four) audio channels is identically but independently processed. The audio input signal is sampled at 48 kHz with a 16-bit quantization. The audio signal is locked to the video. The audio data is processed in frames of 1602 (or 1600) samples per frame for NTCS and 1920 samples per frame for PAL. The duration of an audio frame is equal to the respective video frame duration. Error detection and correction based on inner and outer parity is defined.

The video coding parts of the standards distinguish between 4:2:2 sampled video resulting in DV at 50 Mb/s (DV 50) and 4:1:1 sampled video resulting in DV at 25 Mb/s (DV 25). The coding process starts with an analog input video signal, which is sampled using 13.5 MHz for luminance and 6.75 MHz for each chrominance component, i.e. originally a 4:2:2 sampling is applied. For DV 25 all 720 luminance pixels per line are retained for processing but of the 360 chrominance pixels every other pixel is discarded, thus filtering the original 4:2:2: sampling down to 4:1:1. For DV 50 a 4:2:2 sampling structure is maintained.

Subsequent to sampling a DCT is applied. For DV 25 four DCT blocks for luminance and two DTC blocks for chrominance form one macroblock. In contrast, DV 50 uses two luminance DCT and two chrominance DTC blocks for one macroblock. For NTSC there are 2700 macroblocks per frame for DV 50 and 1350 macroblocks for DV 25. PAL uses 3240 macroblocks per frame for DV 50 and 1620 macroblocks for DV 25.

After sampling and DCT, the weighted DCT coefficients are quantized to 9-bit words and divided by the quantization step. The resulting so-called video segment comprises five compressed macroblocks. Subsequently a video segment is compressed into a 385-byte data stream. During compression, entropy encoding with variable run-length coding is applied.

For transmission of DV 25 encoded video 10 DIF sequences are defined for NTSC and 12 DIF sequences for PAL per frame. DV 50 is transmitted using two channels with 10 (or 12) DIF sequences each.

3.3.4 VIDEO FORMATS IN THE CONTEXT OF CMS

Within a CMS a number of different video formats have to be managed. In future these should be largely digital, based on encoding schemes such as DV or the MEPG standard suite. However, it should be noted that at present there are still a number of analog or proprietary tape-based formats in use within video production. These formats also might have to be administered even if they are not going to be integrated in a fully tapeless workflow. Current video recording formats include:

- analog component formats such as Betacam, Betacam SP and M-2;
- digital composite formats D-2 and D-3;
- digital component format Digital Betacam (using a proprietary compression format with a native data rate of approximately 90 Mb/s);
- Digital component format D-5, which transparently (without video compression) records 10-bit digital video according to the digital studio standard ITU-R BT 601-5 (International Telecommunication Union, 1995)

New digital video tape recording formats are usually based on either DV or MPEG-2 compression algorithms. Existing implementations are mainly represented by:

- the digital component format DVCPRO (using 4:1:1 sampling and a DV-based video compression scheme with a net data rate of 25 Mb/s);
- the digital component format Betacam SX (using 4:2:2 sampling and a MPEG-2, IB-frame video compression scheme with a net data rate of 21 Mb/s);
- the digital component format D-9 (Digital S) (using 4:2:2 sampling and a DV-based video compression scheme with a net data rate of 50 Mb/s);
- the digital component format DVCPRO50 (using 4:2:2 sampling and a DV-based video compression scheme with a net data rate of 50 Mb/s, identical to the one used in D-9);
- the digital component format DVCAM (using 4:2:0 sampling and a DV home video compression scheme with a net data rate of 25 Mb/s);
- the digital component format D-10 (MPEG IMX^{TM}) (using 4:2:2 sampling and a video compression scheme based on MPEG-2 4:2:2P@ML, I-frame with a net data rate of up to 50 Mb/s).

In order to understand the demands video places on a CMS the actual formats used within a specific system have to be considered. In general there will be different formats for various purposes. The low bit rate browse format is used to sift through the material to get an overview. However, the more sophisticated browse applications get, the higher will be the demands of the users. At present dual channel audio is considered sufficient. However, in the context of multiple language programs a number of different language audio tracks are desirable. Hence these kinds of formats will have to support such a feature as well in the near future.

The bit rate of digital broadcast formats can vary depending on the transmission channel. Bit rates between 4 Mb/s and 8 Mb/s are common. These formats are currently often based on MPEG-2. Within production there is a wide range of formats providing different bandwidth and quality levels. The choice of format depends on the kind of production area and the requirements of the producers. In general the bandwidth of these formats can vary between 18 Mb/s for news production to 50 Mb/s for features. A CMS has to be able to handle these formats and to provide access as quickly and easily as possible. From a quality point of view uncompressed (or lossless) compressed video would be preferable. However, the storage and communication requirements in this case are excessive. Table 3.6 gives an overview of some common formats, their transfer

Table 3.6 Video formats and quality levels

Quality level	Format	Transfer rate	Storage requirement, 1 h	Storage requirement, 100 000 h
Browse	Real Video MPEG-4 Advance Simple Profile	128 Kb/s	58 MB	5.8 TB
Preview	MPEG-1	1.5 Mb/s	680 MB	68 TB
Broadcast	MPEG-2 MP@ML	4 Mb/s	1.8 GB	180 TB
News production (Every other frame accessible)	MPEG-2 4:2:2P@ML GOP: IB	18 Mb/s	8.1 GB	810 TB
Production (Every frame accessible)	MPEG-2 4:2:2P@ML GOP: I DVCPro 50	50 Mb/s	23 GB	2.3 PB
Uncompressed 4:3 (Video signal only)	SDTV, ITU R BT 601-5	166 Mb/s	75 GB	7.5 PB
Uncompressed	SDTV, ITU R BT 601-5 (inclusive blanking)	270 Mb/s	121.5 GB	12.15 PB

rates, and storage requirements. The storage requirements have been also calculated for 100 000 hours of video since this is a common range for a medium size broadcast archive. In this context it also has to be considered that a number of formats might have to be kept in parallel.

An issue that can have a serious impact on the preservation of content is the so-called *generation loss* problem. This effect refers to the fact that the quality of the essence can deteriorate with every production step when this involves the decoding and re-encoding of the material. With lossy compression methods a quality loss occurs every time (i.e. in every generation) the material is decoded, processed, and re-encoded. This even happens with high-bandwidth formats such as MPEG-2 4:2:2P@ML and DVCPRO 50 at 50 Mb/s. In the context of a heterogeneous production and communication infrastructure this is a serious problem. Another format-related problem mainly occurring in the archive is obsolete formats that cannot be processed anymore. Together with storage carrier deterioration this is the reason for considerable loss of content that still occurs every year.

There are a number of proprietary video formats where the employed encoding and compression techniques are not (fully) disclosed. For instance, RealVideo as a browse and Internet-capable format or Motion JPEG-based production formats are popular. These formats have to be managed within a CMS; however, the functionality and interaction with these formats in the system is restricted. It is, for instance, only possible to stream, manipulate, and process proprietary formats using native tools. Without knowing the format and structure of the encoding schemes, essence encoded in proprietary formats can only be treated as files. More sophisticated interaction is only possible if the providers of such formats also offer tools that can be easily integrated into a CMS as service or application components. Hence, when selecting video formats one selection criteria has to be how open the formats are and how easily they and any supporting tools can be integrated into other systems.

3.4 AUDIO

Audio (like video) is an essence type that can be categorized as continuous media, i.e. the factor time has to be considered as part of the presentation semantics. Similar to video, audio has to be managed in a CMS and is also used to represent (audio) content using low-resolution copies. It shares many properties and requirements with video but in contrast to video its quantitative aspects (such as bandwidth and storage requirements) are much less demanding. A number of different audio formats exist to satisfy the requirements of different application areas. Within a CMS this data has to be managed, transferred to third-party systems, and (at an application level) presented to the user. Audio can be part of a video or an independent component within the system.

In this section the principles and basics of audio encoding and the major compression standards are introduced and discussed.

3.4.1 AUDIO ENCODING: BASICS AND PRINCIPLES

Sound is produced through the vibration of matter, which creates pressure variations in the surrounding matter (usually air). The vibrations generate a waveform, which is repeated at regular intervals called periods. It is transmitted through the air and when it reaches the human ear a sound is heard (Steinmetz, 2000). The frequency of a sound is defined as the reciprocal value of the period. It is measures in hertz Hz (cycles per seconds) or kilohertz ($1\,kHz = 1000\,Hz$). The human hearing frequency range lies between $20\,Hz$ and $20\,kHz$. The loudness of a tone depends on the amplitude (which is the measure of the displacement of the wave from its mean).

In order to represent the sound waveform digitally it is sampled by an analog-to-digital converter (ADC^3). The sampling rate is also measured in Hz. The standard CD sampling rate is $44.1\,kHz$ (i.e. the wave is sampled at $44\,100$ samples/s). The reverse process to ADC is digital-to-analog conversion (DAC), which translates the digital representation back into a waveform. After the ADC quantization is applied and subsequently the digital data can be compressed. There are different audio formats for different purposes. For example telephone quality audio is sampled at $8\,kHz$ and quantized using 8-bit μ-law encoded quantization, whereas CD quality audio uses 16-bit linear pulse code modulation (PCM). As mentioned above, the two factors determining the quality of digital audio are the sampling rate and the sample quantization.

The resulting data rate of a CD quality digital stereo audio stream is:

$$2 \times 44\,100\tfrac{1}{s} \times 16\,bit = 1\,411\,200\,bit/s$$

A further reduction without loss of quality can be achieved by using difference pulse code modulation.

The waveform format (WAVE) is often used as a reference format for uncompressed digital stereo audio (WAVE, 1991). WAVE is basically a file format that specifies the kind

---

[3] ADC stands for analog-to-digital converter and for analog-to-digital conversion

of encoded audio that can be part of a file. However, more importantly, it defines how the data is packed into a file and the kind of metadata that can be transported in the file. The standard explicitly refers to two encoding schemes suitable for WAVE files (MPEG and PCM). The WAVE encoding format for uncompressed audio is the above described 44.1 kHz sampling and PCM-encoded dual stereo audio in CD quality. Another common uncompressed format is 48.0 kHz PCM-encoded uncompressed audio as used on digital audio tapes (DAT).

3.4.2 MPEG-BASED AUDIO FORMATS

MPEG not only defines standard video with multiplexed audio but also deals with audio only. All three relevant MPEG media coding standard (i.e. MPEG-1, MPEG-2, and MPEG-4) deal with audio.

3.4.2.1 MPEG-1 Audio

The specified MPEG-1 audio coding is compatible with the specifications for compact disc digital audio (CD-DA) and digital audio tape (DAT), i.e. sampling frequencies of 44.1 kHz or 48 kHz with 16 bits per sampling value. In addition a sampling frequency of 32 kHz has also been defined.

In MPEG-1 three different layers are defined. Each layer represents a different level of encoder and decoder complexity and performance. The layers are downward compatible, i.e. a higher layer implementation must always be able to decode the MEPG-1 audio signal of the lower layers. Audio files are often encoded in MPEG-1 Layer III, also referred to as MP3.

The compression starts with a frequency transformation using fast fourier transformation (FFT). During this procedure the spectrum is split into 32 non-interleaved sub-bands. Subsequently the amplitude of the audio signal per sub-band is calculated. In parallel to the FFT the noise level per sub-band is determined by using a psycho-acoustic model. The result of this process controls the quantization. The higher the noise level the bigger are the quantization steps applied to a sub-band. For the quantization of Layers I and II PCM is used whereas Layer III uses Huffman coding (Effelsberg and Steinmetz, 1998; Steinmetz, 2000). As a final compression step entropy encoding can be applied.

The audio coding in MPEG-1 is defined for one channel, two independent channels, two channel stereo, and joint stereo. In the latter case redundancies between channels can be exploited to achieve a higher compression ratio. For each layer 14 different bit rates for encoded data streams are defined (referred to by their bit rate index). The minimum value for each layer is 32 Kb/s while the maximum values per layer differ. Layer I allows a maximal bit rate of 448 Kb/s, Layer II 384 Kb/s, and Layer III 320 Kb/s. Layer III can also support a variable bit rate. Layer II has some restrictions for the different channels (Steinmetz and Nahrstedt, 1995).

3.4.2.2 MPEG-2 Audio

The MPEG-2 standard subsumes and improves the MPEG-1 audio standard. For instance it also allows sampling rates of 16 kHz, 22.05 kHz, and 24 kHz (i.e. half the sampling rates specified for MPEG-1). This results in a bit rate of less than 64 Kb/s. Additionally, the audio part of MPEG-2 allows multiple channels with relatively low bit rates. There are up to five full-bandwidth channels for left, right, middle, and two surround channels, plus one channel to improve the low frequency quality. Further, there can be up to seven channels for different language versions.

3.4.2.3 MPEG-4 Audio

MPEG-4 defines an extensive tool set for speech coding, general audio coding, and synthetic audio and audio composition (Pereira and Ebrahimi, 2002). MPEG-4 audio is (like the video part) object oriented. It allows the composition of audio scenes from multiple (natural or synthetic) audio objects. As defined for MPEG-1 and MPEG-2, natural audio objects are derived from sampled waveforms. The resulting different audio streams are composed into an 'audio scene' in a similar way as described for video.

The speech coding part has been specifically developed for the efficient encoding of the spoken word to save bandwidth and capacity in communication networks and storage systems. Since only speech has to be encoded more efficient coding techniques can be applied. A number of coding techniques are adopted. Most prominent is the Code Excited Linear Prediction (CELP) scheme and the MPEG-4 Harmonic Vector Excitation Coding (HVXC). The produced bit rates of CELP can range between 4 and 6 Kb/s. The HVXC encoding produces bit rates between 2 and 4 Kb/s (Pereira and Ebrahimi, 2002).

The general audio part of the MPEG-4 standard is concerned with the faithful reproduction of the (natural) audio input signal. The MPEG-4 codec is partially based on the MPEG-2 advanced audio coding technology extended by a better compression performance and error resilience. Further, it allows very low bit rates and very low delays. Additionally it provides bit rate scalability (i.e. the ability to decode a subset of the bit stream while still receiving meaningful information) using large-step scalable audio coding and bit-sliced arithmetic coding.

MPEG-4 also defines profiles, levels, and object types for audio. There is an extensive list of object types for general audio, speech, speech and general audio, synthetic audio, and synthetic speech (Pereira and Ebrahimi, 2002). Table 3.7 gives an overview of some selected audio object types.

The MPEG-4 audio profiles define the conformance criteria to which bit streams and decoders have to adhere. There are much fewer profiles than object types defined for MPEG-4 audio. The levels defined for MPEG-4 audio are defined in terms of complexity units (processor and RAM complexity). Further, the number of objects of a specific type can be restricted. The authors have the freedom to use different object types with varying complexity within an audio scene as long as the overall complexity defined by the respective level is not exceeded. Table 3.8 gives an overview of profiles for general audio, and speech and general audio.

3.4.3 AUDIO FORMATS IN THE CONTEXT OF CMS

The standard audio formats (i.e. 44.1 kHz and 48 kHz PCM-encoded audio and MPEG audio) are in terms of bandwidth and storage requirements much less critical than their video counterparts. Thus, they are usually regarded as easy to deal with. However, these

Table 3.7 MPEG-4 audio object types

Audio type	Object type	Description
Speech	*CELP*	Based on CELP. Supports scalable coding. Supported sampling rates: 8–16 kHz; bit rates: 4–16 Kb/s
	Error-Resilient CELP	Error-resilient version of CELP object type. Provides silence compression for greater efficiency
	HVXC	Based on HVCX coder. Provides parametric representation. Supported sampling rate: 8 kHz; bit rates: 2–4 Kb/s
	Error-Resilient HVCX	Error-resilient version of HVCX object type
General audio	*AAC Main*	Based in MPEG-2 AAC Main Profile enhanced with perceptual noise shaping. Five full channels plus one low-frequency channel.
	AAC Low Complexity (LC)	Low complexity version of AAC Main
	Error-Resilient AAC (LC)	Error-resilient version of AAC LC
	AAC Scalable Sampling Rate (SSR)	Base on MPEG-2 AAC SSR Profile
	AAC Long-Term Prediction (LTP)	Based on AAC Main, extended with long-term predictor
	Error-Resilient AAC LTP	Error-resilient version of AAC LTP
	Error-Resilient AAC low-delay	Provides low-delay. Supports error-resilience. Intended for low bit rate, low-delay audio coding
	TwinVQ	Uses fixed-rate quantization. Operates on lower bit rates than AAC
Speech and general audio	*AAC Scalable*	Supports scaling combinations, including TwinVQ and CELP. Supports mono and stereo audio (two channels only)
	Error-Resilient AAC Scalable	Error-resilient version of AAC Scalable

formats also have to be integrated into a CMS, and tools and applications that support them in production and transmission have to be supported.

The MPEG audio coders support the entire range of sound compression. Of major impact was MPEG-1 Layer III (or MP3) since it is the preferred format for audio and music files on the Internet. The relevance of MPEG-4-based audio systems remains to be seen. Other proprietary audio codecs such as RealAudio or Liquid Audio also exist. However, with the rising popularity of MP3 their relevance has decreased.

An additional challenge for CMS can be new formats with much higher sampling rates and quantization intervals. In professional systems sampling rates of 96 kHz might become

Table 3.8 MPEG-4 audio profiles

Audio types	Audio profiles	Description
General audio	*Mobile Audio Internetworking*	Allows ER AAC LC, ER AAC LD, ER AAC Scalable, etc. object types but does not contain speech coders. Intended for high-quality audio in addition to existing speech codecs. Number of levels: 6
Speech and general audio	*Scalable*	Allows AAC LC, AAC LTP, AAC Scalable, TwinVQ, CELP, HVCX object types. Intended for good-quality, low-bit-rate Internet applications but also to support broadcast applications. Number of levels: 4
	High-Quality Audio	Allows AAC LC, ER AAC LC, AAC LTP, AAC Scalable, ER AAC Scalable, CELP and ER CELP object types. Intended for high-quality natural audio coding. Number of levels: 8
	Natural Audio	Allows AAC Main, AAC LC, ER AAC LC, AAC LTP, ER AAC LPT, AAC SSR, AAC Scalable, ER AAC Scalable, ER AAC LD, TwinVQ, CELP, ER CELP, HVXC object types. Comprises all natural audio coding tools. Number of levels: 4

the norm. A CMS used for audio then might have to cope with as wide a rang of audio formats and similar requirements as is already the case for video.

3.5 IMAGES, WEB, TEXT, AND OTHER ESSENCE FORMATS

Apart from video and audio there are a number of other media types (mainly discrete media types) that have to be managed in a CMS. As in the case of video and audio there is a plethora of different formats for images, graphics, documents, etc. Documents for instance can be encoded in plain ASCII text, Microsoft® Word, FrameMaker®, RTF, PDF, etc. Further, there are documents such as presentation slides, project plans or calculation sheets that are encoded in specific (generally proprietary) formats that also have to be managed in a CMS. The main functionality a CMS has to provide in this context is the indexing of these documents to make them searchable, and the integration of native application programs or application views to present the documents to the users in their original form. The former problem is currently addressed by fulltext search engines that can deal with various formats. The latter is part of the integration of third-party applications.

Other important essence types handled by a CMS are images and structured documents, in particular Web pages.

3.5.1 IMAGES

Images are visual, photographic or graphic representations of objects or scenes. They depict a certain situation within a two-dimensional space without any temporal restrictions (i.e. they belong to the discrete media class). Within a CMS an image is represented in a digital image format or as a link to an external image object. In the context of this book only digital image formats are relevant. On a technical level an image can be regarded as a function with resulting values of light intensity at every point of the planar region (Steinmetz and Nahrstedt, 1995). In order to represent these values digitally this function has to be sampled at discrete intervals with subsequent quantization of the sampling values. The digital image itself is then a matrix of numerical values of which each represents a quantized intensity value. The sampling points of an image are (as in the case of video) the picture elements, i.e. pixels. The digital representation of an image can become very large. An image of a size comparable to an NTSC television picture is represented by a 640×480 pixel matrix. With an 8-bit integer representation of 256 discrete gray levels the storage requirement of this monochrome image is around 300 Kbyte. Thus, as in the case of video, compression has to be applied to reduce the number of bits required to represent an image. The most common image formats are JPEG, GIF, TIFF, and BMP.

3.5.1.1 JPEG

The *JPEG* standardization was a joint activity of ISO and CCITT within the *Joint Photographics Experts Group* (JPEG, 1993). JPEG specifies an image encoding and compression scheme for color and monochrome images. JPEG also defines an exchange format that contains the actual image data plus coding tables and parameters. The latter two are not required if the coder and decoder are used within the same context.

The JPEG standard specifies four basic modes (where each has further variants):

1. The Lossy Sequential DCT Base Mode is the so-called baseline process mode that must be supported by every JPEG implementation.
2. The Expanded Lossy DCT Base Mode provides enhancements to the baseline process mode.
3. The *Lossless Mode* allows an exact copy of the original source image to be reconstructed but has a low compression ratio.
4. The *Hierarchical Mode* can contain images of different resolutions and applies algorithms from the other three JPEG modes.

The basic encoding and compression steps (as depicted in Figure 3.6) are the same for each of these modes. Not all the techniques are used within all the modes; the Baseline Process Mode, for instance, uses block, MCU, FDCT, run-length, and Huffman coding.

JPEG specifies a very general image model. The source image for the image preparation step consists of at least one and at most 255 components or planes. Each of these components may have a different number of pixels. These components can for instance represent the different colors (RGB) or brightness and chrominance information (e.g. YUV). The pixel representation is also flexible. A pixel is represented by p bits with values ranging from 0 to $2^p - 1$. All pixels of all components of one image have to be coded with the same number of bits. The JEPG lossy mode uses either 8 or 12 bits per pixel. The processing order for non-interleaved data units is left to right, top to bottom. The uncompressed image samples are than grouped into data units of 8×8 pixels.

The picture processing of the baseline mode uses forward discrete cosine transformation (forward DCT) on the 8×8 pixel groups to map the two-dimensional image values into

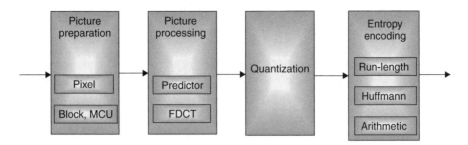

Figure 3.6 JPEG compression steps

the frequency domain. This transformation has to be carried out 64 times. Following this step quantization of all DCT coefficients is performed using a quantization table with 64 entries. Each entry will be used for the quantization of one of the DCT coefficients. Both DCT and quantization are lossy processes. Subsequently entropy encoding is used. In the base mode Huffmann coding is used as entropy encoding method.

The expanded lossy mode allows a higher sample precision (of up to 12 bits). Further, it uses expanded quantization, which enables a progressive rather than a sequential image representation. With the former an image becomes clearer during the presentation whereas the latter technique builds up the image sequentially from top to bottom. Further, arithmetic entropy encoding can be used with the expanded lossy mode.

In the lossless mode data units of single pixels are used for image preparation with a precision between 2 and 16 bits per pixel. The processing and quantization steps are based on a predictive technique where a value is predicted from already known adjacent samples.

3.5.1.2 GIF

The *Graphic Interchange Format* (*GIF*) was originally developed by CompuServe to support platform-independent exchange of images (Steinmetz, 2000). GIF uses a lossless compression scheme. Further, GIF allows interleaving of multiple images within one file.

A GIF image is always encoded as a bit stream. The *logical screen descriptor* specifies the size, position, and kind of the color table used for an encoded image. Further, optional global and local color tables, and the color of pixels that are pointers to such tables are also defined in the logical screen descriptor. The pixel colors are compressed using a special algorithm (called Lempel–Ziv–Welch) that enables the discovery and processing of variable-length bit patterns. The bit patterns are represented in a table in which they are replaced by short bit representations. The bit patterns that occur most frequently are represented by the shortest bit representation.

A GIF image is composed of a number of sectors:

- *header* contains the GIF ID and version number of the algorithm used;
- *application* allows coding of the version and name information of the application that created the image;
- *trailer* marks the end of a GIF stream;
- *control* controls the presentation of a subsequent image block;
- *image* contains an image header, an optional color table, and pixel data;

- *comment* allows the addition of (textual) comments for each image block;
- *plain text* allows ASCII-encoded text to appear in an image.

The control, image, comment, and plain text fields can be repeated and arranged within a GIF image. GIF also allows the encoding of short animations and image sequences. Since it only uses 8-bit color tables it cannot be used for high-quality images.

3.5.1.3 TIFF

The *Tagged Image File Format (TIFF)* is a joint development of Microsoft and the Aldus Corporation (1992). The aim of TIFF is to provide portability and hardware-independent coding of images. TIFF has two components, namely the baseline part and the extensions part. The baseline part specifies those features that have to be supported by every decoding and presentation application. The extensions part defines additional features. TIFF supports a wide range of color models ranging from binary (i.e. black and white), over monochrome to sophisticated color palletes, RGB, etc. Similar to GIF, a TIFF images has a number of fields:

- *header directory* specifies the byte order and version number, and references sub-directories that contain other fields or images;
- *structure* specifies the coding technique and number of tag fields;
- *fields* defines the image coding blocks (e.g. rows, objects, cells, and blocks) and their characteristics (i.e. compression technique, alignment, and resolution);
- *data fields* are graphical objects that are not specified in advance.

The compression techniques supported by TIFF include run-length coding, Huffmann coding, Lempel–Ziv–Welch compression, and JPEG compression schemes. TIFF is a universal format that also allows the encoding of images in different resolutions, e.g. preview formats and high-quality images.

3.5.1.4 BMP

The *bitmap format (BMP)* is a generic image format based on the RGB color model (Steinmetz, 2000). It can also be used for the encoding of monochrome and black and white images. The BMP format defines two main parts, namely the header and the data part. The former is also known as *BITMAPINFO* and specifies the image size, color depth and color table, and the compression technique. The data part holds the pixel value of each point of a row. Possible values for the color depth are 1, 4, 8, and 24. The compression scheme for images with a color depth of 4 and 8 bits per pixel is run-length coding. A special algorithm is used to encode the different reference values in the color table.

3.5.2 STRUCTURED DOCUMENTS

The structured document is another media type that is particularly relevant within a CMS. Their development has been mainly driven from two sides, namely the print media industry and the Web domain. Further, the hypertext and hypermedia initiatives have been instrumental in developing languages and standards for structured documents. In contrast to common document formats such as RTF, MS Word or PDF, structured documents are characterized by the use of mark-up languages and links to external documents and information. For a CMS this poses special challenges.

The major standards to consider in this context are SGML, HTML, and XML. The latter will be discussed in Section 4.5.2 in connection with metadata encoding, transmission, and exchange since it is currently often used in this context.

3.5.2.1 SGML

The *Standard Generalized Markup Language* (*SGML*) (International Organization for Standardization, 1986) is a development mainly driven by American publishers. The basic idea is that a text is written without any formatting using *tags* to mark certain text elements such as titles or paragraphs. With SGML the layout can be determined flexibly at the processing stage of a text. The tag structure also facilitates automatic processing of the text document. SGML defines a framework in which the syntax of the tags is defined; their occurrence and semantic interpretation is left to the applications processing an SGML document. SGML is object-oriented with classes, objects, hierarchies of classes and objects, inheritance, etc. It also allows the specification of processing instructions within an SGML document.

The parsing and formatting of a document are separate processes. The tags determine the structure of a document. However, parts of the layout are also often associated with the structure. Thus, the context the document has been designed in has to be considered when a document is formatted.

Within SGML four tag categories are distinguished:

- *descriptive mark-up (tags)* determine the structure of a document, in the form of `<start-tag> document element </end-tag>`;
- *entity references* are references to other elements that replace the entity reference during the presentation of the document;
- *mark-up declarations* define the elements that can be referenced by an entity reference;
- *processing instructions* are instructions for programs such as formatting instructions. Further, they allow the inclusion of other media types such as video and audio at the presentation of the document.

As SGML defines only the syntax, so-called *document type definitions* (*DTD*) are necessary to define the semantics. Further, a document style semantics and specification language (*DSSSL*) is defined to standardize the layout semantics.

3.5.2.2 Web Pages and HTML

With the World Wide Web becoming more and more popular, there is a much higher demand to manage Web documents within a CMS. Web pages are electronic documents combing text, images, graphics, sound, and video elements, and even small executable programs. These documents are stored on a server from which the users can retrieve them. The different elements of a Web page do not have to be encoded within the page itself but might be referenced by links. Thus, from one Web page a number of other documents or essence elements might be referenced. In order to manage Web pages within a CMS it

Table 3.9 Example of HTML tags

Tag	Description
`<HTML> ... </HTML>`	Declares an HTML document
`<Head> ... </Head>`	Encloses document header
`<Title> ... </Title>`	Defines the document title
`<Body> ... </Body>`	Encloses the document body
`<H`n`> ... </H`n`>`	Defines heading at level n
`<B> ... </B>`	Display as **bold**
`<I> ... </I>`	Display in *italics*
`<LI> ... </LI>`	Start/end of item in list
` `	Carriage return
`<P>`	New paragraph
`<HR>`	Include a horizontal line
`<IMG SRC = "imagead">`	Loads an image with the given address
`<A HREF="URL"> ... </A>`	Includes a link to another document using the given description
`<!-- ... -->`	Comments

is therefore necessary to define the extent of a document. For instance, images and audio or video files referenced by a Web page can be considered part of the document whereas the link to another page, which might even be outside the organization, can be considered as a citation only. Hence when handling Web pages within a CMS it is important to specify the *archiving depth*, i.e. the extent to which linked objects should be included in the storage and management process.

Web pages are written using the *Hypertext Markup Language* (*HTML*). HTML is a simple mark-up language that defines how documents should be presented on screen. Comparable to SGML documents, HTML documents are simple ASCII texts that are structured with tags. These tags can also be regarded as control commands. An HTML document is structured in three parts, namely the document type definition, the document header, and the document body. Within an HTML document meta-information can be encoded that is not presented to the viewer on screen but can be used to state authorship and copyrights issues, or simply to provide Web search engines with information about the page. Table 3.9 shows a number of commonly used HTML tags (Steinmetz, 2000).

As can be seen from the table, some tags require a start and end tag to enclose an element. If the start or end tag is not there the document has a syntactic error.

So-called style sheets can be used in HTML to declare what recurring elements should look like. Using style sheets separates the layout from the actual content. With *cascading style sheets* (*CSS*) the look of a Web page can be exactly determined. Style sheets also have to be managed in a CMS handling Web pages. References from HTML documents to style sheets have to be administered correctly.

3.6 ESSENCE PROCESSING

Essence processing refers to the automatic processing of audiovisual, visual or audio objects to retrieve information that can be used as additional metadata or to facilitate input and retrieval processes. Tools are, for instance, video analysis tools, audio analysis tools (such as speech-to-text and keyword spotting) or image similarity retrieval tools. Other programs in this category are transcoding tools or special programs that segment essence.

3.6.1 ESSENCE PROCESSING APPLICATIONS AND SERVICES

A number of tools exist for the processing of essence and the automatic retrieval of semantic information. Each tool concentrates on specific aspects of essence processing. Not only their processing accuracy is important in the context of content management but also if and how they can be integrated into the CMS architecture.

In general, relevant tools have either application or service character. They process essence fully automatically and can be classified according to the specific characteristics and features of certain media and essence types they exploit in order to extract semantic information:

- *Content segmentation (temporal and spatial) tools and applications* segment a continuous media object considering certain properties such as shots (temporal segmentation) or certain regions or objects in an image (spatial segmentation).
- *Metadata generation tools:* metadata is generated out of the specific properties of certain essence objects or analyzable characteristics (such as, for instance, motion detection).
- *Automatic content description tools* are, for instance, speech recognition tools that transcribe the spoken word or detect keywords to generate explicit descriptions.
- *Indexing tools* classify essence objects according to well-known patterns. Examples are face recognition tools or program classification tools.
- *Content-based retrieval tools* use essence characteristics for the retrieval of content such as image similarity retrieval.

3.6.2 ESSENCE PROCESSING: BASIC PRINCIPLES AND METHODS

Most essence processing tools use the mathematical, statistic or stochastic analysis of certain computational features and properties of the different media. They operate on a file, sound, image or audiovisual signal as a whole, or on a specific component of it at a given time. The basic building blocks of essence processing processes are:

- feature extraction;
- feature interpretation;
- query engine.

Figure 3.7 shows these building blocks as part of a workflow. The feature extraction block includes all the tools that process the raw audiovisual material and extract low-level features by means of an objective measurement (e.g. color histograms, spectrum, dominant motion vector, etc.). Normally these features do not convey any information that could be interpreted by humans. Thus, further processing is needed. This is done by the feature interpretation block, which interprets (with some error margin) the value of one feature (or a combination of several features), trying to match the objective observation with some logical concepts (e.g. the dominant color of the upper part of the picture is light blue, that of the lower one is green, therefore the picture is taken outdoors). This

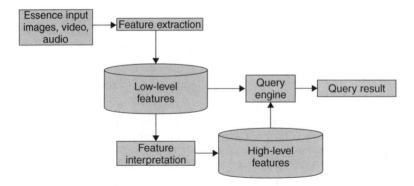

Figure 3.7 Feature extraction workflow

operation will produce a human-understandable description of some characteristic of the content, but as the interpretation must be based on some kind of heuristics this process is prone to misclassification. More advanced schemes try to include media semantic aspects into the consideration.

Another way to directly use low-level features of visual, audio or audiovisual essence is *similarity retrieval*. In this case the user asks the query engine to retrieve items similar to some other item already available to him/her. The engine then extracts on the fly the low-level features relevant to the user query from the example item and matches the values obtained with the stored values obtained with the same analysis process applied to the content of the archive. This also involves a process of abstraction from *comparison of object measurements* to *similarity*. A combination of low-level tools with more high-level tools that also consider the semantic of the content could help to achieve more accurate results.

Sound and audio analysis operates on sound effects, spectral envelopes, and phonemes using stochastic analysis. In general it can be distinguished between general sound processing (which tries to discover and classify the kind sound within a stream), music classification (for instance instrument categorization and theme recognition), and speech analysis (providing speaker detection and recognition, language identification, keyword spotting, and transcription). For speaker recognition acoustic fingerprints of persons are used, for example. In principle the speech analysis is based on the comparison of different characteristic properties. The features extracted from an audio stream are compared to reference examples to identify a speaker or word. There are three steps in the speech analysis process (Steinmetz, 2000):

- an acoustic and phonetic analysis;
- a syntactical analysis to correct errors from the acoustic analysis;
- a semantic analysis which also helps to correct errors.

The speaker-dependent speech analysis engines usually produce better results than speaker-independent ones. This is due to the fact that an additional parameter for comparison can be used. In general there is no speech analysis tool that provides 100% accuracy.

Image analysis of still pictures uses color characterization, region segmentation, texture features, and face detection (which can be regarded as a special kind of texture analysis) to retrieve information about the image content. Image recognition techniques are widely deployed, for instance in OCR tools or handwriting recognition. The three main characteristics used in image recognition technology are color, texture, and edges in an image. Using these features it is possible to find images with similar features, or to classify images according to the occurrence of certain features using statistical and stochastic methods. However, these purely technical analysis methods do not reveal anything about the actual content or semantic of an image. This requires knowledge about the position and orientation of an object within an image.

There are six steps in the image recognition process: *formatting, conditioning, labeling, grouping, extracting*, and *matching*. Conditioning is a process that estimates the informative pattern and suppresses noise. The subsequent labeling labels the informative pattern into units that form a visual entity (mainly using edge detection). The grouping step then forms larger entities. In the extracting step a list of properties (e.g. area, orientation, spatial moments, circumscribing circle, etc.) is extracted for each entity identified during grouping. The matching step matches the identified entity with objects that are known and recognizable. This final step determines the interpretation of the different related objects within an image.

Video analysis tools are based on analysis mechanisms developed for sound and image analysis to perform some basic analysis. In addition, it uses motion information from a video sequence to determine certain events within a video. For instance cut, dissolve, wipe, and effects detection are used to determine shot changes within a video. Using this information allows recreation of the original cut list. Some effects such as dissolves and wipes are hard to distinguish from other visual effects. Thus, the accuracy with which they can be discovered is much lower than for simple cuts. Motion and trajectory estimation can be applied to determine the movement of objects in a video. All this information can be used to create a visual abstract of a video. Keyframes are extracted whenever there is a significant change in the image content. Significant in this context does not necessarily mean relevant to the human viewer. For instance crowd scenes in a sports stadium contain a lot of movement and hence represent a significant change in image content. However, this information is not necessarily relevant. The extracted keyframes can be arranged in various ways to give a quick overview of the video content. Storyboards or mosaicing techniques are often used in this context.

Another visual abstract resulting from video analysis are so-called skims. These are short video summaries based on video analysis results; i.e. short, representative pieces of videos are edited together to give a quick overview of the video content. Sophisticated tools even consider the accompanying audio and edit the audio track separately to create a consistent video and audio abstract.

3.6.3 ESSENCE PROCESSING TOOLS IN CONTENT MANAGEMENT SYSTEMS

The amount of content a content-rich organization has to deal with is growing faster than ever before. Thus, it becomes increasingly important to support the identification and documentation of content by automatic processes. Audio, video, and images have features that allow automatic processing by computer-based tools and applications to extract syntactic and even basic semantic information. The extracted information, however, is not 100% accurate. Therefore it has to be considered carefully in which context essence processing tools are deployed, how accurate the information they provide is, and what additional manual work and procedures have to be implemented to effectively deploy such tools and applications. Currently they are only auxiliary devices that provide input to the documentation process, which still has to be performed by skilled staff.

The retrieval of content can also be supported by automatic processes such as image similarity retrieval or query by humming. In this context it is important to take the user expectations into account. If an editor or journalists queries an image database for a picture of a politician using a sample picture s/he expects only pictures of this person taken at different times, from different angles, etc. However, most tools at present will return all pictures that contain similar shapes, textures, and colors. These are sometimes not even remotely related to the person they were looking for. In order to avoid disappointment, user expectations have to be managed and these tools should only be deployed in a context where they really add value.

Another issue that has to be considered is the integration of automatic processing tools and application into the CMS infrastructure. They require files, data streams or video and audio signals as input. The results they produce are sets of keyframes, audiovisual material or meta-information. This has to be associated with the actual content object and it should be possible to process this information in the CMS. Further, tools such as image similarity retrieval engines produce hit lists of content objects. These hit lists have to reference the actual content items stored within the CMS. Thus, automatic processing tools and applications can only be exploited to their full potential when they are fully integrated into the CMS.

4

Content Representation and Metadata

The core entity of a CMS is the content object that is managed. The main objective of content management is to facilitate content handling and exploitation during an object's entire life cycle from capturing and production over transmission to archiving and preservation. How a content object is represented in the system is of key importance to search and usage processes. The representation is, however, not a neutral description of the content but also depends on the context in which it is used. Thus, content representation models have to take this into account in order to provide a useful abstraction for the various processes in which content data and information is handled.

Metadata is one of the key enablers for CMS; it is generally defined as 'data about data.' Metadata describes a content object considering different viewpoints, aspects, workflows, processes, and existing information models. It is essential to manage, search, find, and retrieve content whenever required. Thus, the description of content and the quality of the description are crucial for the capability of a CMS to provide quick and easy access to content. Metadata has to consider the different roles and perspectives of the users interacting with the system and support the different views resulting from this. Ideally a CMS supports all processes from the creation and ingest of the raw material into the system, over its description and all production-related work steps to content distribution and archiving. Aspects that have to be considered in this context are not only related to the direct interaction with the content but also to all associated areas such as accounting, rights management or program planning.

Apart from pure description and retrieval metadata also governs the inter-organizational and intra-organizational exchange of content. Thus, standards are required to facilitate the actual exchange but also the interpretation of the information being exchanged. Since a standard always has to provide a general and common denominator view, it cannot, by nature, offer all the specific support required by an individual organization. Thus, standards can only cover certain aspects important in a specific functional or thematic context.

A number of data models, content description schemes, and reference models for the representation and description of content have been developed. They usually look at the

Professional Content Management Systems: Handling Digital Media Assets   A. Mauthe, P. Thomas
© 2004 John Wiley & Sons, Ltd   ISBN: 0-470-85542-8

content from different viewpoints and are therefore not fully comparable. A universal or generic content description scheme does not exist. Within a CMS different schemes might actually be needed depending on the emphasis of a specific task or component. This chapter outlines different concepts to represent content depending on the context. It further introduces and discusses the structure of metadata and the kind of information a metadata scheme should capture. Also relevant in this context are data models and meta-data exchange. A number of prominent representation models and description schemes as well as metadata encoding and exchange protocols are introduced in this chapter. This gives an overview of the current state of development in this area. Further, it shows the different approaches currently taken to represent metadata in different contexts. The appropriateness of a specific scheme or model for a particular problem has to be assessed on a case-by-case basis. The discussion in this chapter can facilitate such an assessment.

4.1 THE REPRESENTATION OF CONTENT

The center of a content object is the essence, i.e. the part that carries the actual message or idea. However, there are instances where the information about content is more relevant than the content object itself. This is, for instance, the case for all administrative tasks related to content. Even within the production workflow certain meta-information (i.e. metadata, abstracts, and low-resolution essence versions) can be more important than the essence itself (see Section 2.2.1). In this context it is essential that the appropriate form and suitable metadata representation are chosen to optimally support all workflow and management processes related to a content object. While details of data models, content representation, and workflows may depend on the content and may differ from organization to organization, the main content characteristics and use cases are generic.

A media object (i.e. video, audio track, image, etc.) can be represented by so-called *proxies*. Proxies correspond to specific views on the content depending on a specific aspect relevant in a certain context. They highlight particular characteristics, represent a certain view or visualize specific properties and hence make it better accessible. Proxies can be of different media types. Audiovisual content can be represented by multimedia proxies (i.e. representations that combine different media types). Other proxies are written abstracts, edit decision lists (EDL), database entries, etc. Figure 4.1 shows some common proxies. The categories proxies can be grouped into are *textual proxies, database proxies,* and *multimedia proxies*.

Pure textual proxies are written documentations about the content. These can be abstracts, production documents (e.g. production sheets), scripts, EDL, log sheets (i.e. text transcripts and annotations that are associated with timecodes), contracts, legal documents, etc. This group of proxies comprises all document types that represent the content or specific parts of it in plain text. These documents can be indexed to make them searchable.

Database proxies are all content representations persistently stored in a database management system (*DBMS*). DBMS are characterized by the ability to access large amounts of data efficiently (Ullman, 1988). The content object is represented in an abstract model called the '*data model*'. The data model is hosted by a database. All relevant characteristics are captured in a structured textual representation that facilitates easy search and retrieval according to the attributes searched on. The textual and database proxies are usually referred to as metadata. Metadata is defined as data about data describing the essential

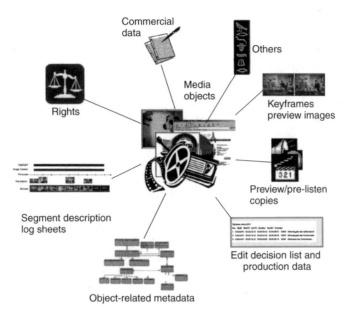

Figure 4.1   Media objects and proxy representation

creative media object. This media object contains the actual message or information that the creator wants to convey to the recipient.

Multimedia proxies are content representations such as low-resolution copies, audiovisual abstracts, and combinations of different media types. Low-resolution copies of the content allow previewing material without accessing the actual essence. Hence, they can be transferred on low-bandwidth links and displayed on low-capability devices. Audiovisual abstracts are combinations of sections of audiovisual content. These abstracts provide a quick overview of the essence by presenting essential parts only. Examples for such audiovisual abstracts are keyframes or skims. Audiovisual abstracts are usually automatically extracted and compiled based on an automatic analysis process. Combinations of different media types can give an even better overview of the actual content by linking written, graphic, visual, and audiovisual information. Examples are so-called storyboards, which are a combination of keyframes and timecoded content descriptions of segments. Others are transcripts linked to low-resolution essence copies. If the content object is of a continuous media type the discrete parts of the content representation also have a time reference that is associated with the timecode of the content object.

In a wider sense multimedia proxies can also be considered as metadata. They are of auxiliary nature and their main objective is to facilitate content representation and retrieval. The textual components of multimedia proxies are clearly part of the metadata description of a content object. Visual abstracts also have a descriptive character. However, low-resolution essence copies are representations of the essence with no descriptive elements.

4.2 METADATA: THE DESCRIPTION OF CONTENT

Metadata is used to describe content and to represent it within systems and system components. It has to cover the entire workflow of media creation from the pre-production stage to archiving. Further, it has to accommodate different user groups with varying rights and interests towards the content. Hence, the description of a content object depends on the different views and purposes.

Metadata is first created during the planning stage and is further amended and used throughout the lifetime of a content object. Stages during which metadata is used are production, transmission, media management, and all other exploitation processes. Metadata has to capture program planning, production planning (including resource material, personnel planning, and references to related contributions), and editorial preparations (drafting, research, collection, and storyboarding) during the planning stage. Other processes and use cases in which metadata is used include documentation, IPR, licences, analysis data, as well as location-related data (such as stock management, location ID, etc.). Another set of metadata is related to the usage and exploitation of the content. Traditionally this includes transmission history, data about the exchange of programs and program material, and information about content sale (including marketing). A new category of metadata is data concerning the use of content on the Internet. Data recorded in this context is concerned with access history, responsible personnel (Webmaster/editor), etc.

Metadata has to capture and described a plethora of characteristics and workflows. Thus, there are different schemes and (sub)systems that deal with metadata. Metadata can be grouped according to the entity it describes. There is *object-related* and *segment-related metadata*. Object-related metadata refers to all metadata that describe certain aspects of a media object as an entity. In contrast, segment-related metadata describes parts of a content object delimited by spatial or temporal boundaries, such as image regions and timecodes.

4.2.1 OBJECT-RELATED METADATA

Object-related metadata describes a content object as one entity. Parts of the object-related metadata are content description (e.g. title, creators, directors, etc.), related organizational data (e.g. program slots, responsible editorial offices, etc.), content identifiers, and involved parties (i.e. persons and organizations handling content, their affiliations, roles, and rights). Further, object-related metadata covers technical metadata and all business processes (including all workflow steps in media production).

Object-related metadata data is usually held in a DBMS. Sometimes object-related metadata is stored in structured files indexed by fulltext search engines to make it retrievable. Documents related to a content object (such as scripts and contracts) can also be stored electronically and indexed to make them searchable. In cases where documents that qualify as metadata have to be kept as physical copies (e.g. legal documents on paper), references to these documents can and sometimes even must be maintained within the CMS data model.

4.2.1.1 Workflows and Content Description

To facilitate the management of content, its characteristics have to be captured as structured metadata. This can be done using an abstract model describing the objects (also called *entities*) themselves as well as links between objects. Workflows are the crucial processes that have to be considered in this context. Along these workflows, metadata is generated. Thus, these workflows define the kind of metadata that is generated at a specific point in the content life cycle. How the metadata is captured and logged, however, is system specific. This section looks at some generic content workflows relevant in the context of content management (such as content acquisition and ingest, production, and archival and documentation). A detailed analysis can only be done when analyzing real-world organizations and their systems and workflows. This has to be carried out individually for each organization dealing with content.

Metadata might already be generated and associated with a piece of essence at the time it is created. Information such as position of the recording (e.g. through GPS positioning), date, time of day, etc. can be automatically associated with the material as it is recorded. Often, however, import and ingest still are the processes where content enters the system and where metadata is first associated with a content object.

Ideally most of the metadata added during this process should be generated automatically. Such information is, for instance, material-related parameters but also data retrieved by automatic analysis processes. The metadata gathered in this process is organized into metadata sets. The structure of these sets depends on the actual data model of a specific system.

Typical metadata gathered during acquisition is mainly material related and may include:

- video source format;
- video compression format;
- audio source format;
- audio compression format;
- recording parameters (such as camera and microphone setting);
- production information.

During ingest metadata is added to identify content but also to support quick retrieval access and to provide an initial rating. This data is usually entered manually into the system. Typical metadata added during ingest is:

- production number;
- title;
- date, time, and location;
- name of source (e.g. agency, cameraman, etc.);
- clip description (textual abstract about the image content);
- mark in/mark out (for clips on collection tapes or files);
- shot quality.

Various systems, departments, and users are involved in the production process. This is also reflected in the metadata that is generated, referenced, and used for retrieval. These different systems are server components (e.g. ingest, production, and playout server), editors and journalists producing an item, craft editors working on the final edit of a content item, and the archive providing material but also cataloging and archiving newly produced items. In all of these processes additional metadata is produced and associated with a content object. The following metadata set shows an example of a set of typical object-related metadata required for post-production and nonlinear editing:

- contribution start/end;
- timecode;

- subtitles;
- commissioning;
- electronic VTR card;
- name of cutter, editor, producer etc.

Additional metadata can be generated and may be collected in other specific metadata sets, e.g. edit set, transmission set, descriptive set, etc. These metadata sets are more explicitly targeted towards a specific task or workflow step.

Once the material has been produced it is ready for distribution. In a broadcast environment the distribution happens according to the program schedule and is recorded in the Program as Broadcast (PasB). Examples of relevant data in this context are:

- date of first broadcast;
- time of first broadcast;
- broadcast channel;
- statistical information (e.g. numbers of viewers and listeners);
- re-broadcasts.

The final step in the life cycle of a content item is the selection and in-depth cataloging. During this step the content is classified and described in detail to make it easier to find, access, and reuse existing material. The metadata added during the production process is amended and consolidated in this step. The applied classification scheme depends on the media type and organization-specific issues. Typical data added in this process is:

- cataloger;
- cataloging date;
- topical classification;
- program affiliation;
- abstract.

During this step a detailed content description is also produced that is part of the segment-related metadata.

4.2.1.2 Rights-Related Metadata

Intellectual property rights (IPR) are a special type of object-related metadata that is closely associated with content exploitation and content management processes. However, IPR are usually managed outside a CMS in a separate rights management system. This is due to complex organizational and legal set-ups that are currently not covered by CMS. Rights management systems keep the IPR status, contractual information, and other legal documents associated with a content object. The information administered in the rights management system should be linked to the content objects managed in the CMS.

IPR describe the rights ownership situation and usage restrictions in relationship to a specific content object. These rights can be very complex, and almost always require interpretation by legal experts. Rights may comprise ownership rights (authors, composers, directors, photographers, etc.), performer rights (actors, musicians, etc.), personality rights, and many more. The specific rights of all the rights holders have to be considered when material is (re)used. Thus, they have to be well documented and kept up-to-date for each content object.

Apart from ownership rights, usage restrictions have to be considered. These comprise, e.g.:

- territorial restrictions (usually geographical territories);
- transmission and delivery methods (e.g. TV, cinema, radio, Web);
- transmission and delivery time (e.g. before or after a certain time of day);
- usage period;
- number of transmissions/usages.

It should be possible to display certain basic rights-related data within a CMS and also highlight simple property-rights-related issues directly in a content object. However, for organizational and legal reasons these can only be indications. A CMS user has to get further clarification on the rights situation from the legal department to ensure that all rights are honored and that content can be used in a specific context.

4.2.1.3 Data Models and Metadata

Data models play a crucial role in CMS; they allow the structured representation of object-related metadata within database systems. A detailed discussion of data models is out of the scope of this book. However, some basic concepts and principles that should be considered in the context of CMS are important.

A data model is a (mathematical) formalization with a notation for describing data and a set of operations to manipulate it (Ullman, 1988). Data models have evolved over the years and there are a number of approaches to how information should be represented in a structured way. There are entity-relationship models, network models, hierarchical models, relational models, and object-oriented models. The entity-relationship model allows the description of an organization in a natural way with a conceptual scheme using an entity relationship diagram. Entities in this context are identifiable objects. The relationships between them indicate how different entities within a model are related. The network data model is comparable to the entity-relationship model with all relationships restricted to binary, many-to-one relationships. Within the hierarchical model the relationships between entities are basically organized in a hierarchy.

The relational data model is value oriented. Operations are defined on relations whose results are also relations. Operations can be combined and cascaded easily in the relational model. The mathematical concept underlying the relation model is the set-theoretic relation that is a subset of the Cartesian product of a list of domains.

Objects in an object-oriented model are records with a unique address. Objects can be complex structures and also define classes and inheritance. Types can have sub-types with special properties. Compared to relational models object-oriented models cannot be composed as easily. This is due to the use of abstract data types and the fact that they do not allow operations on the result of other operations.

The different advantages and disadvantages of data models for content representation have been assessed. Stonebraker (1996) suggests the classification shown in Figure 4.2. This classification takes the data complexity and likely complexity of queries to find the right data model for a specific task as reference coordinates. For data with a simple

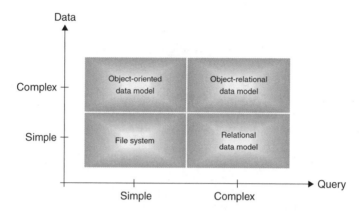

Figure 4.2 Data model classification according to Stonebraker (1996)

structure and simple queries, indexed files stored in a file system are deemed suitable. Complex data that can be queried using simple queries suggests an object-oriented data model. According to this classification the relational data model should be used for simple structured data but complex queries. Where both the data and queries are complex an object-relational data model should be used. The latter is the case for content; its structure is relatively complex and because of the multitude of workflows and use cases that have to be supported the queries are also complex.

However, at the selection of an appropriate data model and the corresponding DBMS other criteria have to be considered as well. There are, for instance, existing databases that might have to be extended or integrated into a CMS. Further, existing expertise within an organization has to be taken into account. Most important in this context is that a candidate data model can capture all relevant data and processes. Thus, an organization designing a CMS should first delineate all characteristics, use cases, workflows, and processes in order to derive the relevant elements and attributes. Further, the links and relationships between entities and processes also have to be considered. Afterwards the appropriate data model (taking into account all the relevant entities and relationships between them) can be developed.

At the heart of a specific data model for a CSM is the content object with all related properties, tasks, workflows, and actors.[1] A data model has to take all the object-related metadata into account. The different relationships and possible interactions with the content objects are mapped onto the data model. Further, the workflow steps and roles of the various users also have to be reflected in the data model for a specific organization. Figure 4.3 shows an example of the stages involved in content creation that can contribute metadata for workflow enhancement. This can be used for the development of a data model covering this specific area.

4.2.2 SEGMENT-RELATED METADATA

Apart from metadata describing an object there is also metadata describing a segment, i.e. part of a content object (conventionally delimited by timecodes or region coordinates).

---

[1] Note, actors in this context refers to all acting entities (human or automatic) that handle content.

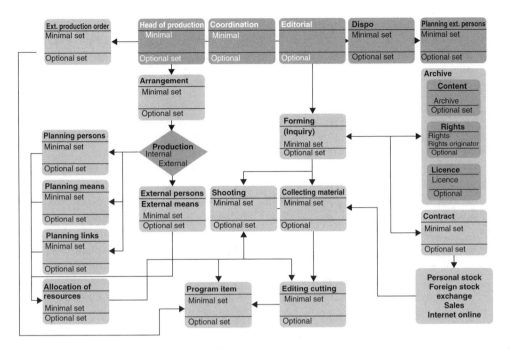

Figure 4.3   Different stages of content creation

The advantage of segment description is the high degree of freedom and flexibility in cataloging specific parts of a content object in depth. This enables catalogers to describe important events in appropriate detail. Further, it allows targeting a search and retrieval operation more specifically and directing it to a particular part of a content object.

Segments can represent different logical parts of a content object. A segment can be spatial or temporal. The former can refer to either a specific region in a content object or (as for instance in MPEG-4) to specific objects present within a content object. In the case of temporal segmentation a segment is identified by the object ID and the start and end timecodes of this segment (or the start timecode and the duration of the segment[2]). A time-line is used as reference scale for the media object timecodes. Figure 4.4 shows the temporal segmentation using a time-line.

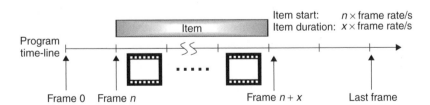

Figure 4.4   Temporal segmentation along a continuous media object time-line

---

[2] Note, items with discontinuous timecodes and possible repetition of timecodes have to be treated separately. However, in this case it should be possible to replace the original time-line with a continuous time-line.

The time-line can be a logical construct to which segments can be associated. The physical time-line of an essence object has to be related to the logical time-line of the documented program. The ability to single out specific segments on any copy should be independent of the storage format of the essence and the integrity and continuity of the content.

The logical decomposition of content using timecodes is traditionally done in log sheets that give an image description of a content item. Timecodes are used to locate a certain event along the time-line of a content object. This is described in detail using either free text or classified documentation schemes and thesauri.

A very powerful concept to add time-oriented annotation is the so-called stratification. Stratification uses the spatio-temporal nature of media objects by introducing anchor points that delimit pieces of the complete media object to which the descriptions can be linked. For example, temporal stratification makes use of time references (associated with the respective time-line) to point to a specific part in the audio or video object. In contrast to traditional log sheets, however, each stratum can be dedicated to specific topical or descriptive themes. The individual descriptions can focus on particular concepts such as image description, keywords, close caption text, persons present, etc. Each of the strata can be segmented individually by associating timecodes with the strata segments. These segments do not have to be related to any other segments or structure (e.g. the shot structure) of the media object. With automatic analysis and feature extraction tools, automatically retrieved information can be associated with a dedicated stratum that uses the automatically retrieved timecodes as segment delimiters. This complements the manual documentation of the catalogers. Hence with stratified documentation human expertise can be combined with automatically generated documentation.

In the case of continuous media the segment description relates time-independent textual information with time-dependent audiovisual information. Some multimedia CMS use a combination of audiovisual and textual information to provide segment-based documentation in so-called storyboards. Here keyframes, segment description, and timecodes are combined to provide an overview of the image content. This is also linked with a copy of the preview video. This technique combines the different media types to form a true multimedia representation of a media object.

Figure 4.5 shows an example of stratified documentation. Metadata such as keywords spotted, closed caption text, copyright or image content description are described on distinct levels. Each level may employ a completely different time-line. It only has to be ensured that a common reference point (i.e. starting point) exists.

During retrieval combinations of strata may be queried to identify exactly the segment the user is interested in. For instance when looking for the segment where a person makes a statement in the presence of another person at a specific location one can query the 'person present' stratum for the name of the involved personalities, the 'closed caption' stratum for the quote, and the 'places' stratum for the location. The combination of the results should point exactly to the desired segment.

The same approach can be applied to spatial stratification. In this case the image is virtually broken into different parts that are then described separately. These parts can be either objects or just regions within a media object.

The costs of cataloging but also retrieval cost in this context are largely dependent on the strategy adopted for the documentation of content within a specific organization.

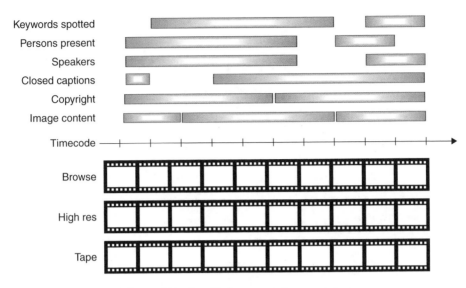

Keywords spotted
Persons present
Speakers
Closed captions
Copyright
Image content
Timecode
Browse
High res
Tape

Figure 4.5   Stratified segment documentation

Effort put into the documentation can improve the reuse quota if it enables users to find the required content item or segment more quickly (or in some cases to find it at all).

4.2.3 LOGICAL CONTENT STRUCTURE AND CONTENT HIERARCHIES

Content objects very often have links with other objects or are part of a hierarchy such as program collections or program series. Apart from representing the metadata pertaining to a specific content object these structures and hierarchies also have to be reflected within a CMS. The following hierarchical structure can be identified:

- *shots* (i.e. sequences of frames between a transition);
- *program items* (single entities that are part of a larger unit, e.g. a news story, an interview, a performance);
- *program* (i.e. logically related chunks of program items that are part of the same transmission or program schedule entity);
- *collections of programs* (grouping of programs that are part of the same product, e.g. a TV series).

Whereas shots are covered by segment descriptions and program items are sufficiently described by object-related metadata, programs and collections of programs need to be considered separately. One way to describe programs is to use links and relationships between the related items. However, this is not sufficient since there is metadata pertaining to a content collection that does not characterize the single items within the hierarchy. Thus, in most cases collections of program items have to be represented as objects in their own right.

The relationship between the different objects within the hierarchy can be expressed using *hierarchical trees* and *relational graphs*. Hierarchical trees as shown in Figure 4.6

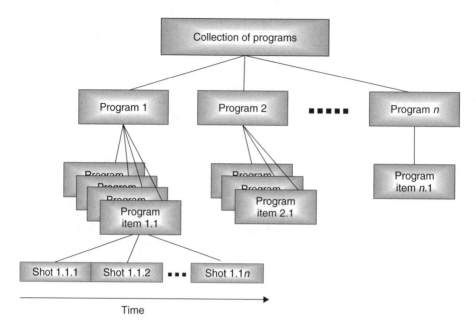

Figure 4.6   Example of hierarchical tree

should be used to express hierarchical relations such as those of program items and programs. Relational graphs are used to express more loose relations such as the grouping of news stories related to a given event or topic. As can be seen from the diagram, there can be more than just two or three levels in the object hierarchy depending on the structure and organization of the different content objects.

Exploiting these hierarchical relationships allows to target a search to a specific level within the hierarchy. For instance if a user wants to retrieve full programs but the original search produces results at item level the relationships can be used to find the object or segment representing the desired granularity.

4.2.4 OBJECT REFERENCES

Object references play a key role in a CMS since they unambiguously identify a content object. They are part of the metadata but also link the metadata to the essence. Apart from identifying a content object as an entity, they can also refer to different content components (such as essence objects) and link the different representations of a content object. Further, they can be used to link related content objects. In order to enable the exchange of content these references should be based on unique identifiers. They have to be at least unique within an organizational context but also might have to be used between organizations. In this case globally unique and possibly registered identifiers are required.

It is important to distinguish between internal system identifiers (such as database keys) and the object identifiers of content objects. The former only identify an object within the context of a specific system. In a content management context this is not sufficient since a content object can be represented in a number of systems, each focusing on specific aspects (e.g. accounting, IPR, program information, etc.). Thus, object references have a much wider implication since they represent the object and not just a specific instantiation.

Apart from uniquely identifying a content object such an identifier can also contain additional information about the source of the content object, its ownership, inception, registration body, etc. encoded within the identifier. A number of different schemes have been proposed. The following kinds of object references can be distinguished:

- *Object unique identifiers* are proprietary references for unique identification required within a local domain or organization. The concept of object unique identifiers can also be used to simplify the integration of legacy systems into a CMS. These identifiers should not be communicated and used outside the scope in which they have been defined.
- *Unique material identifier* (UMID) is proposed for the identification of essence items. UMID is standardized by SMPTE (see Section 4.4.3.3). The UMID format allows the automatic generation of globally unique identifiers based purely on local information. This enables the generation of UMID during recordings in the field. The granularity of identification (how many video frames or audio samples are identified by the same UMID) depends on the one hand on the anticipated usage and on the other on the capability of the storage media to hold metadata. UMID also allows the expression of relations between essence items and between programs and the corresponding essence.
- *Unique program identifier* (UPID) is proposed for tradable program items with associated copyright statements. UPID can have legal implications and hence it is preferable to choose formal, registered identifiers. For instance ISAN (International Standard Audiovisual Number) is registered with the International ISAN Registration Agency, which assigns and maintains the unique identifiers permanently. In this process a set of additional information (e.g. about the registering organization) can also be recorded with the registration body.

Other references deal with specific aspects, in particular time. This is important to register timing, creation, production, and transmission events. Time-related references are, for instance:

- *Time references* are general structured time-stamps that are used to express relationships of the content object to real-world or media stream time coordinates.
- *Real-world time representation* is necessary when there is the need to synchronize metadata and essence items based on real-time events, for instance when several essence versions are independently captured. They can be synchronized by logging the capture time associated to the original source. Thus, indices within a particular version can be automatically propagated to all the others by using the real-world time representation. Several formats are available to express real-world time that are variants of either time unit count since a reference date or the Gregorian date and day time format. The former is often implemented by computer operating systems since it allows easy computation of time spans. The latter is the humanly readable time and date format.
- *Media stream time* is a relative time reference of a specific content object with the start time (presentation of the first frame) as reference point. On the resulting time-line events can be mapped such as time-related logical descriptions of content, temporal segmentations or composition metadata. Unless UMID have been associated to frames,

the media stream time is essential to identify clips within an item (i.e. for segment documentation). Media stream time can be expressed as standard SMPTE timecode or as offset and duration in frame/sample units. Neither of the two can be considered a true representation of the effective time elapsed since the base on which the time is computed is relative to the internal clock of the equipment. This does not consider drift and tolerance of the respective equipment. Mapping the time-lines of different essence versions of the same program can be a complex problem if the essence versions have not been generated in a controlled way. For example, if one essence version has been recorded during broadcast, its duration will probably differ from that of the master, due to fade in/fade outs, commercial insertions, etc.

There are a number of other object references and internal identifiers for content objects. Important in the context of CMS are also *media locators*, which are used to specify where a given item (essence or metadata) can be found. In IT-based systems they can be expressed as uniform resource locators (URLs) according to the W3C definition. When referring to conventional media the physical location has to be addressed according to a unified syntax. In complex systems such as CMS it is recommended that appropriate location services should be implemented. These services have to trace the movement of a content object from one location to another.

4.3 METADATA ACCESS AND EXCHANGE

The purpose of metadata is to make content accessible, easy to find and to exchange. Though metadata in some cases contains enough information to stand by itself, in general it is used to support transactions and processing of the actual media/essence. Hence an integrated search, retrieval, and exchange of metadata is vital for the interaction within a CMS.

4.3.1 SEARCH AND QUERY FOR METADATA

The CMS has to offer unified search capabilities to enable access to the different databases and information systems the CMS interfaces to. Most users will not be experienced database users and will also not be familiar with data models or the representation of content in databases. On the other hand, there are expert users such as archivists and media managers who would like to have instant access using native database queries. Thus, the system has to allow different ways of searching for metadata. The following query forms can be distinguished:

- *Fulltext query* allows querying the system in natural language. The search is usually performed over indexed (sometimes structured) files or using fulltext search capabilities of the DBMS.
- *Query for labels* structures the query according to concepts such as names, places, dates, etc. These concepts are mapped onto suitable attributes of the different involved databases. If metadata is stored in structured files tagged elements can also be included in the search.
- *Query for segment* only searches in the segment description. If a layered documentation model is used it can also be restricted to concepts represented at a specific layer (e.g. persons present or places).

- *Native query* is a search directly applied to a specific database taking into account the data model, query language, and any database-specific restrictions and capabilities.

Queries over a number of databases and information systems are actually common practice. Thus, *federated search* and the consolidation of results have to be supported. If content objects are documented in multiple systems the consolidation of results is achieved via ID matching. How the results are presented to the user is dependent on the role of the user and the application. Databases and fulltext search engines can be queried using their native interfaces. A sensible approach for interfacing to such information systems is using XML messages for information transfer.

4.3.2 METADATA EXCHANGE

Apart from direct access metadata is exchanged on an intra-organizational or inter-organizational level. Within an organization metadata is exchanged to support workflows in media production and other business processes. With a higher level of system integration this kind of metadata exchange will become less important since databases and information systems can be accessed directly from the various applications. Business-to-business metadata exchange occurs between collaborating organizations and in media sales. Here metadata is exchanged at the delivery, reception, and interaction between the business partners. Due to the use of proprietary data models and information systems the overhead in the exchange of metadata using standard data models, metadata dictionary, exchange protocols, etc. is considerable. Figure 4.7 shows the steps involved in exchanging metadata between organizations. In order to exchange metadata it is necessary to encode it in a non-proprietary (preferably standard) data model. However, the common data model is no guarantee that both sides exchanging metadata have a common interpretation of the

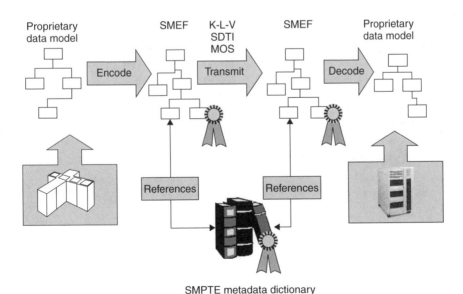

Figure 4.7 Inter-organizational metadata exchange

elements and values. This is ensured by referencing a standard or mutually agreed dictionary that comprises all relevant concepts, elements, and their attributes and values. For transmission the metadata has to be serialized and encoded using a standard transmission protocol or container format.

On a business-to-consumer level metadata has only very recently been introduced. It provides additional information at the delivery and reception of media (for instance in digital broadcasting). The use of metadata in this context also supports some limited interaction.

4.4 METADATA DESCRIPTION SCHEMES

In order to describe a content object it is necessary to consider its usage as well as its structural and content properties. An objective, comprehensive description that comprises all features and supports the workflows of all content-rich organizations (such as an enterprise-wide unified data model) does not seem to be feasible. A number of attempts to define such comprehensive description schemes and data models have been made but none has been found to fully satisfy all requirements of the various organizations and humans that deal with content.

However, these generic description schemes and data models are excellent reference models for the analysis of workflows as well as the design of new or validation of existing data models or schemes for enterprise specific applications. In addition, they can be applied to support content exchange between systems or organizations, i.e. they facilitate the exchange of essence and the descriptive metadata (as shown in Figure 4.7). In the following the currently most important description schemes and data models are reviewed to provide an overview of the different metadata models proposed. Their relevance in the context of specific systems, projects or workflows has to be assessed individually.

4.4.1 SMEF AND P/META

The Standard Media Exchange Framework (SMEF^{TM}) (BBC Technology, 2002) is being developed by the British Broadcasting Corporation's (BBC) Media Data Group (BBC, 2002). SMEF was originally conceived as a data model for the BBC but has later been developed into an *exchange* data model. Hence it does not represent a standard data model applicable to the entire BBC, but is a reference model that attempts to capture all relevant entities, workflows, and processes within the BBC. SMEF covers the entire content life cycle, including media production and all processes involved in handling media at a broadcaster. More specifically, the areas covered are:

- plan, draft, and commissioning;
- content creation;
- promos and programme scheduling;
- playout and transmission;
- media management and archive.

Although SMEF is not deemed to provide the 'required comprehensive level of detail' (BBC Technology, 2002) it is probably the most extensive data model at present that systematically looks at all the aspects involved in content creation and management within a broadcaster. SMEF is an ongoing project and considers input from projects within the BBC but also from other interest groups outside the BBC. The definitions within SMEF are organization independent. Thus, it should be suitable as a reference model for any content-rich organization. SMEF covers all the basic information requirements of new or existing systems. Thus, SMEF provides an initial set of data definitions for content centric projects.

P/Meta is a European Broadcasting Union (EBU) PMC Project intending to facilitate the exchange of metadata of program material (European Broadcasting Union, 2001). The goal of this working group is to develop an EBU standard exchange data model (called Euro SMEF). The BBC's SMEF data model is providing the core information architecture in this context. Euro SMEF also considers additional contributions from other EBU members. The working group takes into account other standardization activities, especially at SMPTE (UMID, Metadata Dictionary), EU INDECES, and the DOI Foundation.

4.4.1.1 SMEF Basics

The core elements of the SMEF data model are media objects described by metadata. The metadata incorporates not only data describing essence properties and related characteristics, but also the use of essence within programs (for instance as part of a service outlet through program schedules), and the organizational entities dealing with it.

The data model is described in terms of *entities* that are defined as 'some thing' an organization maintains information about (BBC Technology, 2001). One or more attributes may be linked to an entity. They describe properties that pertain to an entity. Key attributes are used to uniquely identify individual entities. Each entity and its subtypes are identified by a unique key.

Relationships represent associations between entities. Thus, SMEF uses an entity-relationship concept as a formalization to describe the information and processes related to content. The relationships in the SMEF data model are defined by business rules derived from the business processes that are modeled. SMEF aims to cover all business processes surrounding content within a broadcast organization comprehensively. It uses entity-relationship diagrams to illustrate how entities are related to each other.

The values of the attributes of the different entities are usually dynamic (i.e. data that is altered and deleted by the users). However, reference data can also be used within SMEF. These are predefined values and codes such as, for instance, the ISO country codes that can be used in a SMEF-compliant system. Reference data complements the dynamic data.

4.4.1.2 The SMEF Entity Structure

Entities in SMEF are the information units that embody a specific concept related to the management and handling of content. All the major concepts required to model the objects and processes at a broadcaster have been captured by the list of entities specified in SMEF. However, new entities may be added in future versions if they are found missing. The actual information is represented by attributes pertaining to a specific entity. Each entity has an individual set of attributes.[3] Entities can have subtypes that are a specialization of a super-type.

---

[3] Note, the unique key is not considered as a descriptive attribute and hence not discussed in this section.

At the center of the data model is the editorial object described by the EDITORIAL_ OBJECT_VERSION. The concept of version is an intrinsic part of the EDITORIAL_OB-JECT_VERSION entity. Versions can for instance be pre- and post-watershed versions of the same program. Sixteen attributes are defined for the editorial object, including a unique identifier, title and subtitle, creation date and time, duration, and a synopsis. Subtypes of the EDITORIAL_OBJECT_VERSION include PROGRAMME_OBJECT_VERSION (for the representation of programs that can be a unit of transmission), MUSIC_SPEECH_ SOUND_ITEM_OBJECT, and OTHER_ITEM_OBJECT. A closely related entity of the EDITORIAL_OBJECT_VERSION is EDITORIAL_OBJECT_VERSION_WORK, which maintains a record of the entire creation and development process of an editorial object. The EDITORIAL_OBJECT_VERSION_INSTANCE represents a permanent or temporary instance of work carried out for an editorial object version. An example is a record of a copy that is taken off-site. Table 4.1 shows the different entities related to the editorial object concept, their attributes, and possible sub-types.

Associations of content objects are important in various contexts in a content workflow. Within SMEF three different associations or groupings have been defined (reflecting the different concepts considered by SMEF). An editorial object in the SMEF data model can be part of a group that represents a hierarchy (e.g. series made up of episodes or program strands). This is represented by the EDITORIAL_OBJECT_GROUP entity. Another kind of grouping considered within SMEF is content collections such as CD or records. The ACQUISITION_BLOCK entity has been specified for these kinds of groups. The EDITORIAL_OBJECT_ASSOCIATION is used to link two editorial objects (e.g. in playlists).

The second important content object-related entity in the SMEF data model is the MEDIA_OBJECT. It holds general and editorial metadata about an object of a *single* media type, such as an audio clip, video segment, text, graphic or still element. Thus, media objects in SMEF can be regarded as basic content elements. The MEDIA_OBJECT entity has nine attributes, mostly referring to date and time concepts (i.e. action start data and time, capture start date and time, and creation date and time), title, and an editorial description. The sub-types of the MEDIA_OBJECT are AUDIO_CLIP, DATA (e.g. captions, Web sites, text, teletext data), GRAPHIC, SHOT, and STILL. Media objects can be grouped (represented by MEDIA_OBJECT_GROUP). This is for instance used for the grouping of events such as several shots of the same event on a football pitch. The possible types of media object groups are action, sensor, perspective, and sound. In this context it is important that these groups do not establish a hierarchy but rather a bracket for related media objects.

The MEDIA_OBJECT is a logical or editorial view of a content object. The physical copy of the object (i.e. the actual essence) is represented by the UNIQUE_MATERIAL_ INSTANCE entity. It has four attributes (UMID, compression ratio, creation date, and creation time) and three sub-types (DATA_INSTANCE, AUDIO_CLIP_INSTANCE, and PICTURE_INSTANCE). SMEF also specifies material or essence-related metadata such as coding standards (e.g. image and audio coding standards, compression schemes, etc.) that can be used for the documentation of the essence part of a content object.

These three entities represent concepts that are closely related and have to be jointly used to fully represent a content object. An example of how they are related is for instance a MEDIA_OBJECT that is (re)used in an EDITORIAL_OBJECT_VERSION. In

Table 4.1   Editorial object entities

Entities	Attribute list	Sub-Types
EDITORIAL_OBJECT_ VERSION	EOV_ID EOV_Billing_Desc EOV_Colour_Indicator EOV_Creation_Date EOV_Duration EOV_End_of_Speech_Duration EOV_In_Credit_Seq_Desc EOV_NCS_Slug_Name EOV_NCS_Slug_Type EOV_Out_Credit_Seq_Desc EOV_Sub_Title EOV_Suspended_Date EOV_Synopsis_Desc EOV_Title EOV_Transmittable_Ind EOV_Working_Title	ITEM_VERSION MUSIC_SPEECH_SOUND_ ITEM_OBJECT OTHER_ITEM_OBJECT PROGRAMME_OBJECT_ VERSION
EDITORIAL_OBJECT_ VERSION_ INSTANCE	EOI_ID EOI_Creation_Date EOI_Creation_Reason_Desc EOI_Technical_Acceptance_Ind	
EDITORIAL_OBJECT_ VERSION-WORK	EOW_ID EOW_Creation_Date EOW_Creation_Time EOW_Desc EOW_Editorial_Acceptance_Ind EOW_Last_Update_Date EOW_Last_Update_Time EOW_Name EOW_Production_Number	
EDITORIAL_VERSION_ CREATION_ REASON	EVR_Name EVR_Description	

this case a copy of the original UNIQUE_MATERIAL_INSTANCE is associated with an EDITORIAL_OBJECT_VERSION to model such a relationship.

Apart from the content object itself all the handling, documentation and management processes are captured by SMEF. The entities STORAGE and STORAGE_TYPE describe the physical carrier of the essence. The storage media can for instance be tape or other carriers. For cataloging and documentation the EDITORIAL_DESCRIPTION_SCHEME entity specifies the description context. Several description schemes exist within the BBC that can all be used with SMEF. Since other institutions should also be able to use SMEF other description schemes can also be used in this context. The EDITORIAL_DESCRIP-TION_TERM is a reference table that specifies all allowed terms for the editorial description. Different categories have been specified as sub-types. A genre scheme is for instance also part of the editorial description concept. Further, there is a wide range of entities that describe specific aspects of the media object, for example language, literary works used, locations, origin of music, youth classification, story, etc. Description terms can also be used to form and describe hierarchies. SMEF also contains entities dealing with copyright issues related to different media types (e.g. music, still, script, software, etc.), and contractual rights (for instance related to a specific outlet).

A number of entities have been defined to describe individuals and organizations that deal with content in various roles. The generic entities PERSON and ORGANISATIONS are used to represent them in the data model. ROLE is used to specify the different rights and responsibilities that specific persons or organizations have towards the content. Contractual information is also part of the SMEF data model.

The different media production workflows within a broadcast organization are comprehensively addressed by SMEF. This is reflected in a number of entities dealing with production-related issues. The commissioning process is for instance considered by the BRIEF, OFFER, and COMMISSIONED_PROJECT entities. The actual production process is implicitly represented by the different versions of the editorial and media objects created during the different production steps. SMEF also provides the notion of media folders that can be used during production to group material. Further, the data model also covers all relevant aspects related to the scheduling and transmission of programs. Examples of entities in this context are DELIVERY_REQUIREMENTS, EMISSION_OUTLET_LOCATOR, CONSUMPTION_LOCATION_TYPE, PUBLICATION_EVENT, PUBLICATION_DE-PENDENCIES, TERRITORY, OUTLET (including the WEB), POPULATION_CATEGORY, and POPULATION_GROUP.

4.4.1.3 The SMEF Data Model

The SMEF Data Model is represented in a diagrammatic fashion in the specification document. Entities are shown as rectangles in the data model diagrams that are joined by lines representing relationships. The lines also show the cardinality of relationships and if they are optional or mandatory. Figure 4.8 shows an extract of the SMEF Data Model from version 1.6. This represents a contract negotiation process.

A solid line indicates that a relationship is used as part of the identifying key for the entity that is described in further detail. In the example the CONTRACT is further detailed by CONTRACT_LINE. The symbols at the end of a line indicate the number of occurrences that may exist in a context. The inner symbols specify if an entity occurrence has to exist in a relationship. In the example the CONTRACT contains one or many CONTRACT_LINE, and CONTRACT_LINE must be contained in one CONTRACT. Further, according to the example a CONTRACT must be negotiated by one PERSON (who can negotiate zero or many contracts). The RIGHT must be contained in one CONTRACT_LINE, and the CONTRACT_LINE may contain zero or many RIGHT.

Figure 4.9 shows a small extract (from the SMEF data model version 1.7) of relationships that exist between EDITORIAL_OBJECT_VERSION, EDITORIAL_OBJECT_GROUP, and EDITORIAL_OBJECT_GROUP_ASSOCIATION. Further, the relationships to PROGRAMME_OBJECT_VERSION and EDITORIAL_OBJECT_ASSOCIATION is shown.

An EDITORIAL_OBJECT_GROUP may contain zero to many EDITORIAL_OBJECT_VERSION. Further, the EDITORIAL_OBJECT_VERSION can be linked to zero or more

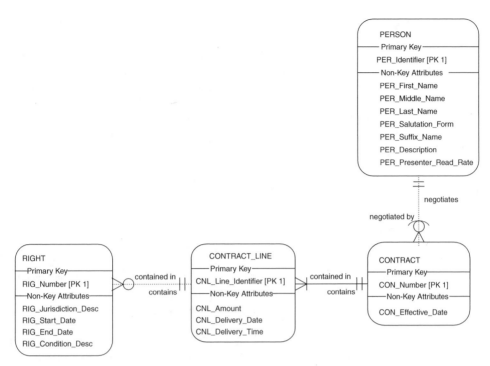

Figure 4.8 Contract negotiation in SMEF data model (v1.6)

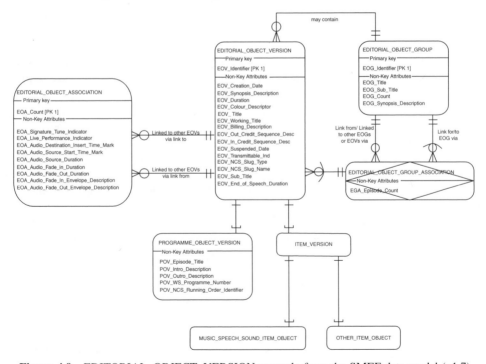

Figure 4.9 EDITORIAL_OBJECT_VERSION example from the SMEF data model (v1.7)

EDITORIAL_OBJECT_ASSOCIATION. Two different kinds of links depending on the origination of the link exist in this context.

The EDITORIAL_OBJECT_GROUP_ASSOCIATION is an entity that allows grouping of editorial objects and enables these groups to be part of other groups. Thus, group hierarchies can be created where each object group association may include one or more editorial groups and/or individual editorial objects. An EDITORIAL_OBJECT_GROUP_ASSOCIATION must be linked from one EDITORIAL_OBJECT_GROUP. Further, it must be either the link for one EDITORIAL_OBJECT_GROUP or for one EDITORIAL_OBJECT_VERSION. Hence, according to this example of the SMEF data model an editorial object representing a content object is part of an editorial object group or editorial object group association. It can reference editorial object associations such as EDL. Further, it is linked to program object version and item version. Thus, this example shows the context for an editorial object within SMEF.

Full discussion of the SMEF data model with all entities and the relationships between them is out of the scope of this book. It is fully documented in the SMEF specification document. All the relevant workflows and relationships identified by the BBC Media Data Group and within projects at the BBC or by external partners (such as P/Meta) have been considered in this model. The entire data model with all possible relationships is depicted in the specification document. Since it is rather extensive it is split into a number of diagrams that show the relationships between all the entities specified in SMEF. The diagrams deal for instance with the relationships between editorial and media object-related entities, contract and role entity relationships, commission and editorial genre-related entities and their relationships, and also all outlet-related entities and their relationships.

4.4.2 MPEG-7

MPEG-7 is a standardization initiative of the ISO/IEC Moving Picture Expert Group (MPEG, 2002b). In contrast to other MPEG standards (MPEG-1, MPEG-2, and MPEG-4) MPEG-7 is not concerned with the encoding of the audiovisual part of the content (i.e. essence) but specifies a Multimedia Content Description Interface. MPEG-7's goal is to provide a standard that is widely applicable and not specific to any application domain or content type. The metadata specified by MPEG-7 is applicable at different levels and defines low-level and high-level features. The high-level features defined by MPEG-7 cover metadata such as title, summaries, events, and usage history whereas low-level features are concerned with concepts such as color schemes, encoding, and region motion tracking. MPEG-7-based descriptions should be able to stand alone, being multiplexed with the essence, or associated with one or more versions of the essence. MPEG-7 descriptions deal with content objects that are independent of the media or carrier, or the encoding format of the essence (Manjuhath *et al.*, 2002).

A key role in MPEG-7 is played by the tool set representing the interface. It supports the description of content by human users as well as automatic systems processing of media/essence. These tools constitute the three main elements of MPEG-7, namely *Descriptors, Description Schemes*, and the *Description Definition Language* (DDL). A descriptor defines the syntax and semantics of audiovisual feature representation within MPEG-7. Building on this, the Description Scheme specifies the syntax and semantics of the relationship between these components, and allows the modeling and describing of multimedia content. The DDL defines the syntax that MPEG-7 compliant description tools have to use. It specifies the use of XML schema in an MPEG-7 enhanced form for Descriptors and Description Schemes (Manjuhath *et al.*, 2002).

The description of a content object itself is organized hierarchically in a tree structure. In this description tree the nodes represent the descriptive information whereas the links represent a containment relationship between nodes.

The MPEG-7 standard consists of eight parts (as shown in Figure 4.10) reflecting the major functionality specified by MPEG-7 (MPEG, 2002b).

4.4.2.1 MPEG-7 Systems

The *MPEG-7 Systems* part identifies the tools that are required to compile and process MPEG-7 descriptions considering efficient transport and storage (also in binary format). It also specifies the terminal architecture and normative interfaces. A terminal in the context of MPEG-7 is an entity that makes use of the MPEG-7 description. It can be a stand-alone application or be part of an application system.

Further, the MPEG-7 Systems part deals with languages for the representation of Description Schemes and of binary and dynamic descriptions. It specifies incremental delivery of descriptions by so-called access units (AU). AUs are structured as commands

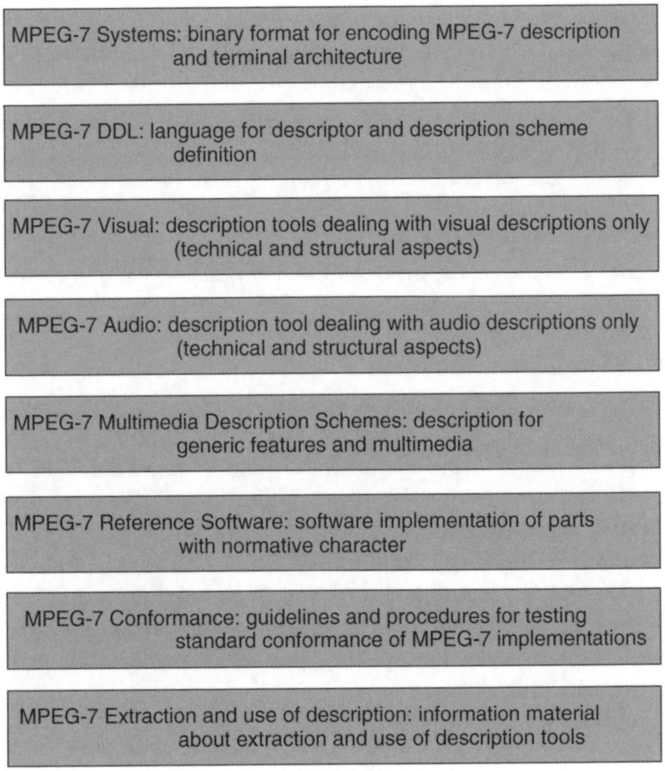

MPEG-7 Systems: binary format for encoding MPEG-7 description and terminal architecture

MPEG-7 DDL: language for descriptor and description scheme definition

MPEG-7 Visual: description tools dealing with visual descriptions only (technical and structural aspects)

MPEG-7 Audio: description tool dealing with audio descriptions only (technical and structural aspects)

MPEG-7 Multimedia Description Schemes: description for generic features and multimedia

MPEG-7 Reference Software: software implementation of parts with normative character

MPEG-7 Conformance: guidelines and procedures for testing standard conformance of MPEG-7 implementations

MPEG-7 Extraction and use of description: information material about extraction and use of description tools

Figure 4.10   MPEG-7 standard parts

encapsulating the MPEG-7 Descriptions (i.e. parts of the description tree). The binary format for MPEG-7 (BiM) has been defined to support compression and the streaming of content descriptions. It can either be parsed and interpreted directly by MPEG-7 tools, or might be mapped onto the DDL textual description (a bidirectional mapping exists between BiM and the DDL textual description). Thus, MPEG-7 descriptions can also be made available in human readable form.

4.4.2.2 MPEG-7 Data Description Language (DDL)

The *MPEG-7 DDL* provides the language for the definition of the structure and content of MPEG-7 documents. Descriptors and Description Schemes must conform to the syntactic, structural, and value constrains specified by the MPEG-7 DDL. In order to accommodate the specific requirements of individual applications, syntactic rules are provided according to which existing description schemes can be combined, extended, and refined. Further, it must be possible to express relationships between Descriptors and Description Schemes (namely structural, spatial, temporal, spatiotemporal, conceptual, and inheritance relationships).

The MPEG-7 DDL is based on XML Schema Language enhanced by MPEG-7 specific extensions. These extensions were necessary to accommodate the special requirements of audiovisual content description. The MPEG-7 DDL can be broken down into three components (Manjuhath *et al.*, 2002):

- *XML schema structural language components,* which include namespaces (for assigning universally unique names) and scheme wrappers around the definitions and declarations, element declarations (specifying type definitions, default information, and occurrences of a schema element), attribute declarations (enabling attribute definition), and type definitions (defining internal schema components).
- *XML schema data type language components,* which include built-in primitive data types (e.g. string, Boolean, decimal, float, etc.) and built-in derived data types (a set of data types derived from primitive data types).
- *MPEG-7 specific extensions*, i.e. array and matrix data types, and *basicTimePoint* and *basicDuration* built-in derived data types (based on ISO 8601).

4.4.2.3 MPEG-7 Visual and Audio Parts

The *MPEG-7 Visual* and *Audio* parts deal with description tools concerned with the technical and structural aspects of visual and audio descriptions. The visual descriptors have been defined to facilitate the identification and categorization of image or video objects using low-level descriptors. The goal is that visual objects can be grouped and classified using this information. Retrieval and filter operations based on these descriptors should enable the retrieval of images and videos with certain properties. In this context it is important to note that the search and retrieval operation does not have to be based on a textual query but can apply search by example mechanisms (e.g. 'give me all images that contain this object' or 'give me all images with a similar dominant texture'). The MPEG-7 descriptors include:

- *color descriptors* describing color distribution, spatial layout, and structure of color;
- *texture descriptors* describing visual patterns such as homogeneity, color multiplicity and intensity;

- *shape descriptors* describing the shape of visual objects (they are based on region and contour descriptors);
- *motion descriptors* describing motion in video objects, for instance camera motion and object motion;
- *face descriptors* describing specific facial characteristics for applications such as face reorganization (they are based on 48 basis vectors that span the space of all possible face vectors).

The Audio part of MPEG-7 specifies low-level features but also includes high-level Description Tools. The audio tools are intended to be used in applications such as query by humming, query for spoken content, and in the support of audio editing.

The generic low-level tools apply to any audio signal. They are part of the audio description framework and comprise the scalable series, low-level descriptors (LLD), and uniform silence segment. Table 4.2 shows the 17 temporal and spectral audio descriptors and the groups into which they are divided. The LLD are all defined referring to two sub-types, namely *AudioLLDScalarType* (for single-value descriptors, e.g. power of waveforms) and *AudioLLDVectorType* (for multi-value descriptors, e.g. spectrum information).

The high-level description tools build on the foundation layer in the standard. They have been specified with specific applications in mind (for instance to facilitate query

Table 4.2   Basic audio descriptors

Group	Descriptors	Description
Basic descriptors	*AudioWaveformType* *AudioPowerType*	To describe instantaneous waveforms and power values
Basic spectral descriptor	*AudioSpectrumEnvelopeType* *AudioSpectrumCentroidType* *AudioSpectrumSpreadType* *AudioSpectrumFlatnessType*	To describe low-frequency power spectrum and spectral features, including spectral centroid, spectral spread, and spectral flatness
Basic signal parameter descriptors	*AudioFundamentalFrequencyType* *AudioHarmonicityType*	To describe the fundamental frequency of quasi-periodic signals and the harmonicity of signals
Temporal timbral descriptors	*LogAttackTimeType* *TemporalCentroid*	To describe log attack time and temporal centroid of a single, well-segmented sound
Spectral timbral descriptors	*HarmonicSpectralCentroid* *HarmonicSpectralDevination* *HarmonicSpectralSpread* *HarmonicSpectralVariation* *SpectralCentroid*	To describe specialized spectral features in a linear frequency space (including a spectral centroid) and spectral features specific to harmonic portions of signals (including harmonic spectral centroid, harmonic spectral deviation, harmonic spectral spread, harmonic spectral variation)
Spectral basis representation descriptors	*AudioSpectrumBasisType* *AudioSpectrumProjectionType*	To describe parameters for the projection into a low-dimensional space. Used in conjunction with sound recognition

operations). The tools for sound recognition and indexing are intended for indexing and categorization of sound and sound effects. This includes a taxonomy of sound classes and tools for specifying an ontology of sound recognizers. Spoken content tools do not directly produce a simple textual transcript but rely on combined word and phone lattices for each speaker in a sound stream. They are aimed at supporting indexing and retrieval of spoken content in audio streams and annotations. Further, there are tools for musical instrument timbre description tools (descriptions related to attack, brightness, and richness) and melody description tools.

4.4.2.4 MPEG-7 Multimedia Description Schemes (MDS) Tools

The *MPEG-7 MDS* tools are concerned with the description of content at different levels. They comprise tools for generic descriptions as well as multimedia descriptions. They cover vector and time parameters but are also concerned with textual descriptions and controlled vocabularies that can be generally used for the description of content. Multimedia description tools are more complex and are used when more than one medium has to be described. Figure 4.11 gives an overview of the MPEG-7 MDS and the structure of this part of the standard. The main components of MDS are *Basic Elements, Content Description, Content Management, Content Navigation and Access*, and *User Interaction*.

The Basic Elements (schema tools, basic datatypes, link and media localizations, and basic tools) are used throughout the definition of the MPEG-7 Description Schemes (DS). Content Description components describe the structure and semantics of content objects. The former includes regions, video frames, and audio segments. It is intended to be used for the description of segments that are part of a content object. These segments are delimited by spatial, temporal or spatiotemporal boundaries. The segment description schemes can be organized sequentially or hierarchically. The latter, for instance, allows building a table of content. The structural features can be described by using MPEG-7 Audio and Video descriptors as well as text annotations.

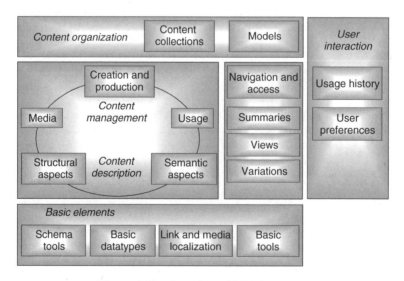

Figure 4.11   Overview of MPEG-7 MDS

The semantic aspects of the MPEG-7 MDS include object descriptions, events, relationships, and abstract concepts. They consider real-world semantics and conceptual notions for the descriptions of content objects. Content semantics are considered as narrative by nature. Thus, MPEG-7 supports the description of narrative worlds that are contained in a piece of content using semantic descriptions. A semantic abstraction model is defined to describe abstract concepts and abstract quantities. There are two types of abstractions, namely media abstractions (i.e. to describe events independent of media) and formal abstractions (i.e. to describe patterns with placeholders and variables for a specific event). The semantic entities described by the MPEG-7 Semantic Description Schemes are directly related to the narrative world. They cover the description of objects (using Object DS and AgentObject DS), events (using Event DS), concepts (using Concepts DS), states (using SemanticsState DS), places (using SemanticPlace DS), and times (using SemanticTime DS). The SemanticBase specifies basic types for these semantic entities. It has to be noted that some of the semantic descriptors are incongruent to the actual event. The narrative of a film, for example, might be set in London (semantic place) in the 17th century (semantic time) but is actually being shoot in Hollywood (creation place) in 2003 (creation time). Semantic attributes can be used to describe content using labels or text. Further, they can consider properties and features of the content object or segment. Semantic attribute tools can also describe abstraction levels and semantic measurements in time and space (Manjuhath *et al.*, 2002). Semantic relationships can be expressed in normative ways and non-normative ways. Examples of the standardized normative relations are for instance *agent, agentOf, annotates, annotatedBy, component, componentOf, key, keyFor, part, partOf, result, resulOf, user, userOf*, etc. Structural and semantic descriptions can also be related using a set of links that allow content to be described on the basis of the structure and semantics together.

For content management purposes MPEG-7 specifies an extensive set of descriptors. Content creation information includes attributes such as title, and textual annotation (including creator information, creation locations, and date). Further, classification information (comprising genre, subject, purpose, language, etc.), review and guidance information, and related material information (describing links with other material) are also part of the creation information. The usage information covers usage records (e.g. broadcasting, on-demand delivery, CD sales, etc.) and financial description (related to production cost, income created, etc). Rights information is not explicitly included but links are considered to the rights holder and existing information schemes. Media description deals with information about the media such as compression, coding and storage format, and the description of the master (or the original source format). A so-called *Media Profile* describes a copy of the content.

Content organization represents the analysis and variations of a number of audiovisual content objects and their organization in collections. The collection tools describe the collection of multimedia content, segments, descriptor instances, concepts or mixed content. The idea is to use the content organization descriptors to describe structures such as albums of songs. The model tools are divided into probability models, analytic models, and cluster models.

The navigation and access tools are intended to provide summaries to facilitate discovery, browsing, navigation, visualization, and providing an audio impression of the content. The navigation support can be hierarchical or sequential. The view descriptions describe different decompositions of the signals considering spatial, temporal, and frequency aspects. They should support the representation of content in different views (e.g. for multi-resolution access and progressive retrieval). Finally, user interaction description schemes deal with user preferences and usage history associated with the consumption of content.

4.4.2.5 MPEG-7 Reference Tools and Relationships

MPEG-7 specifies an extensive set of low-level and high-level descriptors also considering abstract models, semantic entities, attributes, and relationship. A list of principle concepts relevant in MPEG-7 has been identified (ISO/IEC, 2000). It contains 183 concepts including Author, Bandwidth, Character, Contract, Copyright, Duration, Edit, Image, Language, etc. MPEG-7 does not prescribe any specific model or scheme but provides a standard framework for the description of content. In this context it has to be noted that MPEG-7 specifies a content description interface that does not include a database model or dictionary description. Despite the comprehensive approach MPEG-7 takes, there are still problems mainly related to fusing the language syntax and schemata semantics, the representation of the semantics of media expression, the semantic mapping of schemata, and the general applicability of MPEG-7 (Nack and Hardman, 2002). The latter point refers to the complexity of the internal organization and the need to represent the description in a hierarchical document structure.

Part of the standard is also the *MPEG-7 Reference Software*. The eXperimentation Model (XM) is a simulation platform with normative character for MPEG-7 Descriptors, Description Schemes, Coding Schemes, and DDL. There are two types of XML applications, i.e. the server (or extraction) application and the client (search, filtering, and/or transcoding) application.

The *MPEG-7 Conformance* and *MPEG-7 Extraction and Use of Descriptions* parts are still being developed. The former will include guidelines and procedures for testing the conformance of MPEG-7 implementations and the latter will provide informative material about the extraction and use of selected MPEG-7 Description Tools.

MPEG-7 takes an all-inclusive approach to content description. It has links with other activities in this area such as EBU/P-Meta (European Broadcasting Union, 2001), SMPTE Metadata Dictionary (SMPTE 2001), and Dublin Core. Other related standardization activities are TV Anytime and W3C.

4.4.3 SMPTE METADATA DICTIONARY

The Society of Motion Picture and Television Engineers (SMPTE) developed the SMPTE Metadata Dictionary (Society of Motion Picture and Television Engineers, 2001a) as a standard for capturing and exchanging metadata. It is stressed that the Metadata Dictionary does not specify a data model, cataloging convention or description scheme. It rather provides a framework that supports interoperability between systems by defining metadata tags and universal labels for metadata elements. Within this framework the various organizations dealing with metadata can map existing schemes onto the framework structure or develop conventions that meet their individual requirements.

The SMPTE Metadata Dictionary defines distinct classes that comprise metadata elements, which share common characteristics or attributes. The organization of classes and within classes is hierarchical. The universal label concept is used to identify the metadata elements in the dictionary.

The Metadata Dictionary is a dynamic, living document that allows new entries to be defined. SMPTE acts as *Registry Organization*, i.e. it guarantees compliance with the general structure of the dictionary and the uniqueness of the entry as long as it is registered. The Standard is augmented by documents describing administrative processes (*MAP*) and engineering guidelines (*MEG*). As a standardization body SMPTE ensures that changes and additions are made in accordance to the standard.

4.4.3.1 The Metadata Dictionary Structure

The Metadata Dictionary comprises a number of sections (or sub-dictionaries) divided into distinct classes. Figure 4.12 shows the class structure of the SMPTE metadata dictionary. Up to 127 classes could be defined to deal with the different characteristics associated with content. At present there are seven standard classes and three classes for organization-specific, internal or experimental use.

The following standard classes have been defined:

- Class 1 *Identifiers and Locators* covers all metadata elements related to identifying information that describes the essence of an overall bit stream or file. Of particular concern is the unambiguous identification of the essence by a single, standardized numbering scheme such as the SMPTE UMID (see Section 4.4.3.3). This class also includes identifying information about metadata elements (so called meta-metadata). Examples for sub-class titles are *Globally Unique Identifiers (GUID)*, *ISO Identifiers, Object Identifiers, Device Identifiers, Unique IPR Identifiers, Local Locators*, and *Titles*.
- Class 2 *Administration* contains metadata representing administrative or business information. This also includes rights information, usage restrictions, encryption information and financial data. Examples for sub-class titles are *Supplier, Rights, Financial Information, Security, Publication Outlet, Participating Parties*, and *Broadcast and Repeat Statistics*.

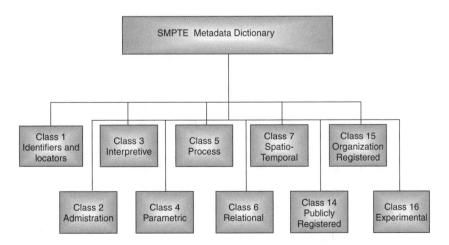

Figure 4.12   SMPTE Metadata Dictionary Classes

- Class 3 *Interpretive* comprises descriptive information. This includes manual, human-generated (i.e. subjective and classified) description, and information retrieved by automatic essence analysis processes. Class 3 information is intended for indexing, cataloging, administering, searching, and retrieving of content (or essence). The interpretive information incorporates textual descriptions (e.g. content and narrative descriptions, scripts, keywords, genre categories, etc.) but also low-level technical descriptors (e.g. color histograms, texture maps, object shapes, facial features, etc.) Examples for sub-class titles are *Fundamental* (e.g. ISO Language Code, length, and time systems), *Descriptive, Categorization, Assessments*, and *Descriptors* (machine assigned or computed).
- Class 4 *Parametric* covers information related to the technical characteristics of the camera, sensor or system that originates the essence or metadata. Information about the technical characteristics of the essence itself and the metadata is also part of this class. This includes creation parameters and the configuration of the originating system. Examples of sub-class titles are *Video Essence Encoding Characteristics, Audio Essence Encoding Characteristics, Data Essence Encoding Characteristics, Metadata Encoding Characteristics, Audio Test Parameters, Film Pulldown Characteristics, Fundamental Sequencing and Scanning, MPEG Coding Characteristics*, and *Timecode Characteristics*.
- Class 5 *Process* contains information that describes the processing (or processing results) of the essence life cycle. This includes for instance EDL parameters, an audit trail of all changes, a record of compression and decompression steps, and any changes in storage media and format. Examples for sub-class titles are *Process Indicators, Manipulation, Downstream Processing History, Enhancement or Modification, Audio Processor Settings*, and *Editing Information*.
- Class 6 *Relational* deals specifically with relationships between objects. The relationships can be between content objects, any combination of essence, objects, and metadata. Examples for sub-class titles are *Generic Relationships, Relatives, Essence-to-Essence Relationships, Metadata-to-Essence Relationships, Metadata-to-Metadata Relationships, Object-to-Object Relationships, Metadata-to-Object Relationships, Related Production Material, Numerical Sequence*, and *Relationship Structures*.
- Class 7 *Spatio-Temporal* includes metadata concerned with temporal, place, and spatial aspects related to content or the originating camera, sensor, or system. The geo-spatial information defines the absolute or relative positions or places of objects, scenes, individuals or any other component of the essence. Further, this class also covers temporal elements such as date, timecode, synchronization marks, temporal keywords, and motion vector parameters. Examples for sub-class titles are *Position and Space Vectors, Absolute Position, Image Positional Information, Rate and Direction of Positional Change, Abstract Locations, Angular Specifications, Distance Measurements, Latency, Delay, Setting Date and Time, Relative Durations, Absolute Date and Time, Operational Date and Time*, and *Rights Date and Time*.

There are also three classes reserved for the organizational or experimental use of metadata registered with SMPTE in the Metadata Dictionary. These classes and their elements are reserved and managed separately from the other seven classes in the dictionary. The three classes are:

- Class 14 *Organizationally Registered for Public Use* is defined for individual elements that have been registered by a specific organization or individual. This metadata is published and can be used by all organizations that use the Metadata Dictionary. It is intended for the exchange of individual metadata between organizations. The information about this kind of metadata appears in the appropriate sections of the published Metadata Dictionary Contents. It is managed by the SMPTE Registration Authority in accordance with the SMPTE Administrative Practices.

- Class 15 *Organizationally Registered as Private* comprises metadata elements registered by a specific organization or individual for internal use. This metadata element itself is not made public but the metadata tag referring to the element is identified in the Metadata Dictionary Contents. It is therefore reserved for use by the registered organization. It is managed by the SMPTE Registration Authority in accordance with the SMPTE Administrative Practices.

- Class 16 *Experimental* has been defined for metadata used in multimedia research, systems with limited access, and experimental environments where the experimentation requires new metadata elements. The definition and use of elements defined in this class does not have to conform to the definitions in the Metadata Dictionary. Class 16 metadata should only be used in an experimental set-up or laboratory environment.

4.4.3.2 Metadata Dictionary Element Structure

The SMPTE Metadata Dictionary defines a combination of attributes, sets, and data types. These are called SMPTE Universal Label and uniquely identify a metadata entry. The dictionary defines the meaning of an attribute or attribute set. Public labels and attributes are registered with SMPTE. Private labels for public use (i.e. Class 14 labels) can also be registered with SMPTE and published. The labels referencing an item consist of a two-part 16-byte numerical value. This value is language independent and unique, i.e. a generic representation of the concepts behind a metadata element. The first eight bytes can be used to label the second as a tag. The data element tag is then used to define a metadata element (or to give it meaning). These identify a specific metadata element in a hierarchical fashion. The element names in the dictionary are in English. Further, an English language definition of what is represented by an element is used. However, this should not restrict the generic nature of the dictionary.

The SMPTE Metadata Dictionary specifies the *type* of an element as information about the required format of a metadata 'value.' This specification also gives the allowable range of a metadata element as a list or bounded range. The value length defines the permitted length in bytes or character. The length itself can be variable. It is important to note that individual data element values can be represented in various ways. For instance a textual value can be represented as ASCII or Unicode. Hence the representation has to be known and registered. The last active word of a tag defines the representation in use.

The Metadata Dictionary is organized in nodes and leaves where the different dictionary classes form the class nodes with the different sub-classes defining the sub-nodes. The nodes are represented by tags that have no values assigned to them. The data elements themselves are represented by leaves to which values can be assigned.

Figure 4.13 shows the example of an SMPTE Universal Label for a keyframe set that is (in hexadecimal notation) 06-0E-2B-34-01-01-01-01-03-02-01-02-06-00-00-00. The first eight octets represent the designator that identify the sequence as a SMPTE Metadata

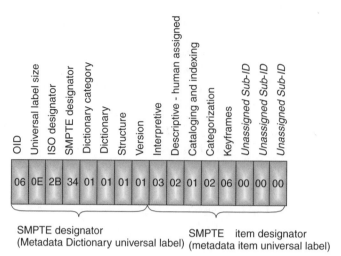

Figure 4.13   SMPTE universal label for start time code

Dictionary universal label. These eight octets are identical for all items in the Metadata Dictionary. The next eight octets then identify the dictionary item along the node and leaf structure of the dictionary. The example in Figure 4.13 resolves the structure as 'Interpretive/Descriptive – human assigned/Cataloging and indexing/Categorization/Indexing/ Keyframes'.

For the transfer of SMPTE Metadata Dictionary elements the Key-Length-Value protocol has been defined (see Section 4.5.1). Alternatively, XML encoding can be used (see Section 4.5.2).

4.4.3.3 SMPTE Unique Material Identifier

The *Unique Material Identifier* (*UMID*) has an important role within the SMPTE Metadata Dictionary Class 1. The UMID is a locally generated identifier that is nevertheless globally unique. It allows an unambiguous identification of content. Extended UMID (64-byte, see below) provides this identification with single frame granularity. It is intended for broad usage within storage and streaming applications. It should be automatically generated with the essence and remain associated with a content object (or parts of it) for its entire lifetime. The main purpose of the UMID is to identify and locate material in storage systems, to provide constant identification throughout production, archival and transmission, and to create links between metadata and essence.

The standard proposes that essence and associated metadata should share the same UMID. This, however, should not be interpreted such that the UMID must also be the unique identifier of the overall content object. This would not be feasible for generic content management applications, since it would require that a content object could only be created in a database when the related essence is already available (the UMID is available to the CMS only when the respective essence exists, since it is generated when creating the essence). Such an approach would prohibit workflows as described above that start with planning and drafting. It is very sensible, however, to store the UMID as an important attribute of the metadata set that describes the content object whenever essence is associated to the object.

The UMID allows SQL techniques to be applied for the retrieval of essence and metadata. The data representation of the UMID can be split into two parts:

1. 32-Byte (mandatory) basic UMID which contains information on:
   - universal label identifier for an SMPTE-UMID
   - length description of the length of UMID
   - instance number identification of a copy (version) of the clip
   - material number identification of a clip, unique number
2. 32-Byte (optional) 'signature' UMID that contains information on:
   - time/date, i.e. time of creation of the clip (to the level of an individual frame)
   - spatial coordinates of the camera location (coordinates) of origin of the clip
   - country code for the country of the producer
   - organization code for the organization of the producer
   - use code for the name of the producer.

The basic UMID operates on a shot-by-shot level. Each time a recording is initiated, a new and unique UMID will be generated and added to the essence. It is important to note again that the UMID must be generated automatically.

4.4.4 DUBLIN CORE

The *Dublin Core Metadata Initiative* (DCMI) was conceived in 1995 as an institution that promotes the widespread adoption of interoperable metadata standards. It involves a number of organizations such as Deutsche Bibliothek, Library of Congress, National Institute of Informatics, National Science Foundation, etc. DCMI develops metadata vocabularies to describe resources and thus facilitates better information discovery and retrieval (DCMI, 2003). Resources in the context of Dublin Core (DC) are addressable entities, namely Web sites, collections of documents, and also non-electronic forms of media (e.g. physical archives).

The goals of DC are (Hillman, 2001):

- *simplicity of creation and maintenance* of a set of essential metadata;
- *commonly understood semantics* to support access and retrieval for different user groups;
- *international scope* by translating the element set into various languages;
- *extensibility* by linking elements of other metadata sets to DC.

Dublin Core specifies a set of 15 metadata elements in the so-called *Dublin Core Metadata Element Set* (DCMES) (DCMI, 1999). This core set of descriptive semantic definitions is deemed to be appropriate for the description of content in various industries and disciplines, and across organizations. Table 4.3 gives an overview of the 15 metadata elements.

Individual institutions or organizations are not able to create new Dublin Core elements that go beyond the scope of the 15 specified metadata elements. Dublin Core does not restrict the length of the fields. Attribute–value pairs are used to represent the properties of content objects. Dublin Core can be represented in different syntaxes, in particular HTML and RDF/XML (the W3C Resource Description Framework using XML; Brickley and Guha, 2003).

Table 4.3 Dublin Core element set v1.1

Element	Definition	Description
Title	*Name given to resource*	Name by which the resource is formally known
Creator	*Entity responsible for the creation of content*	Name of persons, organizations, and services creating content
Subject	*Topic of the covered by the content*	Keywords, key phrases or classification codes to describe content
Description	*Account of content*	Abstract, table of contents, graphical representation or free-text content descriptions
Publisher	*Entity responsible for making resource available*	Name of persons, organizations or services publishing content
Contributor	*Entity making contributions to the content*	Name of persons, organizations or services contributing to content
Date	*Data associated with an event in the life cycle of the content*	Data values that can be freely defined. Recommended best practice for data values according to ISO 8601, namely YYYY-MM-DD
Type	*Nature or genre of the content*	Terms describing general categories, functions, genres or aggregation levels
Format	*Physical or digital manifestation of the essence*	Type and format of the essence. Also to determine software, hardware, and other equipment to display the content
Identifier	*Unambiguous reference*	Formal ID including Uniform Resource Identifiers (UCI) and URL, Digital Object Identifiers (DOI) and International Standard Book Number (ISAN)
Source	*Reference to original source*	Reference to the original source or master using formal identifiers
Language	*Language of the intellectual content of the resource*	Language elements defined according to IETF RFC 1766
Relation	*Reference to the related resources*	Reference to related material using formal identifiers
Coverage	*Extent or scope of the content*	Spatial, temporal, and legal coverage of the content using controlled vocabulary
Rights	*Information about rights held in and over the content/essence*	IPR, copyright, and property rights statements or links to this information

Dublin Core also specifies a list of qualifiers that refine the meaning and use of metadata elements. For instance they indicate if a value can be compound or structured but also allow the use of controlled vocabularies such as thesauri. The qualifiers allowed within Dublin Core can be grouped into two classes (DCMI, 2000):

- *Element Refinement Qualifiers* make the meaning of an element narrower or more specific. This restricts the scope of the meaning compared to an unqualified element. Element refinement should not restrict the usage of a metadata element since clients who are not aware of a specific element refinement term should be able to ignore the qualifier and treat the metadata values as if it was an unqualified element. The definitions of element refinement terms for qualifiers must be publicly available.

- *Encoding Scheme Qualifiers* identify schemes that help in the interpretation of element values. Such schemes include controlled vocabularies (where the value can be regarded as tokens or reference) and formal notations on parsing rules (e.g. date parsing rules to interpret 2003-02-05). The definitive description of an encoding scheme for qualifiers must be clearly identified and available to the public.

The qualifiers are non-normative. The qualifiers specified by DCMI form a foundation for a large body of qualifiers that can evolve by including additional qualifiers that have been developed by various communities and approved by DCMI. Qualifiers have to conform to the 'dump-down' principle, i.e. a qualified element may be dropped but useful information for discovery still has to be maintained. Qualifiers should only facilitate the precision of metadata, which still has to be useful without them.

Examples of qualifiers are:

- element refinements such as *Created, Valid, Available, Issued,* and *Modified* for the Date element;
- element refinements such as *Is Version Of, Has Version, Is Replaced By, Replaces, Is Required By, Requires, Is Part Of, Has Part, Is Referenced By, References, Is Format Of,* and *Has Format* for the Relation element;
- element refinements and encoding schemes such as *Spatial* with *DCMI Point, ISO 3166, DCMI Box, TGN, Temporal with DCMI Period,* and *W3C-DTF* for the Coverage element;
- element encoding schemes such as *ISO 639-2* and *IETF RFC 1766* for the Language element;
- element encoding schemes such as *Library of Congress Subject Headings, Medical Subject Headings, Library of Congress Classification,* and *Universal Decimal Classification* for the Subject element.

The Coverage element allows giving time periods and hence could be used to create segments. However, this was not the intention and although the W3C-DTF qualifier allows specifying also fractions of seconds there are no qualifiers for relative times. Still, in general Dublin Core considers content objects and not segments.

4.5 STANDARDS FOR METADATA TRANSMISSION AND EXCHANGE

In order to exchange metadata between organizations or within an organization it has to be transferred between systems and organizational units. The different systems between which the data is exchanged might use different metadata representations for a content object. Thus, it is important that the meaning of the metadata is preserved during the transmission and that the different entities handling it have the same reference document or specification to interpret the data. Standards are essential in this context to ensure interoperability between systems and organizations. A number of standards and transmission protocols and encoding schemes for metadata exchange have been defined.

The encoding schemes and standards introduced in this section have all been defined to facilitate the exchange of metadata between systems. However, their background and the main requirements they consider are different. Whereas the key-length-value protocol aims to specify an efficient, storage- and bandwidth-saving metadata encoding protocol, XML has been defined for a structured and flexible representation of content that is also suitable for its exchange. The Media Object Server (MOS) protocol has been specifically defined for the exchange of messages in a newsroom environment. The Simple Object Access Protocol (SOAP) is specified by W3C for the transmission of structured message. Both protocol are XML based. It is important to understand the structure, objectives and features of these protocols to assess their suitability in a specific system context. Other technologies such as CORBA (Object Management Group, 2002) or the TCP/IP protocol suite (Stevens, 1994) can of course also be used for the transmission of metadata. However, these are pure transmission protocols and the metadata still has to be encoded according to a well-defined metadata encoding scheme or standard.

4.5.1 THE KEY-LENGTH-VALUE (KLV) PROTOCOL

The *key-length-value* (KLV) data encoding protocol (Society of Motion Picture and Television Engineers, 2001b) is being standardized by SMPTE to encode metadata elements for transmission between systems. It allows the mapping of metadata onto different transport media and provides a common interchange point for all compliant applications. KLV takes into account the specific requirements of the SMPTE Metadata Dictionary (see Section 4.4.3) and has been especially devised for the production environment in television.

The standard defines a protocol that uses octet-level data encoding (i.e. a byte-oriented data encoding) to represent metadata items and data groups. It uses a key-length-value triplet where the *key* identifies the data type, the *length* specifies the length of the data, and the *value* is the actual data itself. It is also possible to combine triplets in data sets. In this case KLV encoding is used for the elements of the set and the set itself.

4.5.1.1 Key-Length-Value Structure

The KLV protocol uses a 16-byte universal label (UL) as an identification key followed by a numeric value that gives the length of the subsequent data value. Figure 4.14 shows the basic structure of KLV encoding. The key is a universal label according to SMPTE 298M (Society of Motion Picture and Television Engineers, 1997a). The header starts with

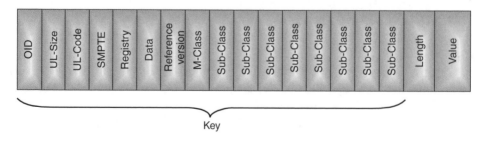

Figure 4.14 Basic structure of KLV encoding

an object identifier followed by the universal label size. The universal label designator starts with a code identifying the standards organization (e.g. ISO, ORG) followed by the SMPTE sub-identifier. The following byte identifies the registry category such as the SMPTE Metadata Dictionary. Bytes 6 and 7 more specifically identify the registry and structure followed by the version number. Unique identification of the particular item within the context of a universal label designator is represented by bytes 9–16.

The length field is encoded according to the basic encoding rules (BER). Either the short or long form encoding of length bytes according to the ISO specification can be used. The data values may either be individual data items or data groups.

4.5.1.2 KLV Data Coding

The coding of individual data items uses the key to identify the reference framework and data type, the length, and subsequently the encoded value. Since there are many dictionary items that allow multiple data representations of the same descriptor it has to be determined which representation is used. A default representation should be used (indicated by a key with at least one trailing zero byte). Alternative representations are marked by a non-zero value in the leftmost byte with the values documented in the dictionary.

Group coding schemes have been defined, which allow logical groups of individual data elements or groups of elements to be encoded together. This can for instance reduce the overhead of repeating redundant information within key units. Group coding can also be used to build logical groups of data elements. The KLV Coding Protocol can be used to support *Universal Sets, Global Sets, Local Sets, Variable-length Packs*, and *Fixed-length Packs*. The coding efficiency increases from the Universal Sets (which have the lowest efficiency) to Fixed-length Packs (which have the highest coding efficiency). In more detail the KLV group codings are:

- *Universal Sets* are used to construct a logical grouping of data elements and other KLV-encoded items. Full KLV encoding is used throughout the universal set.
- *Global Sets* are defined the same way as Universal Sets but offer more coding efficiency by sharing a common key header. This is equivalent to a lossless compression where the original key can be fully recovered.
- *Local Sets* are defined the same way as Universal Sets but have an increased coding efficiency through the use of short local tags. These tags only have meaning in the context of the Local Set. The structure of the KLV data construct is maintained but for the local tags a separate specification or standard to define their meaning is required. This also has to specify how local tag values are mapped onto the universal label key value.
- *Variable-length Packs* are defined as a further grouping of data elements that eliminates the use of UL Keys and Local Tags for all individual elements within the group. Variable-length Packs therefore rely on a standard or specification that defines the order of data elements within the pack.
- *Fixed-length Packs* eliminate the use of both UL Keys and Local Tags and also remove the length for all individual elements within the group. Thus, Fixed-length Packs rely on a standard or specification that defines the order of data elements and the length of each data element within the pack. This makes Fixed-length Packs the most efficient but also the least flexible grouping of data elements.

Sets and packs consist of a number of individual data elements coded as a group by the KLV set or pack data construct. They are defined by a full universal label key registered with the SMPTE Registration Authority. The data elements encoded by a set may be not only individual dictionary items but also other sets or packs. However, a pack shall only encode a group of individual dictionary items, i.e. they cannot use recursive coding. The presence of sets or packs is indicated in byte 5, with a specification of the type of set or pack in byte 6 of the key.

Labels are a special type used to identify objects that have a meaning (or value) that is determined by the label itself. Thus, a label does need a length field or a value field. They have to be defined in a special label dictionary.

4.5.2 EXTENSIBLE MARKUP LANGUAGE (XML)

The *eXtensible Markup Language* (XML) has been defined by W3C (2003) as a very flexible text format derived from SGML (ISO 8879; see Section 3.5.2.1). XML was originally intended for large-scale electronic publishing and is now increasingly used for a wide variety of data (mostly on the Web). It defines a container format for data content and a structure that allows automatic validation, i.e. XML describes a class of data objects (called XML documents) and partially describes the behavior of computer programs that process them. There are two major documents in which XML is specified, namely Bray *et al.* (2000) and Cowan (2002). The design goals of XML were that:

- it is straightforwardly usable over the Internet;
- it shall support a wide variety of applications;
- it shall be compatible with SGML;
- it facilitates the development of programs to process XML documents;
- optional features are kept to an absolute minimum;
- XML-encoded documents are human-legible, and quickly and easily created;
- XML documents shall be easy to create.

XML defines how documents are created, outlines their structure, and specifies the entities they may contain. Key elements of an XML document are the so-called Markups, which have the form of tags. Most commonly known are the start tags and end tags that enclose text fragments. They are used to structure documents and part of documents. Further tags comprise entity and character references, comments, document type and XML declarations, and processing instructions, inter alia. Figure 4.15 gives a small example of an XML structured document. It starts with a header giving the version of XML, the start

```
<?xml version="1.0"?>
<note>
<to>Andreas</to>
<from>Peter</from>
<heading>Reminder</heading>
<body>Don't forget the Content!</body>
</note>
```

Figure 4.15   Sample XML code

tag of the document (called note), and further start and end tags defining the different elements of the note.

By the definition of tags and the Document Type Declaration no specific structure is prescribed, i.e. it is not defined what exactly an XML document has to look like, how many elements it has to contain, what format these elements have or which exact form the document has to take to be considered a well-formed XML document. XML 1.0 defines that a document has to contain one or more elements, which are delimited by start and end tags. Hence, it provides a syntactic framework to create well-formed XML documents that can be parsed to retrieve information. In addition to simple elements an XML document can contain attributes, i.e. entities where a value can be assigned in a document.

Without any further definition and qualification of elements within an XML document, their characteristics and relationships, the number of occurrences of elements in the document, and their location within a given document structure an XML document cannot be interpreted. Since there is no reference point it is not possible to determine if a document is not only well formed but also valid. Further, in order to exchange information formatted in XML it is necessary that all involved parties have the same understanding and interpretation of the structure and content of the documents. A so-called *XML schema* is used to specify the characteristics and relationships of XML elements and attributes for a certain class of XML documents (Ahmed *et al.*, 2001). The classification of information and documents and their exchange is facilitated by referring to a *specific* XML schema or document type definition. This basically defines a template for a particular document class. A number of XML schemas that can be used to define XML document types and classifications have been specified. The most relevant ones are Document Type Definitions (DTD) and the W3C XML Schema.

Important in the context of content management is that XML can be applied not only to documents but also to streamed data. Thus, it is suitable for the transmission of content and not only for basic file transfer.

4.5.2.1 Document Type Definitions (DTD)

One form of XML schema that can specify the structure of XML documents is the DTD (Ahmed *et al.*, 2001). DTDs are written in a non-XML syntax. A DTD gives a description of the structure and allowed content that a document of a particular document type has to adhere to. DTD specifications are based on SGML employing the Extended Backus–Naur Form (EBNF). The DTD gives the document type declaration, and the element and attribute declarations for an XML document. Since it has been intended for document definitions (i.e. has a document centric rather than a data centric view) the support for different data types is not incisive. A DTD can basically only state that an element contains parsable character data but is not able to specify any data types. Hence it cannot specify what data types are allowed or should be expected for a certain element. It only gives the structure and elements that a document of a certain type should contain.

4.5.2.2 W3C XML Schema

The W3C initiative specifies an XML schema using XML syntax that not only specifies the structure of XML documents/document types but also provides the integration with XML namespaces, the integration of structural schemas with data types, and inheritance. The three main documents specifying the W3C XML Schema Definition (XSD) are XML

Schema Part 0: Primer (Fallside, 2001), XML Schema Part 1: Structures (Thompson *et al.*, 2001), and XML Schema Part 2: Datatypes (Biron and Malhotra, 2001). Part 1 specifies the XML schema definition language. This is used to describe the structure and constraints the content of XML documents conforming to XSD have to adhere to. Part 2 describes a number of built-in, basic data types (such as Integers, Dates, Binary Data) and defines how additional data types can be defined (by schema designers) using the XML Schema Definition Language. XSD provides similar mechanisms for Element Type Declarations and Attribute Declarations as in the case of DTD. Additionally, it allows specifying cardinalities for the definition of minimal and maximal occurrences.

The greatest difference between DTD and XSD is the list of data types defined for XSD. It distinguishes between primitive data types (i.e. data types that exist *ab initio*) and derived data types (i.e. data types that are defined in terms of other data types). The latter case distinguishes further data types derived by restriction and derived by list. Table 4.4 gives an overview of the built-in primitive data types.

Apart from the primitive data types a number of derived data types (such as normalisedString, Name, ID, long, short, byte, etc.) are also defined in Part 2. The namespaces defined by XSD use the namespace prefix *xsd*.

4.5.2.3 XML Processing

An important aspect of XML is that it can be parsed and processed automatically as well as contain instructions to control the processing and execute commands on a computer.

Table 4.4   XSD primitive data types

Primitive data types	Description
String	Character strings in XML
Boolean	Binary value logic: true, false
Decimal	Decimal numbers in arbitrary precision
Float	IEEE single-precision floating point (IEEE 754–1985)
Double	IEEE double-precision 64-bit floating point (IEEE 754–1985)
Duration	Duration of time with value space yyyy/mm/hh/minmin/ss according to (ISO 8601)[a]
dateTime	Specific instant of time
Time	Instant of time that recurs every day
Date	Calendar date
gYearMonth	Specific Gregorian month in a specific Gregorian year
Gyear	Gregorian calendar year
gMonthDay	Recurring Gregorian date in a specific month
Gday	Recurring Gregorian date for a day (e.g. 5th of a month)
GMonth	Recurring Gregorian month
hexBinary	Arbitrary binary encoded data
Base64Binary	Base64 encoded arbitrary binary data
AnyURI	Uniform Resource Identifiers according to RFC 2396 and RFC 2732
Qname	XML qualified names as set of tuples {any URI, local Part}
NOTATION	XML 1.0 NOTATION Attribute type with a set of QNames defining the value space

[a]The seconds part may have a decimal fraction.

The querying process is essential to retrieve information about content objects. Links are used to provide information about relationships of XML data items or sets of data items (Ahmed *et al.*, 2001). Querying and linking are both fundamental processes to deal with (complex) information. A special *XML Linking Language* (XLink) has been defined that allows inserting elements into XML documents that can be used to create and describe links between resources. The linking functionality provided by XLink is defined via global attributes that can be attached to link elements from other namespaces. An XLink associates resources (i.e. addressable units of information or services). Links are traversed to go from a starting resource to an ending resource. The so-called arc provides information about how to traverse a link (e.g. direction). It distinguishes between simple links (providing similar functionality to hyperlinks as used by HTML) and extended links. In the latter case their elements can be nested within the element that has the extended attribute. It may have an arbitrary number of extended resources that participate in the link.

The *XML Path Language* (XPath) has been defined to query and traverse XML structures by addressing parts of an XML document. It provides basic facilities for the manipulation of strings, numbers, and Booleans (Clark, 1999). With XPath an XML document is modeled as a hierarchy of nodes with a root node that represents the document entity as the root of a tree. The different types of tree nodes include element nodes, attribute nodes, and text nodes. XPath defines a way to compute string values for the different nodes types.

A query starts at the root and follows according to the 'directions' given in the XPath statement. It operates on the abstract logical structure of the documents rather than its surface syntax. Important in the context of queries is the functionality to match patterns (i.e. to test if a node matches a given pattern).

Apart from XPath the *XML Query Language* XQuery is being defined to allow search operations on information stored in XML documents. It is intended for documents that do not fit the relational model and dynamically created documents. The query language used within XQuery is going to be declarative and should be able to be expressed in XML syntax.

4.5.2.4 XML and Metadata

XML and its extensions provide powerful concepts and tool sets for the structuring, manipulation, and handling of metadata. However, XML does not provide a data model or even a metadata reference model. In order to use XML to handle and exchange metadata these models have to be defined. MPEG-7, for instance, uses an extended version of XSD. The Resource Description Framework (RDF) (Brickley and Guha, 2003) has been defined to represent information about resources in the World Wide Web and to facilitate the exchange and interpretation of documents and hence of metadata. It provides a common framework to express such information but does not, however, define the vocabulary of descriptive properties (e.g. title or author). It rather specifies mechanisms that may be used to name and describe properties and the classes of resources they describe. Dublin Core is one application field where RDF can be applied to a specific metadata model.

Information can be flexibly encoded in XML; however, it is not sufficient for the interpretation and exchange of metadata. It requires specific XML schemas, description frameworks, and metadata models that can be used as reference for the encoded information. Only if two systems use not just XML but also the same XML schema, description framework, and metadata reference model is an exchange of metadata between these systems possible.

4.5.3 THE MEDIA OBJECT SERVER PROTOCOL

The *Media Object Server* (MOS) protocol (MOS Consortium, 2001) is being developed by an industry consortium led by Associated Press (AP). MOS was conceived in conjunction with AP's newsroom system ENPS. Members of the consortium are mainly companies involved in the development of broadcast systems. The objective is to develop an open industry standard to support integration between systems related to media production, particularly in the area of news. Such systems are video servers, newsroom systems, studio automation systems, and CMS, inter alia.

The protocol is basically a tagged text data stream. It specifies a set of classified, well-defined XML messages for the exchange of information between MOS capable systems. Each MOS message begins with the root tag 'mos' followed by the ID and message type. The subsequent data is also in tagged form. Apart from the specified message types the protocol also allows the addition of XML tags (<mosExternalMetadata>tags) for the exchange of additional information. The data tags and constants are formatted in English with some descriptive data fields that can contain other languages. The object description is restricted to Unicode UCS-2 text; formatted text is not allowed in the unstructured description area.

MOS messages are exchanged in a unidirectional manner between an entity called Newsroom Control System (NCS) and a Media Object Server (MOS). Messages have to be acknowledged and a device should not send another message until it receives a positive or negative acknowledgement. In order to allow bidirectional communication between two systems both have to implement an NCS and MOS server. TCP/IP is used for the transmission of messages between NCS and MOS.

Essentially the MOS protocol covers a set of very specific requirements in the context of information exchanged between production systems. It provides a basic subset with some potential for extension but does not go beyond the special needs of the (news) production domain.

4.5.4 THE SIMPLE OBJECT ACCESS PROTOCOL (SOAP)

The *Simple Object Access Protocol* (SOAP) is a standardization activity of the W3C (Gudgin *et al.*, 2002). It is a lightweight protocol for exchanging structured information especially in decentralized and distributed environments. SOAP was developed having Web services in mind. Web services are Internet services that can be accessed using standard Internet and Web protocols and technologies (such as HTTP, XML, SMTP, etc.). SOAP has been conceived as a standardized packaging protocol for the exchange of messages between applications. It uses XML technologies to define an extensible messaging framework. SOAP is independent of any particular programming model and implementation-specific semantic. The protocol itself does not deal with issues such as reliability, security, and routing. The motivation for using a simple XML-based messaging approach was to support platform- and system-independent information exchange.

The SOAP standard specifies an XML-based envelope format for information exchange and a set of rules to translate application- and platform-specific data types into XML representation (Tidwell *et al.*, 2001). The envelope contains an optional header and one (and only one) body. The header is made of one or more blocks that contain information regarding the processing of the messages (including routing and delivery settings, authentication, and transaction context). The message body contains the actual message encoded in XML syntax. The body can have as many child nodes as required. The XML message needs to be well formed and namespace qualified. It should not contain any processing instructions or DTD reference.

SOAP supports two basic kinds of message exchanges, namely request response messages and one-way notification messages. The former is, for instance, used for remote procedure calls (RPC) whereas the latter is suitable for applications such as document transfer. For RPC style messages the called method is represented in a single structure with the parameters modeled as a field (in the same order as they appear in the method to be invoked). The response is also modeled as a single structure with a field for each return parameter. Conventionally the response is named after the invoked method appended by *Response*.

Essentially, at the core of the SOAP exchange model is just a one-way transmission of an envelope from a sender to a receiver. Intermediate systems in the message path might process the message on its way from the sender to the receiver. However, SOAP does not specify routing itself and proprietary standards such as the Microsoft WS-Routing protocol have been proposed for establishing message paths.

The encoding styles specified in section 5 of the SOAP standard define how applications (possibly running on different platforms) can communicate although they might not have common data types or representations. These encoding rules are optional and have been defined to facilitate the exchange of information without a priori knowledge about the types of information to be exchanged. The two concepts important in this context are *value* and *accessor*. A value represents a single data unit or a combination of data units that are enclosed in XML tags. An accessor is an element with a reference to a value. This reference can be direct or (as in the case of multi-referenced accessors) via IDs. The data type of accessors can be determined in three different ways, namely by referring explicitly to XML Schema data types, by referring to an XML Schema document that defines the data type of a specific element, and by referring to other types of schema documents. The XML Schema data types are explicitly supported by SOAP (Biron and Malhotra, 2001). Although SOAP is called an object protocol it does not have the concept of an object or object reference. The SOAP encoding rules rather specify how to represent objects in XML.

As a packaging protocol SOAP uses transport protocols for the transmission of SOAP messages.[4] Protocols used in this context are HTTP, FTP, TCP, SMPT, POP3, etc. Hence there are various possibilities for the transfer of SOAP messages. However, because of its pervasiveness in a Web environment, HTTP turns out to be currently the most widely used transport protocol for SOAP. The mapping of SOAP onto HTTP is given special consideration in the SOAP specification. SOAP over HTTP is considered a natural match since HTTP is a request–response base protocol. The SOAP specification outlines how the semantics of the SOAP message exchange model map onto HTTP. For instance the *SOAPAction* HTTP header is defined within the SOAP specification. It indicates the objective of the SOAP HTTP request. However, there are also concerns regarding the SOAP over HTTP mapping. One issue is that for security reasons the use of port 80 (as commonly used by HTTP) is considered problematic. Further, error handling and addressing are being discussed as problematic in this context.

---

[4] Note, transport protocol in this context is not used in the way of the ISO seven-layer model. Transport protocol here refers to transmission schemes that allow the transfer of data between networked entities.

5

File Formats

In content production an increasing trend towards integrated production supported by a plethora of IT-based devices can be observed. In this context files are becoming more and more important for the storage and exchange of content. Files can be more flexibly accessed and transmitted than streamed media. Thus, the structure, features, and properties of the formats used in this context are also becoming increasingly important. For a CMS it is crucial to be able to handle and manage the relevant formats alongside metadata and pure essence.

File formats specify how essence and metadata can be stored and exchanged. A number of proprietary file formats have been defined. In the context of CMS files are the items that are handled, managed, and stored in the systems. A CMS should be able to manage all kinds of files (i.e. it should be format-agnostic). However, in the case where the CMS has to process files, retrieve information from a file or perform any other operations on them it needs to know the file formats it is dealing with.

File formats are designed as wrappers encapsulating essence and metadata. The essence components in a file are atomic parts such as instances of video clips, audio (as audio tracks), images or text. These components can be structured and hierarchically ordered. Metadata is present in the file header and might be repeatedly listed throughout the file. Depending on the file structure it can be also co-located with the essence.

A file might contain a single program with various different components (e.g. video and a number of foreign language audio tracks). It might also comprise a more complex structure such as a program series or hierarchical structured media.

In general, file formats are specified to support different essence formats (such as MPEG-1, MPEG-2, and DV-based formats for video, uncompressed PCM-encoded audio and MP-3, and JPEG, GIF, TIFF, etc. for images). The metadata can also be structured according to different metadata standards. Figure 5.1 shows the generic SMPTE/EBU Content Package Model, a reference model for file formats (SMPTE/EBU, 1998).

Depending on the purpose and type of media stored in a CMS different file formats are relevant. For instance in authoring systems the file format has to support the creation process whereas in delivery systems the characteristics of a file format have to primarily support transport and delivery of complete multimedia programs. Apart from proprietary formats proposed by certain companies, a number of standardization bodies and interest

Professional Content Management Systems: Handling Digital Media Assets   A. Mauthe, P. Thomas
© 2004 John Wiley & Sons, Ltd   ISBN: 0-470-85542-8

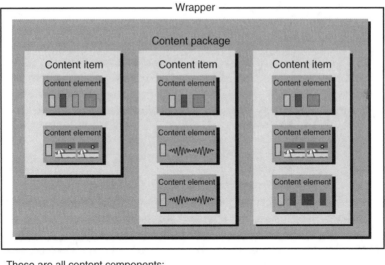

These are all content components:

    Essence component (video)   Metadata item

    Essence component (audio)   Vital metadata (e.g. essence type)

    Essence component (other data)   Association metadata (e.g. timecode)

Figure 5.1   SMPTE/EBU content package model

groups have been specifying file formats. In the following a number of open file formats are reviewed that have been specified with the special requirements of the content industry (in particular broadcast) in mind.

5.1 MEDIA EXCHANGE FORMAT (MXF)

The Media eXchange Format (MXF) (SMPTE/Pro-MPEG Forum, 2002) is an industry initiative that has been developing a file format for the exchange of program material. MXF is a joint development between major standardization bodies and interest groups, namely the Pro-MPEG Forum and the AAF Association. Further, the EBU P/PITV (PMC Project on Packetized Interface in Television Production) group and G-FORS members have been actively involved in the development of MXF.

MXF specifies wrappers around multimedia containers that can store different essence formats (for instance MPEG-2, DV-based formats or ITU 601.5 uncompressed video). It supports the exchange of content (i.e. essence and metadata) between servers but also tape streams and archive operations. MXF allows partial file transfer and supports random access of essence. Further, it also allows the recovery from interruptions that might occur during transmission. Thus, MXF specifies a file format that is useful in a number of contexts and does not only focus on a specific aspect.

The essence containers of an MXF file can be structured. They can contain a complete material sequence or clips and may contain segments of the same or different essence types. MXF fully implements the SMPTE 336M key-length-value (KLV) data coding (see Section 4.5.1). All data (i.e. metadata and essence) within an MXF file has to be

KLV encoded. MXF and AAF (see Section 5.2) are interoperable, and AAF-to-MXF export and MXF-to-AAF import operations are defined.

5.1.1 THE MXF FILE STRUCTURE

An MXF file is structured into a file header, followed by the file body, and completed by a file footer as depicted in Figure 5.2. Metadata can be carried in the file header, footer, and partition elements (which can also be part of the file body).

The file header of an MXF file is a mandatory element that has an optional run-in, the header partition packet, the header metadata, and an optional index table. Unique identifiers in the form of SMPTE Universal Labels are used in the file header and throughout the file to facilitate the identification of the MXF Operational Patterns, descriptive metadata, and the essence containers in the file body.

The file body comprises the essence elements stored in an MXF file. Within the body there can be one or more essence containers. If there is more than one essence container they are multiplexed together using partition elements. Besides the actual essence, essence containers may also contain related metadata. The file body is not a mandatory element since MXF metadata-only files may not have a file body.

The file footer is located at the end of the file. A mandatory part of the footer is the Footer Partition Packet. Further, in the footer the header metadata can be repeated and index information for random access can be kept.

Apart from the basic file structure MXF defines five structural elements:

- *Sector* is a bit aligned block;
- *Partition* defines the division between header, footer, and body partitions;
- *Container* specifies the encapsulation of different essence types;
- *Track* represents sequences and their temporal relationships;
- *Package* defines a container for related tracks (defined into Composition, Material, File, and Source Package).

The partitions of an MXF file are required to construct files of different complexity. Files with multiple partitions can for instance be used to interleave essence streams or to allow partial file restore.

Index tables are used to allow fast and non-sequential access to essence. The idea is to use a byte offset value to get to any point within an essence stream from a defined address. This facilitates for instance partial file retrieval. The referenced essence may be interleaved or can be part of a single essence element. An index table indexes one essence container only. Different essence types such as picture, sound, field coded, and

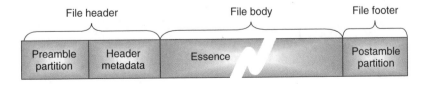

Figure 5.2   Simple MXF file

frame coded interleaved and variable bit rate essence can be described by MXF index tables. An index table can be placed in all the different partitions of an MXF file (i.e. header, body, and footer partition). If it is created on the fly it will be placed in the footer of the file. Index tables also allow the temporal re-ordering of content as used with Long GoP MPEG.

The Random Index Pack placed in the footer is used to find partitions distributed throughout an MXF file. It gives the first byte of the partition pack key and is used by decoders to access index tables and to find the partitions to which a specific index table points. The Random Index Pack is an optional element within an MXF file.

Figure 5.3 shows an example of an MXF file with multiple partitions and optional components. It has two essence containers. The header metadata is repeated in the file footer, which also contains a Random Index Pack. The run-in sequence is used to camouflage bytes or as synchronization bytes. A run-in is for instance used with BWF files (see Section 5.3.1) where the BWF header precedes the MXF header to make it look like a BWF file. A run-in sequence is ignored by MXF decoders.

5.1.2 METADATA IN MXF

Metadata is specified in the MXF file header, it is split into two categories, i.e. structural and descriptive metadata. The structural metadata is defined by the MXF standard whereas the descriptive metadata is defined as a plug-in in order to accommodate different descriptive metadata schemes.

The header metadata can be repeated in the body and footer partitions. The repetition of metadata supports the recovery of critical metadata, for instance when transmission was interrupted or where an application joins the transmission in mid-transfer. The KLV encoding of MXF facilitates the easy repetition of metadata elements throughout the file. The metadata repeated in the different partitions is usually identical with the original metadata specified in the header. However, in the case of an open header (e.g. when an MXF file is created during recording) each repetition of the header metadata is an updated copy identified by a higher generation number. The highest generation number within a file represents the master version.

5.1.2.1 Structural Metadata

The structural metadata part always occupies the first part of the header metadata. Structural metadata defines the capabilities of an MXF file and how it is constructed. Apart from the structural metadata that has to be known by all decoders it is possible to place metadata into an MXF file that is specified outside MXF. This is known as dark metadata

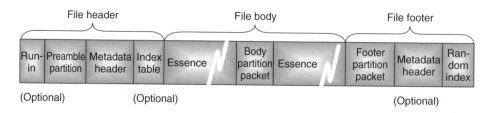

Figure 5.3 MXF file with partitions and optional elements

and does not have to be interpreted by all MXF-compliant decoders. Dark metadata has to be ignored by those decoders that do not recognize it.

The different components of the structural metadata set are placed sequentially into the header metadata and are linked using strong (i.e. one-to-one relationships) or weak (i.e. one-to-many relationships) references to construct a logical data model.

The structural elements of an MXF file are reflected in the structural metadata. It distinguishes between Material, File, and Source Packages. These are constructs that allow flexible creation of a content item out of existing components or the use of different essence parts to compile a content object. The Material Package is an element that represents the actual item by defining the output time-line of the file. It owns a Track defining its origin (i.e. it specifies the start and so-called Edit Unit Rate of the output) and a Sequence that specifies the duration of the output. A Sequence can have one or more SourceClips associated with it. SourceClips link the packages to the actual source material. They reference other packages, essence descriptors, and essence containers (i.e. the actual essence). The File Package and Source Packages are structured in a similar way (i.e. they also have Tracks, Sequences, and SourceClips associated with them). The idea of this package hierarchy is to enable the creation of content out of existing content elements. Take the example of a Material Package that defines the output of a promotional program of a television broadcaster. The Material Package represents the entire promotional clip and governs the synchronization and play order of the SourceClips defined by the Source Package. It references the File Packages that represent the material from the different programs that form part of the promotional program. The Source Packages finally describe the MXF files originally used for the different television programs. Source Packages not only reference essence containers but can also be used to store information about how programs have been captured, e.g. the tape format used to originally record an item.

Certain metadata elements can appear in different packages and the context of the occurrence has to be considered in addition to the actual value. For instance the timecode in the Material Package is continuous and defines the timecode for the playout of the file. In contrast, the timecode track in the File Package represents the timecode in the associated essence container.

SourceClips are the basic entities within the packages. SourceClips in a sequence are contiguous since MXF supports cuts-only edits (MXF does not include metadata descriptions for video and audio effects). Thus, when a sequence set references more than one SourceClip the access order of the SourceClips is important.

5.1.2.2 Descriptive Metadata

Descriptive metadata carries mainly content-related user-oriented metadata. An MXF file can contain more than one descriptive metadata scheme. These metadata schemes are defined as plug-ins, hence various different description schemes that have been defined outside the MXF standardization process can be supported.

In the Geneva Scheme of MXF, collections of descriptive metadata are defined. In particular these are production collection (containing descriptive metadata sets and items that provide identification and labeling information), scene collection (describing events and actions in the content material which are segment oriented), and clip collection (metadata related to the production of the content such as clapper-board information).

5.1.3 OPERATIONAL PATTERNS

MXF enables the composition of content files with almost arbitrary complexity. This approach accommodates all the requirements that can be placed on content files in terms of composition out of various other file components. However, this can result in very complex files. In order to deal with this complexity operational patterns are defined that represent a certain complexity class. The actual operational patterns are not defined in the standard but in separate documents if and when they are required.

The standard only defines generalized operational patterns that consist of two components, i.e. the operational pattern axes and the operational pattern qualifiers. The former has two dimensions, namely the item complexity and the package complexity. The latter defines the file parameters that are common to all operational patterns.

The two dimensions of the axes used to classify operational patterns have three complexity levels each. The different kinds of item complexity defined are:

- *Single item,* which are files that contain one Material Package Source Clip only that has the same duration as the File Package
- *Playlist item,* which are files that contain multiple items butted one against the other (with optional audio fade out/in). There is a one-to-one relationship between the different Material Package Source Clip duration and the respective File Package duration.
- *Edit item,* which are files with several items with one or more cut edits. Material Package Source Clips can come from anywhere apart from a respective File Package.

There are also three different package complexity levels defined:

- *Single package* represented by a single Material Package that accesses a single Source Package at a time.
- *Ganged packages* represented by a single Material Package that can access one or more Source Packages at a time that share a common synchronization time-line.
- *Alternate packages,* represented by two or more Material Packages of which each can access one or more File Packages at a time. An example given for this case is different Material Packages for different language versions.

The combination of item and package complexity leads to a matrix that gives the possible cases characterizing the generalized operational patterns defined in the standard. Figure 5.4 shows the generalized operational matrix with the operational pattern axes and nine different classes into which actual operational patterns defined outside the standard can be grouped. Applications can use this classification to state the operational pattern complexity they conform to. An application should always report the simplest operational pattern it conforms to.

The different operational patterns are currently being specified. A proposal for the MXF Operational Pattern 1a (Simple Item, Single Package) is given in Devlin (2002). Figure 5.2 shows the file structure of a simple MXF file equivalent to the MXF Operational Pattern 1a for Simple Programs. It comprises the Preamble Partition that represents the start of the file with a KLV-encoded partition pack. The Header Metadata contains Structural Metadata to directly access the file and Descriptive Metadata. The file body contains the essence within a single essence element (which might also be interleaved essence elements). In

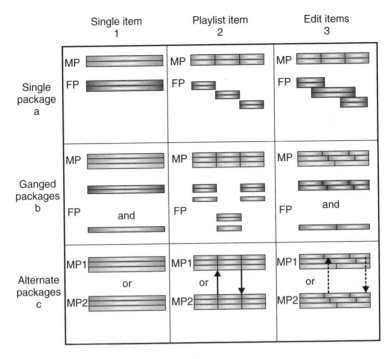

Figure 5.4   Generalized operational pattern axes

the file footer the Postamble Partition terminates the file with a KLV Partition Pack. The file footer can also contain a repetition of the File Metadata and an optional Random Index Pack that defines pointers to different partitions in the file.

Apart from MXF Operational Pattern 1a operational patterns up to Operational Pattern 3c are being specified. Operational Pattern 2a is defined to combine essence objects for edited programs in a playlist like manner, while Operational Pattern 3a allows cuts-only edits to be applied to the Source Package. In these cases video and audio are always interleaved. Operational Patterns 1b to 3b are similar to Patterns 1a to 3a. The difference is that multiple Source Packages can be accessed. This allows, e.g., non-interleaved video and audio. Operational Patterns 1c to 3c enhance this by also allowing choice between alternate Source Packages. This allows, e.g., provision of multiple language versions in a single MXF file.

5.2 ADVANCED AUTHORING FORMAT (AAF)

The Advanced Authoring Format (AAF) is also an industry-driven initiative for the development of a cross-platform multimedia file format. The AAF Association Inc. has been formed to promote and develop AAF technology. Founding members of the association are Avid, the BBC, CNN, Discreet, Matrox, Microsoft, Sony, Turner Entertainment Networks, inter alia.

AAF is a universal multimedia file format proposed for the exchange of digital media (i.e. essences) and metadata across platforms, and between systems and applications that are AAF-compliant (AAF Association, 2000). It is not designed as a streaming essence format. It considers the native capture and playback process and also supports the flexible storage of large data objects. AAF specifically considers the requirements of the content creation and authoring process. Important in this context is, for instance, that one or more source essence files can be accessed concurrently, that the essence can be manipulated or edited easily, and that the result is subsequently saved in an AAF-compliant file. During this process metadata is generated that has to be stored together with the created multimedia item within an AAF file.

The target applications for AAF are TV studio systems (including NLE, server, effects processors, automation systems, and CMS), post-production systems (including offline editors, graphics and rendering systems, image manipulation applications, and audio production systems including multitrack mixers and samplers), and advanced multimedia production tools (such as 3D rendering tools, content repackaging applications, etc.).

AAF supports the interchange of essence and (certain types of) metadata. AAF therefore specifies a structured container representing a content object. The essence part of the container can consist of audio, video, still images, graphics, text, animation, music, and other multimedia data. The metadata part of AAF is concerned with information on how to combine and modify the included essence components, and how the essence has been created but also includes additional information about the essence characteristics. The compositional information describes exactly how sections of audio, video or still images are combined or modified for the presentation of the content object. This is considered as part of the creative metadata. A history of the creation process can be kept within an AAF file (including times, dates, and versions). Within an AAF file it can also be specifically stated how a certain kind of essence was derived from other (original) material. In this context it is also important that within an AAF file it is also possible to keep references to other external files.

Particularly relevant for the creation of multimedia content and films is the extensible video and audio effects architecture. It includes an extensive set of built-in base effects commonly used in the editing process of audiovisual material.

5.2.1 AAF FILE STRUCTURE

An AAF file contains a header object (with related objects), packages (comprising the different objects they have) and the essence data. Figure 5.5 shows an example of an AAF file.

There is exactly one header object per file that specifies the byte order used to store data in the file, date and time of the last modification, the version number of the AAF standard referred to, a dictionary object (for all definitions), and a content storage object.

The dictionary holds class definitions, property definitions, type definitions, data definitions, parameter definitions, and effect definitions. Any of these definitions that are not defined by the AAF specification have to be defined within the dictionary. This also allows the definition of extensions, for instance for new effects, new kind of metadata, and new kinds of essence.

A package within an AAF file is an object that has a unique identifier and describes the essence and the relationship between essence components, i.e. it consists of metadata related to essence.

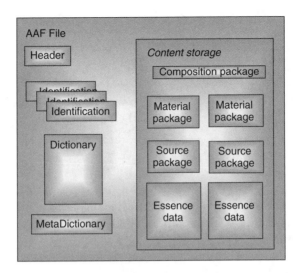

Figure 5.5 AAF file including packages and essence data objects

Packages have one or more slots of which each refers to one type of essence; each slot is separately identifiable. Each kind of slot defines the relationship between a specific kind of essence data and the time. There are Static Slots for discrete media (such as static images and text), Time-line Slots for continuous media with a fixed reference to time (such as audio, video, timecode), and Event Slots that describe irregular relationships with time (e.g. for GPI or interactive events).

Different package types have been specified within AAF. Physical Source Packages describe the original source of the essence (i.e. the source an essence component was derived from). File Source Packages describe the digital essence data (e.g. MPEG-2, WAVE, JPEG). The Material Packages are required to locate File Source Packages and also describe how individual essence elements should be synchronized or interleaved. Composition Packages contain the creative decisions on how the essence should be presented. This includes, for instance, the order and placement of essence components, and effects that modify or combine essence components. For instance transitions between two essence segments can be specified in a Composition Package.

5.2.2 ESSENCE IN AAF

An AAF file can store multiple types of essence. These essence types comprise:

- video encoded in different formats, e.g. MPEG, RGBA, YCbCr;
- audio encoded in different formats, e.g. Broadcast WAVE;
- static images in different formats, e.g. JPEG, GIF, TIFF;
- musical instrument digital interface (MIDI) music essence;
- text in different formats;
- compound formats, e.g. DV, MPEG transport streams.

The Essence Descriptors (within the Source Packages) describe the details of the essence format. An Essence Descriptor is an abstract class that describes the format of the essence data. Essence Descriptors describe essence stored in Essence Data Objects or referenced non-container files (so-called File Descriptors), but are also used to describe the physical media source.

For continuous media, timecodes are particular important for the description of essence. Within an AAF file a timecode can either be specified by using a starting timecode that serves as a reference point, or by including a stream of timecode data. This also allows non-contiguous timecodes, which are particularly important in the context of compound media objects.

5.2.3 AAF CLASS MODEL AND CLASS HIERARCHY

The AAF specification is object-oriented. Within the AAF specification (AAF Association, 2000) a class hierarchy and class model are defined to describe multimedia compositions and data. The main goal for this is to provide mechanisms to encapsulate multiple types of essence and metadata.

An AAF class specifies an object by defining the kind of information it may contain, and how it should be used and interpreted. Inheritance is defined for AAF classes, i.e. a class inherits from one but only one immediate super-class.

Two root classes have been specified in the AAF class hierarchy, namely Interchange Class and the MetaDefinition Class. Extensions to the base AAF classes can be defined in the file's AAF Header object's *ClassDictionary* and *Definitions*.

5.3 BWF AND OTHER MULTIMEDIA FILE FORMATS

Within the multimedia domain there are a number of (mainly proprietary) file formats. Most of them focus on the way essence can be stored in a structured way within a certain system context. Examples are AVI (for the storage of video), Quick Time® (for playback and streaming media), the Adobe Photoshop format (mainly for the storage of still images), MS DirectX® (optimized for 3D images), AVID's Open Media Framework® (OMF), Advanced Streaming Format (ASF), etc. There are currently no formats that could be used as a generic file format suitable for all multimedia applications. AVI for instance does not support the storage of compositional information or ancillary information (e.g. timecode information). Others do not allow storing compressed essence data, etc. The disadvantage of proprietary formats is that they have been specified considering only certain requirements and that they are sometimes not open. A CMS has to be able to store and handle them but processing and the extraction of information stored in them is a specific task that has to be implemented individually for each format.

Apart from the already introduced formats that are mainly used in a professional context there is also the Broadcast Wave format for audio and a number of other open multimedia file formats that have to be handled within a CMS. Microsoft in cooperation with IBM has defined the *Resource Interchange File Format* (RIFF) as a basic tagged file structure upon which many multimedia file formats can be based (WAVE, 1991). It specifically considers the requirements for recording and playing back multimedia data, and for exchanging multimedia data between platforms. Basic building blocks of a RIFF are chunks that contain metadata about the encoded media data and the actual encoded essence itself. Further RIFF forms are defined with a list of registered forms. This list includes PAL (a file format that represents a logical color palette), RDIB (the RIFF Device Independent

Bitmap format), RMID (the RIFF MIDI format), RMMP (the RIFF Multimedia Movie File Format), and WAVE (the original Waveform Audio format). The rich text format (RTF), which defines a standard method of encoding formatted text and graphics using only 7-bit ASCII characters, is also included within the set of RIFF-compliant file formats.

Most multimedia file formats carry metadata alongside the actual essence. However, in most cases this is a very specific set of metadata characterizing the actual essence encoding or processing. Thus, many of these formats may be only useful for the specific purpose they have been defined for but not as generic multimedia file formats for the exchange of essence and an extensive set of descriptive metadata.

5.3.1 THE BROADCAST WAVE FORMAT

The *Waveform Audio File Format* (WAVE) was initially defined by Microsoft and IBM in (WAVE, 1991). The *Broadcast Wave* (BWF) format as defined by EBU is based on this original definition (European Broadcasting Union, 1997a). WAVE is a file format for audio data derived from the Microsoft RIFF. BWF defines some broadcast-specific extensions to the original format.

Two encoding types have been explicitly specified to be carried within a BWF file, namely pulse code modulation (PCM) audio and MPEG-1 audio. The former is lossless encoded CD quality audio whereas the latter is standard MPEG-1 encoded audio. It is possible to use other encoding types within a BWF file. Therefore an extended wave-form format structure is used that allows certain parameters and characteristics for these encoding types to be specified.

The basic building blocks of a BWF file are chunks containing specific information, an identification field, and a size field. There are different kinds of chunks for the various bits of information carried in a BWF file. Figure 5.6 shows the structure of a BWF file.

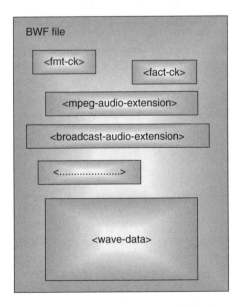

Figure 5.6   Broadcast wave format file structure

The <fmt-ck> is the mandatory WAVE format chunk that defines the format of the <wave-data>. In this chunk the format category, number of channels, sampling rate, buffer estimation, and data block alignment are specified. Two format categories are defined within BWF, namely PCM and MPEG-1.[1] There are up to two channels; channel 0 represents the left channel of stereo audio and channel 1 the right. The buffer estimation specifies the number of bytes per second at which the waveform data should be transferred.

The <fact-ck> and the <mpeg-audio-extension> are MPEG-specific chunks. The former is actually required for all WAVE formats other than PCM. Together with an extension of the <fmt-ck> it states information that is required to specify the MPEG encoding options. The <fact-ck> contains the file-dependent information about the contents of a <wave-data chunk>. For MPEG it specifies the length of the data samples. The <mpeg-audio-extension> chunk gives additional information about the sound file such as the MPEG-1 audio layer and ancillary data length. Within the MPEG (European Broadcasting Union, 1997b) encoded data it is possible to state certain kinds of metadata on a frame-by-frame level. This for instance includes copyright and protection information.

BWF files also contain the <broadcast-audio-extension> chunk. This has been specifically defined for the exchange of material between broadcasters. It contains:

- a 256 character long description field;
- an originator field (for the name of the originator or producer);
- an originator reference (which should be a globally unique ID);
- the creation date and time;
- a time reference (i.e. a timecode of a sequence);
- the version of the referred BWF standard;
- the coding history (which specifies the type of sound with its specific parameters).

Additional chunks can be added but these are optional and do not have to be supported by all applications. The <wave-data> chunk contains the encoded audio data.

---

[1] Note, with the original Microsoft WAVE format other categories are registered as well.

6

Content Management System Architecture

A CMS is made up of two major elements, namely the hardware infrastructure and the software components represented by the unifying system architecture. The former provides the basic resources for the storage, processing, and transmission of content. However, assembling these components is not enough to manage content since they only provide the infrastructure. Even enhanced by their standard software these components are just basic parts that only contribute to the overall system. Out of these components and a number of software systems that fulfill specific tasks the CMS is developed. In order to do this a unifying architecture is required that defines the different modules and how they interact in a system context.

The system architecture has to take into account the user requirements and workflows in a content-rich organization (see Chapter 2). Further, it has to consider the characteristics and specific features of the various content components. Essence can be encoded in various formats that might serve different purposes and is used in various contexts. Multiple copies of essence might have to be managed and sometimes there will be versions that are not electronically available but nevertheless are part of the system. The different ways to represent content also need to be supported by the CMS. There is no standard metadata model or even universally acknowledged framework that could be used. Moreover, long-established organizations have also developed specific workflows and systems to represent content and administer it in their organizations. Thus, a CMS has to be able to integrate existing systems and/or provide a way to replace them. This implies that a CMS cannot just offer a standard metadata representation scheme on top of a suitable information system or database. It has to be open to the integration of multiple systems and yet provide a generic service via a unified interface to the user.

However, just supporting current work processes is not sufficient in today's content production. New ways of processing content and supporting production, transmission and archiving are becoming more and more important. One example is the move towards a fully digital, tapeless, file-based production environment. This is still very much in a development stage and no standard ways, modules or even best practice schemes have

Professional Content Management Systems: Handling Digital Media Assets   A. Mauthe, P. Thomas
© 2004 John Wiley & Sons, Ltd   ISBN: 0-470-85542-8

been established. Even small developments such as automatic essence processing can have a profound effect on how content is handled although this is also still in an experimental phase.

All this places a number of requirements on the underlying system architecture that can only be supported if the system is flexible and can easily be adapted to change. In general, a CMS has to operate in a heterogeneous environment. This is not only true for the demands coming from the users and various systems that have to be integrated, but also for the software components and hardware systems that are part of the overall software and hardware infrastructure.

In this chapter an exemplary CMS architecture is introduced that has been developed taking the key requirements of modern content production and management into account. First, the basic design principles for such a system are introduced. Subsequently, the model software architecture is presented and discussed in detail. The different modules, their functions, and characteristics are examined. Application and system integration aspects are not part of this chapter. They are introduced separately in Chapters 8 and 9.

6.1 SOFTWARE DESIGN PRINCIPLES

Modularization is the key design principle of the proposed system architecture. It provides the flexibility that is crucial to master the inherent heterogeneity and complexity of a CMS. The following sections provide some insights into these principles, which guide the development of the subsequently introduced architecture. It is important to understand these principles to see how central concepts and requirements are realized within the different parts of the system in order to provide the necessary support.

6.1.1 SCALABILITY BY DISTRIBUTION

A CMS in an enterprise-wide context typically serves large numbers of users, manages enormous amounts of assets, stores huge amounts of data, and has to do so with a very high level of performance and reliability. Using large central servers and databases to accomplish this intrinsically introduces bottlenecks. Hence, the basic philosophy behind designing enterprise CMS is scalability by distribution. This allows adding resources to the system whenever there is a need for more performance or more storage space.

To do this, the system design must fulfill some basic requirements:

- Each component of the architecture must be able to run on a dedicated server platform. This means that the communication between all components must be fully network based.
- Since CMS components can be distributed over servers that may be physically far apart, this means that the network protocol employed must be able to establish the communication between components via Wide Area Networks (WAN).
- Wherever possible, CMS components should be designed in a way that they can be launched in multiple instances at the same time, improving performance and availability by load distribution.
- Wherever possible, CMS components should not require the availability of specific hardware; it should be possible to launch these components on any server platform within the CMS hardware architecture.

- CMS components should be resilient against unavailability or failure of other CMS components. This means that they have to include reconnect strategies and recovery procedures as standard functionality.

When a CMS meets these requirements, it is possible to install (software) components that are required for the operation of the CMS on more than one server, and launch these components on any of these additional platforms whenever there is a need to do so (e.g. to recover rapidly from a server hardware failure). This allows flexible failover scenarios that can be handled either by human operators or by advanced cluster software packages to be built.

6.1.2 SERVICE GROUPS

Services are self-contained modules in the overall architecture that are dedicated to perform a specific task, thus enhancing the functionality of the CMS. Examples of such functionality modules are ingest, import or export, essence analysis and indexing, conversion and transcoding, and trimming and splicing of essence.

In general, services are job-oriented background tasks that access existing essence and/or metadata and modify it or create new essence and/or metadata. In line with the philosophy of achieving scalability and redundancy by distribution, many services can employ a peer service group concept as shown in Figure 6.1. In this model multiple instances of a service join a group and offer their joint resources to the rest of the system via a single interface, i.e. they appear as if they were a single service.

Within a service group, each instance of a service runs on a dedicated server. Depending on the service and the server platform, the service offers a certain number of resources (i.e. jobs it can process concurrently). For example, a Video Analysis Service instance

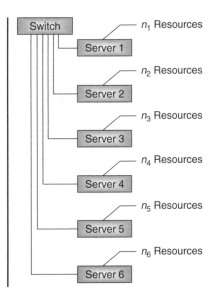

A service runs on a dedicated server
A service offers n_i resources
A service is restart capable
m services running on m servers join a services group
A services group looks to the outside world like a single service with:

   Σn_i resources
   A transaction secure job list
   An event and status log functionality
   A processing time estimation
   A resource reservation

The internal organization of these central tasks is managed by the services group
Each service can register and un-register with the services group
One service acts as coordinator of the group (manager)
In case of failure of the manager another service automatically registers as manager

Figure 6.1 Service group concept

installed on a specific server hardware platform may be able to run two or three analysis processes concurrently and still finish them faster than real-time.

The service has to fulfill certain requirements, e.g. it must be possible to restart a service at any time, and after failure the service has to recover and get back into a consistent state.

Using this concept the number of available resources is easily scalable by increasing the number of service instances. Depending on the server platform an instance runs on, each instance again offers a certain number of resources. All services of the same kind join a service group. This group presents itself to the remaining system as a single service offering all resources available from the services within the group. Via this unique interface the group accepts jobs, offers access to status and event log information, and provides an interface for querying for the estimated remaining processing time as well as for an advance reservation of resources. The group itself is responsible for dispatching jobs to the single servers, managing its internal communication, and offering the aforementioned interface.

From an organizational point of view, any service can register and un-register with a service group. One service acts as the manager of the group, handling the joint interface. Typically, the manager is determined as the one being the first to register with the system's naming authority (i.e. the Naming Service). All subsequently launched services then recognize the manager from the Naming Service and register directly with the manager. While the manager is responsible for periodically checking the existence of the Naming Service and the validity of its registration with this service, its peers in the service group have to periodically check the existence and availability of the manager.

Should the Naming Service fail, the manager registers with another Naming Service instance that has taken over. Should the master fail, all services within the group connect to the Naming Service in order to negotiate a new manager (again using the *first come, first served* principle) and register with the new manager.

This procedure allows for dynamic adding and removing of services in order to configure available resources, and for stable operation in case of failure of system components. Thus, the required level of redundancy concerning resources and service instances can be flexibly configured.

6.1.3 BROKER–MANAGER–SERVER CONCEPT

The CMS should provide applications and services with transparent access to distributed essence, and device and data management systems, even on remote locations via WAN. That is, to the service user (be it an application, service or other CMS module) the service is seen as one instance, accessed via a unified, standard interface. To do so, a broker–manager–server concept as depicted in Figure 6.2 may be employed. The idea is to allow:

- structuring of multiple instances of specific server components for optimum performance and availability, on the *single server level*;
- structuring of multiple storage clusters for maximum flexibility and scalability at a *single site of an organization*;
- integrating of *multiple sites of an organization* so that they feel like a single unified system;
- interfacing of *multiple organizations* allowing them to share selected items seamlessly.

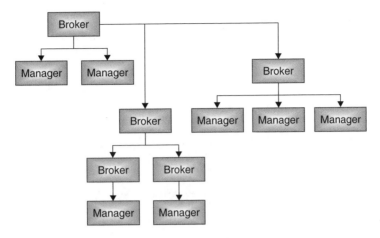

Figure 6.2   Brokers and managers

Brokers basically offer two kinds of functionality, depending on the service they support:

- They may act as distribution and collection node. An example is a database broker:
  - accepting queries from a client;
  - federating these queries to multiple systems and existing databases;
  - collecting the results from these databases;
  - passing the consolidated result back to the client.
- They may act as a factory. An example is a stream broker:
  - accepting requests from a client to stream essence to the client;
  - selecting a suitable stream server from a pool of servers known to the broker;
  - creating an instance of a control interface on this server;
  - passing the interface back to the client, which then takes direct control over the server without involving the broker any longer.

Brokers replicate the interfaces provided by managers, thus forming a superset of the interfaces provided by them. In fact, the brokers really hold the complete data model or the full APIs available in a specific installation. It is transparent to applications whether they access a manager or a broker, i.e. to them each access is equivalent to accessing a broker providing certain functionality.

Brokers can relay requests to subordinate managers or to other brokers. Thus, requests can be distributed over a broker/manager hierarchy. Brokers collect and assemble responses to requests and submit the collected answers to the superior broker instance. This way the result of a distributed request is finally assembled by the top-level broker and then submitted to the requesting application. It is also possible to configure the response flow in such a way that all incoming responses are immediately forwarded to the superior unit.

6.2 SOFTWARE ARCHITECTURE

Before suggesting a more detailed architecture proposal for a content management system, Figure 6.3 introduces the basic components of a CMS. These components are concerned

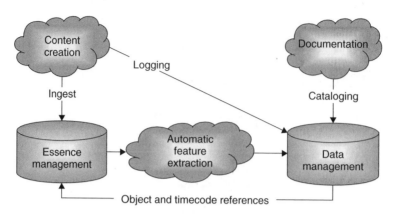

Figure 6.3 Basic components of a CMS

with the major management aspects of a CMS (i.e. the management of data and essence) and the contribution or access of the content. Whereas access is largely passive (i.e. essence and metadata are not changed), the contribution (either during content creation or documentation) is an active process. As a consequence of the search and retrieval there might be an export of content but this interaction does not actually change the content.

For simplicity, there are three major functionalities noted in this figure, i.e.:

- content creation, which comprises all kinds of work associated with production, such as program planning, recording of feeds or signals, editorial work, editing and post-production, scheduling, and playout;
- documentation, comprising all kinds of manual cataloging and indexing work;
- automatic feature extraction, which is a functionality very popular in CMS since this comprises all kinds of technologies that automatically derive indexing information directly from the essence (e.g. shot detection, keyframe extraction, speech recognition, speaker recognition, sound analysis, etc.).

It is interesting to note that CMS tend to provide real-time annotation capabilities for manual indexing while recording ('logging'), which brings elements of the cataloging work into very early stages of content creation.

Besides the functional elements, there are two major building blocks that form the foundation of each CMS, namely essence management and data management.

6.2.1 ESSENCE MANAGEMENT

The essence management is responsible for all operations where the primary objective is storage, provision, and manipulation of essence. Major components of the essence management are:

- storage management services that automatically and efficiently migrate essence items within the storage hierarchy according to various configurable storage policies;
- transcoding services for file format and encoding format conversions;

- delivery services, such as file transfer and streaming services, for ingest and provision of essence;
- a database that holds technical metadata (formats, bit rates, color depth, etc) as well as management information (locations, file sizes, etc.).

The essence management system may manage multiple versions of an essence object. Each of these versions may serve a specific purpose. Here are some examples:

- *Original tape copy:* the original copy ingested into the CMS. Usually, the carrier is a conventional analog or digital videotape. A reference to this copy has to exist within the essence management.
- *High quality file copy:* the file version of the original copy. The signal quality should be indistinguishable from that of the original tape copy. It typically is generated via encoding at ingest, but may also be imported as an already encoded file. If a state-of-the-art digital compressed videotape format is ingested, the high quality file copy should preserve the original encoding format (algorithm, parameters, bit rate).
- *EDL browsing copy:* a mid-quality file copy that can be used within the CMS environment for clipping and hard-cut offline editing, wherever the necessary bandwidth is available.
- *Content browsing copy:* a low quality version of the essence used in both LAN and WAN environments for viewing and selection purposes.
- *Keyframes:* still frames, typically automatically extracted from the essence.

6.2.2 DATA MANAGEMENT

Being classic IT, the data management basically consists of databases and search engines. It is responsible for the organization and management of the metadata describing the essence. This descriptive metadata is the key to the exploitation of programs or content items.

The relation between programs or items and the corresponding essence can be one-to-many, as one program may be archived in several different essence versions, all of which may share the same program identification and description. Metadata may describe the program or item as a whole, or may reference a segment or sequence via links to timecodes (see Section 4.2).

While the data management in itself stores and provides metadata, a CMS typically also provides special applications to support certain indexing workflows. Examples are:

- *Real-time annotation:* annotating incoming material while it is being ingested. This is required, for example, for incoming news feeds, during live sports events, or in some feature production scenarios, where the footage is classified during ingest. Real-time annotation typically provides a limited set of metadata, sufficient to identify the material and to allow successful retrieval even while the material is still being recorded, but lacks the depth of full cataloging.
- *In-depth cataloging:* typically is a longer-than-real-time process. Applications that support in-depth cataloging tend to support features like thesaurus support and legal lists, and typically provide more complex user interfaces for detailed annotation of content.

- *Automatic feature extraction:* these tools are a rather new field of technology. While shot detection and keyframe extraction are already state-of-the-art in video indexing, advanced tools like speech or speaker recognition or audio classification still require improvements to deploy them productively. However, it is this kind of technology that has the potential to realize the vision of automatic content classification in the nearer future.

6.3 CMS COMPONENT SOFTWARE ARCHITECTURE

Comparable to an operating system a CMS provides the platform for the integration of services and applications. However, a large CMS is a complex federated architecture where the key to scalability is distribution. Into this platform additional components can be integrated, thus increasing functionality and enabling additional workflows. These additional components can be quite different in their design, structure, and implementation since the CMS architecture cannot enforce a common design onto third-party components. Therefore it is crucial that all interfaces are clearly specified and possibly standardized. Hence, each CMS module needs to be defined by a:

- complete *interface design specification* (IDS);
- detailed *functional design specification* (FDS).

Any implementation that at least meets the requirements set forth in these specifications can be integrated into the CMS as a module.

Figure 6.4 shows a generic CMS architecture following the principles of separating basic functionalities, represented by components and modules within the architecture. The overall system is divided into three planes, each of which hosts a defined part of the overall system. These system parts are the:

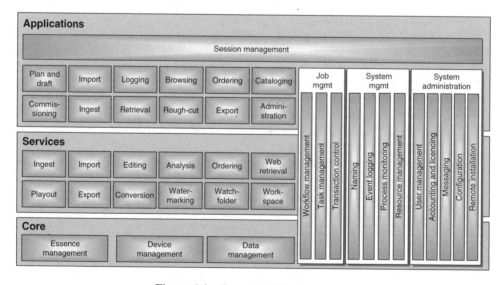

Figure 6.4   Generic CMS architecture

- *Core*, hosting all components concerned with the management and administration of content and the control of the related tools and devices;
- *Services*, encompassing all components that enhance the functionality or the workflow support of the system;
- *Applications*, comprising all components relevant for interaction with the user.

Each plane hosts certain components or modules that will be discussed in detail in the following sections. It is important to note that this 'layered' view is logical rather than functional. Applications may directly access core components or may interact with services. There is no prescribed way of interacting. Therefore the term plane rather than layer was chosen deliberately.

In addition there are three major groups of vertical services that are available throughout the system:

- *Job Management*
- *System Management*
- *System Administration.*

The system works primarily in a job oriented way, i.e. clients (being applications or other system components) request the processing of jobs. The job management functionality offered by the system processes these jobs asynchronously. The requesting entity of the process is notified of success or failure of a job. By employing a flexible and modular job management scheme it is possible to define new complex jobs based on basic modules (or primitive jobs) supported by the system. Thus, the system offers extensibility in its processing capabilities. Augmenting the job-oriented approach, parts of the system (e.g. streaming servers and automation systems) rely on the classic client–server approach where the client manages the delivery process by controlling the server.

The system management provides the internal monitoring and management of the system, which is very important for a distributed CMS. The system administration deals with administration issues that either affect the entire system or are common for all system components.

There are different solutions for communication between the several client and server components within the CMS. One of them is the Common Object Request Broker Architecture (CORBA), proposed by the Object Management Group (OMG) (2002). The CORBA Interoperability Platform has lately been accepted by ITU as a standard (ISO/IEC 19500-2); the complete CORBA framework is also to be submitted to ISO for standardization. Other communication methods that can be used here comprise for instance Microsoft DCOM on Windows platforms Microsoft Corporation (1996) and Remote Procedure Call (RPC) implementations available for all common hardware platforms.

6.4 THE CORE

The core plane (or layer) provides the core functionality of the CMS. It facilitates the control of storage devices, which can for instance be disk-based or tape-based solutions. Further, it is concerned with the control of input, output, and transport devices such as VTR, crossbars (or matrix switches) or other studio devices. Other tasks of the core plane are import of content, finding content, and export and delivery of content to other components related to the CMS or using its services. The main task of this part of the system, however, is the permanent storage (archiving) of content. This includes the organization of content and its basic components (i.e. essence and metadata).

Basically, the core plane encapsulates the actual essence, device and data management subsystems as 'black boxes', together with the interfaces to get content in and out and to control external devices.

6.4.1 ESSENCE MANAGEMENT

The *Essence Management* is responsible for storing, managing, and providing essence objects in a distributed storage environment. It takes requests from clients, manages all queues necessary to handle these requests, and provides requesting applications with statistical information about the status of the request and how and when the request will be processed. It handles:

- access to mass storage systems by addressing the Archive Management Server;
- file migration between online and near-online storage by controlling the Archive Transfer Servers;
- the creation of multiple copies of essence files in the archive;
- the contents and integrity of the online storage systems by addressing the Cache Servers;
- the delivery of content to clients by linking them to the Streaming Servers;
- accepting essence from or delivering essence to remote storage systems by accessing the Transfer Servers.

The Essence Management is involved in various tasks within the systems. For instance it can provide access to Streaming Servers to playout preview quality copies and images via remote communication from the server to the client. For the purpose of streaming audio and video a CMS should be able to integrate third-party streaming servers controlled by client components that are provided by the respective vendors. In such a configuration the client can use all functionality provided by the respective streaming server (such as read-while-writing, split audio editing, multiple playback speeds, etc.). The Essence Management ensures timecode synchronization between different copies of a single content object, independent of the format of the copies.

Besides such streams of continuous media the streaming of structured data should be supported. This does not necessarily refer to the streaming of continuous media objects only. The Essence Management can for instance also use such a capability to transfer sets of images, for example keyframe sets, between server and client.

To support the workflow in news operations the Essence Management has to allow access to streams while they are ingested. In this context it is crucial that the latency is kept minimal. This capability enables other system components to work on a stream while it is being recorded.

As shown in Figure 6.5, its fundamental building blocks are:

- *Archive Manager*
- *Archive Transfer Manager*
- *Archive Transfer Servers*
- *Cache Servers*

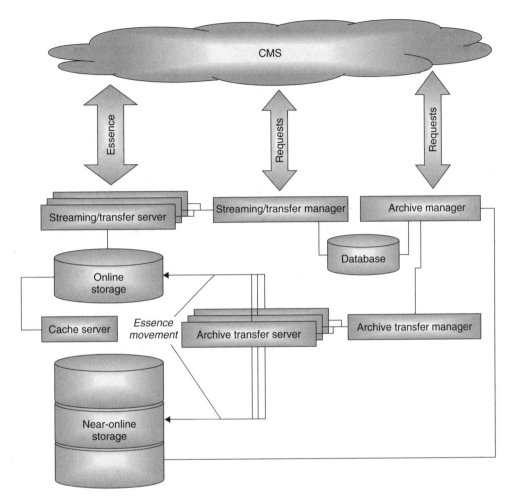

Figure 6.5   Essence management

- *Streaming* and *Transfer Manager*
- *Streaming Servers*
- *Transfer Servers.*

These components together provide all the functionality expected from an Essence Manager where each of them addresses a specific aspect. The following sections describe the functionality of these components.

6.4.1.1 Archive Management Servers

In most cases the core of the mass storage system of the CMS is a near-online storage system. An example of such a system is a robotic tape library connected to one or more hard-disk based staging areas (online storage). The storage medium of a near-online system is data tape. Popular alternative near-online systems are hard-disk systems based on technologies such as serial ATA. Up to a certain size these systems can economically compete with data-tape-based systems.

The Archive Management Server is the 'brain' of the archive management part of the Essence Management. It keeps track of all storage units in the mass storage system (such as tapes and tape pools) or disk pools. It receives requests from the CMS to archive files or to restore files. In case of data-tape-based near-online storage systems the Archive Management Server relays commands to move the tapes between slots and tape drives to the library or the library management software. When the mass storage system is ready to fulfill archive or restore tasks the Archive Management Server passes the respective request to the Archive Transfer Manager.

When the near-online storage solution is based on data tape the Archive Management Server should also provide means to ensure the overall integrity of the data stored on the data cartridges in the automated cartridge system. This includes a series of tasks to check the content on a tape and take appropriate action if required. The following procedures are performed in this context:

- Mounting tapes just written by one tape drive in another tape drive and re-reading the data in order to eliminate the possibility of write errors due to misalignment of heads or other technical defects in the writing unit.
- Rewinding and re-tensioning tapes that have not been accessed for a given amount of time in order to avoid sticking of tapes.
- Reading tapes at regular intervals in order to check the bit error rate.
- Copying tapes that have a bit error rate higher than a given safe bit error rate to new tapes.
- Copying tapes that have reached end-of-life to new tapes. A tape has reached end-of-life when its overall age due to its time spent inside of a controlled environment and the time spent outside of a controlled environment is larger than a given safe age. Tapes age faster when they are outside of a controlled environment.
- Migrating data to new tape formats when new tape drives are introduced into the library.
- Deleting objects and thus releasing available space on tapes by reading the tape and writing the data not deleted to a new tape. This process is governed by configurable watermarks.

It is important that the system is configured in a way that these processes do not obstruct the productive work. They are important in the context of content preservation but other, high-priority tasks always have precedence. Thus, these procedures are automatically performed in the background when the overall load on the system allows performing them without affecting the productive work process.

Instead of using standard IT mass storage solutions it is also possible to use automated studio tape libraries in place of a data tape library as storage vaults for high quality material. In this configuration an automation or media management system can be used to move essence from tape via studio (or video) lines, matrix switches (or crossbars), and routers to a receiving device and back can be used. Ideally, in this scenario the automation or media management system also controls the tape library. However, this depends on the capabilities of the automation system.[1] Therefore it is also possible to use an Archive

---

[1] Section 8.3.2.1 discusses the integration with studio automation and media management systems in more detail.

Management Server to control the library and to mount and unmount studio tapes in recording/playback devices attached to the library, so that the media management system can focus on the essence transfer.

6.4.1.2 Archive Transfer Manager

Archive Transfer Managers receive requests from the Archive Management Server for archiving and restoration of files, which actually means copying files between the online storage and the near-online storage. Since the number of requests that are passed from the Archive Management Server to the Archive Transfer Manager may exceed the number of concurrent requests that can be handled by the available Archive Transfer Servers, the manager has to provide a transaction secure request queue. Ideally, the request queue provides means to assign priorities to requests. Requests that are kept in this queue are then passed to the Archive Transfer Servers on a first come, first served (but higher priority first) basis.

6.4.1.3 Archive Transfer Servers

Archive Transfer Servers handle the migration of data between near-online and online storage. Typically, an Archive Transfer Server has a connection to at least one tape drive (which may or may not be attached to a robotic tape library) or to an archive disk pool. It copies data from online storage to tapes mounted in this drive or to the disk pool, or reads the data back from tape or disk pool and writes it to the online storage. The online storage can be a hard disk system connected to the Archive Transfer Server (local storage) or a shared storage environment (Storage Area Network; SAN). Further, it can also be storage that is accessible via a network (for instance as Network Attached Storage; NAS), or a remote storage system accessible via File Transfer Protocol (FTP, RFC 959) (Postel and Reynolds, 1985). Typical storage systems accessible via FTP are disk recorders or video servers.

Archive Transfer Servers can provide certain optimization strategies for writing to and reading from the archive storage. These strategies for instance include batch write, which is collecting files until a configurable number of files or a minimum total size is reached. Further, operators, archivists, or media managers might select items for transfer. These selectable collections can be defined by a list of file names the users select from the appropriate applications. Another strategy is partial file restore. In this case only a relevant part from an archived file is read. Note that partial file restore has to be supported by the essence file format used for archiving.

Archive Transfer Servers may also be used to integrate optical media into the storage strategy or, by addressing tape drives that are external to the tape library, to make offline copies accessible to the system. However, Archive Transfer Servers are *not* used to transfer essence via studio connections such as SDI.

With the advent of integrated SAN management solutions providing server-less data transfer mechanisms it is predicted that in the long term Archive Transfer Servers will become less important (especially in pure SAN environments).

6.4.1.4 Cache Servers

Cache Servers manage the online storage systems, i.e. the caching area where essence is kept for online access. A Cache Server always attempts to keep the most frequently accessed material online since it is expected that this will be frequently accessed for production. Cache Servers manage their caching area autonomously. User-definable storage rules are used by the Cache Servers to decide on what to keep and what to remove to clear the cache for new material. Since material might be of relevance although it has not been used frequently according to the defined rules, applications or services must be able to lock essence on the online storage during access. The Cache Server must respect this lock to prevent unwanted release from the cache. This for instance may be the case for impending and foreseeable events such as anniversaries, jubilees, sports championships, etc. Material might be prepared and made available on caches in advance to ensure that all relevant content is immediately available if and when required.

Cache Servers provide information on what essence is available online. One permanent management policy is to keep the amount of data stored between configurable high and low watermarks. For this purpose a Cache Server autonomously deletes essence from the online storage when the high watermark is exceeded and continues doing so until the low watermark is reached. When selecting essence to be deleted, at least the following conditions have to be considered:

- the essence already must have been successfully transferred to the near-online system;
- the essence must not be locked by an application or a service (i.e. it may not be in use);
- prefer deleting essence that has not been touched for the longest time (i.e. 'oldest essence first');
- Prefer deleting essence that occupies a large amount of disk space (i.e. 'largest essence first').

In addition, various additional options should be configurable, such as marking a certain class of objects to be undeletable or specifying a minimum time frame for which essence is to be kept online after ingest.

6.4.1.5 Streaming Managers

Since the number of concurrent requests for the streaming of content may exceed the maximum number of streams that can be handled by one Streaming Server, a CMS should allow the installation of groups of Streaming Servers that share access to the same content. In such a configuration, a Streaming Manager can be employed to select a Streaming Server suitable to fulfill an incoming request, and to negotiate the connection between the client and the selected Streaming Server.

An appropriate streaming server is selected according to the load situation and physical and logical proximity of the client and the respective server. Physical proximity refers to the location and connection features (i.e. QoS parameters such as bandwidth, jitter, and delay) between the client and server. Logical proximity is found in cases where the requested content is already served to other clients. Thus, synergies could be exploited.

6.4.1.6 Streaming Servers

Streaming Servers grant streamed access to media content. In most cases the Streaming Server is built around an existing, proven audio or video server that has to have direct access to the essence's online storage. The Streaming Server then negotiates and establishes the connection between a client and this streaming product.

The streaming solution interfaced by the Streaming Server accesses essence that resides on the online storage system and streams it to the client. If necessary the essence is staged from the near-online storage tape to the online storage via an Archive Transfer Server. Multiple Streaming Servers may share the same online storage. This allows scaling the number of concurrent streams by adding additional Streaming Servers while minimizing the required disk capacity. In the case where several Streaming Servers share the same online storage, one of the Streaming Servers acts as the manager of this Streaming Server Service Group, providing a single point of contact for the group.

Streaming is either done by using connections with controlled bandwidth (this includes dedicated connections) or by employing best effort and buffering strategies. The actual transmission may be either synchronous and controlled by the application (e.g. real-time video playout) or asynchronous. Asynchronous streaming is a best-effort download via streaming protocols, but without any control from the side of the application. In contrast to standard download, the essence is still streamed and not transported via a file transfer.

In the synchronous case a bandwidth less than or equal to a maximum allocatable bandwidth for this connection is used (and preferably reserved), whereas in the asynchronous case a bandwidth equal to the maximum allocated bandwidth is possible (but not necessarily used). This makes the behaviour of a Streaming Server to some extent deterministic.

A special class of Streaming Servers are Keyframe Servers. Keyframe Servers stream keyframes to clients. However, typically streaming of content (especially continuous media content) is a real-time or near real-time process (as is the case for audio or video playback). In contrast, in the case of keyframes the user wants to see the result 'as fast as possible' but there are no inherent time restrictions. Thus, Keyframe Servers always stream essence by using the maximum allocated bandwidth, which is comparable to a download process.

6.4.1.7 Transfer Managers

Like streaming requests, the number of concurrent requests for file transfer of content may exceed the maximum number of transfers that can be handled by a single Transfer Server. Thus a CMS should allow installing groups of Transfer Servers that share access to the same content. The Transfer Manager then is used to select a Transfer Server suitable to fulfill an incoming request.

As is the case with Streaming Managers, physical and logical proximity of client to server is a selection criterion alongside the actual load situation of a candidate server.

6.4.1.8 Transfer Servers

Transfer Servers are used to exchange files with remote content servers that are not managed by the CMS. Such a content server can be any kind of system providing or accepting content or essence. Examples are encoders or decoders, production systems, media jukeboxes, cart machines, playout servers, or nonlinear editing (NLE) systems.

Transfer Servers may be used to perform the physical file transfer within export and import processes. Nonlinear editing systems for example are considered to be export targets when content is placed on them for modification. This is the case when selected material is prepared for further production in the pre-production process. When content is brought back into the system after modification they are import sources. This step can involve a versioning process. Another example where Transfer Servers might be used is the file exchange via file transfer between CMS instances that are not integrated. This is a valid way of information interchange between such systems even within a single organization.

The Transfer Server software components are either installed on the content server itself or on a proxy server that has a suitable control and media exchange connection to the content server. They provide a control API that is suitable for content exchange. In addition, they offer network transport mechanisms for content to facilitate the physical exchange of media objects.

6.4.2 DEVICE MANAGEMENT

The *Device Management* enables control of external devices via dedicated *Device Servers* granting an IT-based access to systems that otherwise may only be accessible via dedicated control connections such as RS-422. As shown in Figure 6.6, its fundamental building

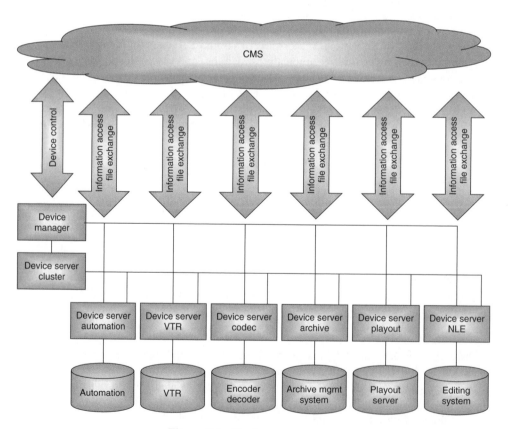

Figure 6.6   Device management

blocks are the *Device Manager,* one or more *Device Server Clusters,* and a number of special *Device Servers* (each dedicated to manage a certain device).

Examples of devices that are integrated in this way are automation and media management systems, special archive systems, NLE, playout servers, etc. In this section the principles of the Device Manager are discussed. In Section 8.3 the integration of certain third-party tools relevant in the context of CMS are presented. System automation and NLE are also part of the discussion there. Depending on the kind of integration it might go beyond the integration via the Device Manager.

6.4.2.1 Device Managers

The *Device Manager* is the entity that the CMS contacts when a service or application demands direct device control over an external device. The physical device control interface to such an external device is implemented by a dedicated Device Server. All Device Servers that provide a device control interface register with the Device Manager. This registration includes all parameters required to control the respective device. The Device Manager establishes the connection between the requesting entity and the device and manages the control connection. The Device Manager should support popular control protocols to maximize the range of systems supported.

When demands on latency and frame accuracy are not of prime concern the CMS can directly control studio devices remotely (i.e. via IT or control networks) using the respective Device Servers through the Device Manager. An example is the generation of a timecode accurate preview copy from a video stored on a studio tape. Here, the CMS prepares an MPEG-1 encoder to start encoding at reception of a certain timecode, and issues a 'start replay from timecode' command to the VTR. The latency involved between issue of the command and start of the replay is not of prime concern. The result will be exactly what is required.

In the case where response time is critical and latency would introduce operative problems (like in a scheduled recording or while working through a transmission list) the Device Manager can allow access to a studio automation system or a media management system that has a direct control connection to the respective devices in order to schedule certain jobs. In this scenario the automation would be in charge of controlling the studio devices, and would receive its jobs from the CMS from the respective Device Server via the Device Manager.

Comparable to device drivers in an operating system, Device Servers (see below) are software components that are used to interface to external devices, e.g. studio devices. Device Servers typically require customization for each new device that needs to be supported by the CMS.

To support the Device Manager, Device Servers abstract from a device-specific control protocol and provide a unified control protocol for certain device classes. Examples of such device classes are:

- video tape recorders or player;
- audio tape recorders or players;
- video disk recorders;
- automation or media management systems;
- encoders, decoders, etc.

Using this approach, a Device Manager can address devices via a well-defined standard interface. The goal is to perform the integration of third-party tools via the Device Managers. However, this might not always be possible. A tighter integration might involve other components such as the Data Managers, etc.

6.4.2.2 Device Servers

The Device Server provides unified control interfaces to various kinds of external devices such as VTR, disk recorders, video servers, archive management systems, studio automation systems or external file systems. For each class, it may offer a number of interfaces and functionalities, such as:

- a file system interface, granting access to the assets present on the device via direct file access;
- a device control interface, enabling remote control of the device via a network;
- inventory management, storing and providing metadata and status information for each asset;
- event handling that can trigger actions depending on certain events, such as appearance, deletion or status changes of assets;
- cache management that keeps the used storage space between high and low watermarks, thereby recognizing side conditions for each asset such as maximum and minimum lifetime on the device as well as status.

A Device Server should be fully self-contained, which means it should be able to manage the device it interfaces without any need for information to be provided by the CMS. In addition, it has to provide the necessary interfaces so that the CMS can take control over the device, retrieve or modify information about assets present on the device, transfer assets to and from the device, or delete assets from the device.

Figure 6.7 shows a possible architecture of such a Device Server. According to this design a Device Server would consist of a core that provides interfaces for file level access and device control, since each device interfaced by a Device Server may have different (and often proprietary) means to implement these interfaces. In Section 8.1 the principles of system integration are introduced. In the context of the Device Server the two most relevant forms of integration are via protocols (or data exchange) and through Application Programming Interfaces (API). The integration on the protocol level includes access at file level, for instance via FTP, via shared file access protocols such as Network File System (NFS) or Common Internet File System (CIFS). Further, it could provide device control, for example, via standard protocols such as Video Disk Control Protocol (VDCP), Network Disk Control Protocol (NDCP) or Video Archive Control Protocol (VACP). Since there are many possible means to exchange information and content elements it is sensible to encapsulate these implementations as plug-ins to the Device Server (e.g. via Dynamic Link Libraries). This allows customizing Device Servers easily to support new devices.

The Inventory Management keeps metadata about assets that may exceed the metadata that can be provided by the device itself (e.g. additional status information). It is reasonable to support the Inventory Management by using a database. Since this database must be synchronized with the actual metadata that the device holds about each asset, the Device

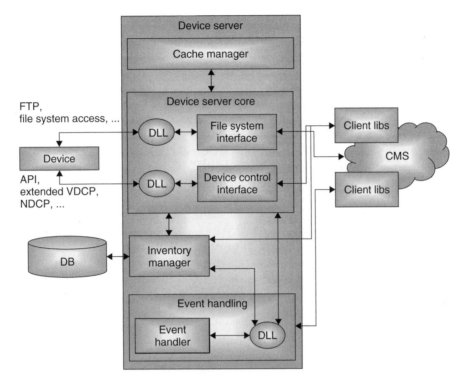

Figure 6.7 Possible Device Server architecture

Server Core must provide means that allow the Inventory Manager to retrieve information about clips from the device.

The Event Handler may communicate with the Inventory Manager or the Device Server Core in order to identify whether a change has occurred that qualifies as an event. This event is then processed by a plug-in that performs the actions required to handle the event.

The Cache Manager manages the storage space available on the device depending on certain rules. Typical rules are:

- allow deletion of content objects that have exceeded their maximum lifetime;
- do not allow deletion of content objects that have not yet reached their minimum lifetime;
- do not allow deletion of content objects that have a certain status;
- prefer deleting large content objects;
- prefer deleting content objects that have not been accessed for the longest time;
- start deleting when high watermark is reached;
- stop deleting when low watermark is reached.

Finally, for easy integration into the CMS the Device Server implementation should be complemented by a software development kit (SDK). The SDK provides the API description together with the necessary client libraries.

6.4.3 DATA MANAGEMENT

The *Data Management* handles access to all databases and information systems that store descriptive metadata or can be used otherwise to identify content during retrieval. Metadata is either introduced automatically by services or entered manually using applications. Figure 6.8 shows a possible architecture of a Data Management integrating distributed heterogeneous information systems.

The major building blocks of the Data Management are database systems. A CMS should not build on a single information system but has to be able to integrate multiple databases and metadata information systems. Especially if the CMS is implemented in an enterprise-wide context it has to be possible to use special, purpose-built data stores for certain metadata components. In general information systems that have to be integrated comprise generic database(s), legacy systems, and specialized databases. The generic databases are used internally by the CMS for the management of essential metadata that is not stored in any other system. Legacy systems comprise databases such as existing bespoke cataloging systems that are in most cases already in use. Rights management systems can also be considered as legacy systems. This is part of the integration of legacy and third-party systems and is described in further detail in Section 8.2. Specialized databases provide dedicated cataloging functionality such as stratified documentation.

Often, a CMS also integrates different kinds of search engines, such as fulltext search engines, image similarity search engines or audio similarity search engines. These search engines support specific searches and facilitate the selective search using specific media features or enable unskilled users to interact with the system.

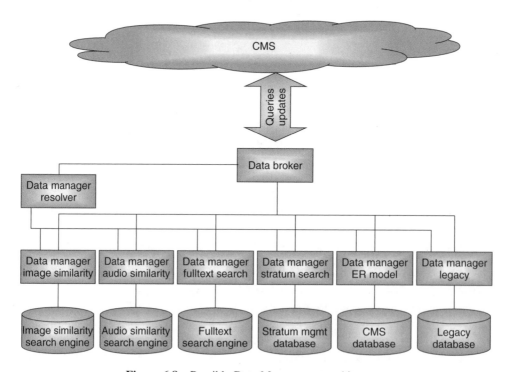

Figure 6.8   Possible Data Management architecture

6.4.3.1 Information Systems

Legacy Databases

In many cases organizations that plan to introduce a CMS already have an existing documentation system or catalog that is widely used within the organization. These systems store metadata describing a large part or even all of the content objects of the organization. It cannot be expected that the organization instantly replaces these systems as part of the introduction of a CMS. Thus, they need to be interfaced to and often even be treated as the master database. Other systems that may contain valuable metadata are:

- newsroom systems, keeping rundowns, scripts, etc.;
- systems supporting planning and drafting (synopses, stories, etc.);
- the management systems of various production systems (EDLs, titles, descriptions, etc.);
- studio automation systems (playlists, as-run logs, etc.).

All these information systems, including the existing catalog, are here referred to as *legacy databases*.

The CMS needs to interact with these legacy systems and may even have to regard certain of these systems as master databases. It is the task of a *Data Manager* to transparently provide this functionality and to integrate legacy systems into the overall CMS solution, while it is the task of a *Data Broker to* handle the distribution of metadata over the various Data Managers. A Data Manager typically falls into two major parts, namely a system-specific part and a generic part. The former implements the specific interface to the legacy system to allow query, update, and deletion of data sets. The latter implements the standard API provided to the CMS.

There are two ways in which legacy systems can be integrated into a CMS. One possibility is to map the relevant part of the data model present in the legacy system to the CMS's internal data model. This data model may be an implementation of a corporation-wide data model covering most or all business processes that involve the CMS. In this scenario, a Data Manager must be implemented that understands what part of this data model is implemented by the database it interfaces to, and that can process the respective requests. Given this Data Manager is present, the legacy system then is regarded just as one of the native CMS databases covering a specific subset of the overall data model, and accesses are automatically federated to this system. While this approach seems to be quite elegant, the implementation tends to create a considerable overhead that may induce severe latencies.

The second possibility is adapting a standard Data Manager to the legacy system in order to support a predefined set of queries and reports and a predefined set of commands that each have a system-wide consistent semantic. In this case the scope of the integration is limited to a relevant sub-domain of the corporate data model. However, it still has consistent meaning within the scope of the CMS. The legacy system is then addressed as a standard information system that can deliver information within this scope. This approach easily allows dedicating specific information systems to deliver specific sets of information (such as a dedicated rights database, a dedicated catalog for video, audio or stills, or a press database). The only thing necessary in this case is a consistent specification of the information a specific system can provide.

Considering this, legacy systems may be used to store parts of the metadata managed by the CMS, or they may represent independent sources of information. In any case a dedicated Data Manager instance is required for each legacy system. Hence, Data Managers are often subject to customization.

CMS Databases

The standard way of representing a corporate data model is developing an entity-relationship diagram to describe the data model and to implement the model accordingly in a standard relational database management system (RDBMS). A good example of such a corporate data model is the BBC's Standard Media Exchange Framework (SMEF™), which also was input to the EBU project P/Meta (see Section 4.4.1).

Since these data models differ tremendously from organization to organization, and since also the workflows that are applied on these data models are quite different, there is no possibility that a CMS can work with a standard custom data model that comes 'right out of the box.' Thus, a typical CMS project includes customization to the required data model and workflows. This implies that the respective Data Manager also needs to be customized.

For persistent storage, a single database or a number of federated databases may be used. How the physical implementation is done is fully transparent. The CMS will access this database via Data Managers and thus is not relying on specific knowledge about the physical representation of Metadata in databases.

Fulltext Search

A user-configurable set of attributes within the corporate data model used in a CMS implementation should be selectable to be indexed for fulltext retrieval. The fulltext search capabilities are provided either by a separate fulltext search engine, or via fulltext search capabilities of the database itself. If a separate fulltext search engine is used either the full-text search engine directly accesses the database and indexes the attributes, or the database writes the selected attributes as database reports into structured files. These files are placed in a protected region of the file system. The fulltext search engine then indexes these files.

Both approaches have their drawbacks. If the fulltext search engine directly accesses the database tables, a strong coupling between the mapping of the data model and the indexer of the search engine has to be established and maintained. In case of using database reports, the database is responsible for keeping up referential integrity between these reports and the database contents.

Which fulltext engine should be used is primarily a question of requested functionality and language support. Integrating a fulltext engine into a CMS should be easy to accomplish, which again means that a clearly defined interface is required.

Some requirements apply that may limit the selection of such fulltext search engines. The engine should provide

- a configurable parser;
- support for attribute-based retrieval and sorting;
- support for limiting searches to ranges of numbers;
- for CMS applications in certain countries, Unicode support;
- a reasonable API that grants access to retrieval functionality and the highlighting of information.

Depending on the capabilities of the selected fulltext engine, the CMS can provide fuzzy search, ranking, and other advanced retrieval features. This approach of separating fulltext search and database can provide very good scalability with respect to the typically highly asymmetric update to query ratio, thus making use of the excellent performance of today's fulltext search engines.

Image Similarity Search

In a CMS video objects are typically represented by a set of keyframes, which are all image objects. Besides keyframes other images may be stored in the essence management. Image similarity engines provide an alternative approach to query and retrieval of image objects. Objects are not retrieved by querying text-based metadata but by comparing image features in order to find images that look like an image that is provided as an example.

The caveat is that searches based on image similarity often deliver unexpected results. For instance a user may be searching for images of a certain person. The user offers a mug shot of the person as an example image and asks for similar images. Most users now expect to find all images showing this person, close-ups and totals, front and side views, with and without glasses, with and without hats, and so on. What they will get, however, will be a lot of images showing faces of different persons, animals or even objects that look similar to humans. Since this is not what they have expected, uneducated users are often disappointed with the search results.

This technology can be useful in other application contexts, however. One possible application is not looking for image similarity, but for image *identity* (or a very high degree of similarity). For example, a user has found an interesting clip during retrieval but it is too short or it has inserts that make it unsuitable for reuse in a new production. So the user wants to query the system and find longer versions of the clip, or versions without inserts. The user could do this by picking a characteristic image or keyframe from the clip and querying all keyframes for *very* similar images (basically almost identical ones). Via the keyframes found as a result from this query the user finds all clips represented by them and thus the clips that he is looking for. In this scenario the result matches user expectations, i.e. the technology can be deployed in a useful way.

With keyframes as the prime source for image similarity search the major problem is scalability. Let us consider an average archive with 100 000 hours of video material. On average, about 1000 keyframes may be selected to represent 1 hour of video. This means that we have to compare one image to 100 000 000 images in order to find all images that look like the example image provided. If we assume that the maximum response time should be less than 10 seconds, it is clear that this functionality is quite demanding.

Recognizing the rapid progress of technology, there may be solutions to this problem upcoming. Therefore a clear interface description that allows integration of an image similarity search engine into the CMS framework at any point in time is desired.

Audio Similarity Search

Another kind of advanced search functionality is audio similarity search. The user presents a sound clip to the search engine, and the search engine returns audio objects that contain the presented clip (audio identity) of a similar audio element. Since the audio object may be an audio track of a video object, this kind of search applies to audio and video content.

These kinds of search engines are at present still in the research stage. However, like image similarity engines that become scalable to extremely high numbers of images, technology advancements will lead to solutions in the near future. Again, a clear interface description that allows integrating an audio similarity search engine into the CMS framework at any point in time is desired.

6.4.3.2 Data Managers

Data Managers are used to integrate information systems into the overall architecture. For each information system that has to be included a dedicated Data Manager has to be provided.

A Data Manager can be divided into two parts, namely a system-specific part and a generic part. The system-specific part is the wrapper around an information system whereas the generic part presents the interface provided to the CMS. Data Managers are responsible for providing any modules requesting information (e.g. services and applications) with a unified interface to information systems that store metadata (not essence). It is important to note that the Data Managers have to fully abstract from the respective data management system's query language and physical data storage characteristics. A Data Manager receives queries as well as requests for creation, updates, and removal of data objects. As a response, it delivers hit lists, reports for details, etc.

All messages passed between a Data Manager and a client (which may be an application, a service or a Data Broker) should be encoded in a standard exchange format such as XML. For unique referencing of attributes, SMPTE 335M (Metadata Dictionary Structure) (Society of Motion Picture and Television Engineers, 2001a), together with SMPTE Recommended Practice RP210, can be used. The attributes can be coded according to SMPTE 336M (Key-Length-Value protocol; KLV) (Society of Motion Picture and Television Engineers, 2001b). For mapping of KLV to XML, the document type definition (DTD) of the messages should comply with the format specified by the Advanced Authoring Format (AAF) Association (2000).

The Data Manager transforms incoming query requests into the native query language of the respective information system (which may, for example, be an ANSI SQL RDBMS) and maps the responses provided by this information system into standard response messages. If necessary it may mediate between the CMS data model, and the data models and data storage representations of the information system it interfaces to.

A possible approach to abstract from native data models is using the concept of labeled queries. A label is the identifier of a specific query to an arbitrary set of attributes within the information system. Labels need to have a consistent semantic meaning throughout the CMS, but it is up to the configuration of the respective Data Manager which part of the data model hosted by the respective information system the label addresses. Examples of such labels are 'What', 'Where', 'When', and so on.

The Data Manager receives a query as the label, together with a number of attributes passed as parameters. It selects the query identified by the label from the list of queries it stores and maps these attributes into this query. This results in a complete query formulated in the native query language of the information system that is interfaced. Up to a certain level of complexity of queries, this approach even allows federating the same labeled query over a number of Data Managers. This is crucial since especially unskilled and casual users expect a single result list that does not contain any duplicates. Thus, queries using labels allow specific search for a common concept known in a number of information systems while abstracting from the implementation details of the individual system.

Via labels or more advanced structures (e.g. encoded in XML) a Data Manager can address a subset of the CMS's data model, which typically is a company-wide data model (e.g. BBC's SMEF^{TM}). The size of the subset is defined by the capabilities of the database system that the Data Manager interfaces; it may provide anything from certain attributes of a single object up to the full data model.

6.4.3.3 Data Manager Resolver

A Data Manager integrates an arbitrary information system, which grants access to a certain part of the metadata available for a content object (or asset if IPR are managed). Other metadata may reside in another information system, interfaced by another Data Manager. Most probably, each of these information systems uses its own proprietary unique ID to address this asset within its domain. Thus, there needs to be the possibility to map the various proprietary ID to a system-wide unique ID that represents the asset within the domain of the CMS. This can be accomplished by a *Data Manager Resolver.*

The Data Manager Resolver is basically a simple mapping table, that contains:

- a row that contains all known IDs for each known asset;
- a column that stores the CMS-wide ID for each asset;
- a column for each Data Manager that stores the ID of the respective object in the system the Data Manager interfaces to.

Queries against this mapping table allow the CMS ID of an object to be found when you know the Data Manager and the respective local ID, or the local ID of an object in a given information system to be found when you know the CMS ID and the respective Data Manager.

Since a Data Manager provides a unified interface to the CMS that fully abstracts from proprietary information present in the respective information system, it is the Data Manager that accesses the Data Manager Resolver. This mechanism allows query for CMS content objects identified by CMS IDs, and delivering such CMS content objects as results.

6.4.3.4 Data Brokers

When a query needs to be federated over more than one Data Manager, the CMS must provide an entity that can distribute the query over the Data Managers that need to be involved and receive the responses from these Data Managers. Further, it has to merge the responses into a unified response set and (if required) can enrich the attribute set to be delivered by again querying specific Data Managers for additional information for each object in the result set.

This component is called the *Data Broker.* It implements the broker manager scheme introduced above on the metadata level. The Data Broker provides a unique query, update, and delete interface to the CMS, so that CMS components have no need to know exactly how the Data Management as a whole is structured. A Data Broker submits requests (queries and database maintenance commands) to the Data Managers that have registered with the Broker. Since the Data Broker itself can be a distributed entity it will also pass on a request to other Data Brokers known and accessible to it. After receiving the results from all the units it called, a Data Broker collects responses and forwards them to the calling unit. This process depends on the chosen calling depth and broker selection.

6.5 SERVICES

Services extend the scope and capabilities of the CMS. Like the daemons or services in an operating system, CMS services typically are background processes that may accept jobs from a client (e.g. an application or another service) and may access the Core in order to perform various tasks accessing essence, metadata or devices. The following sections discuss a number of services that have been identified as being useful in CMS that are managing content on a enterprise-wide level. A CMS may comprise these services but neither is required nor limited to provide the services described.

6.5.1 INGEST

The *Ingest Service* is responsible for ingesting audiovisual signals into the CMS. A signal may be an incoming feed from satellite or cable, or a signal from a studio device, like a tape, a DAT or a CD recorder/player. The signal is digitized into multiple formats via suitable encoders and written to an online storage area.

The major tasks of the Ingest Service are:

- identifying the object that the essence to be ingested should be associated to (if applicable);
- generating unique identifiers when necessary;
- enforcing entry of minimum metadata;
- monitoring and supervising the encoding process;
- identifying and handling files resulting from the encoding;
- registering essence with the essence management system while it is being recorded;
- introducing relevant metadata into the data management.

6.5.2 PLAYOUT

The *Playout Service* is used to replay audiovisual signals at signal level. It uses suitable decoders or video severs to accomplish this task. While a CMS typically is not directly involved in on-air procedures there are may workflows (such as re-encoding a browse copy from signal, or playing archived essence to tape) that require provision of file-based essence as baseband signal. Thus, the playout service handles all streaming requests that are not an inherent part of the content handling process.

6.5.3 IMPORT

The Import Service is invoked whenever an application or a workflow processor requests import of assets into the CMS. The Import Service should allow import of single objects as well as batch-controlled mass import. What exactly needs to be done to complete an import process strongly depends on the workflow requirements, the business rules to be applied to the installation, and the condition of the assets to be imported.

Figure 6.9 proposes an Import Service architecture that is flexible enough to handle diverse requirements and conditions that may apply to imports. This Import Service consists of a workflow processor that manages the import procedure and a set of processors that can be added as plug-ins. Each of these processors may be invoked along the workflow, and each processor may again use external systems, external code or CMS services

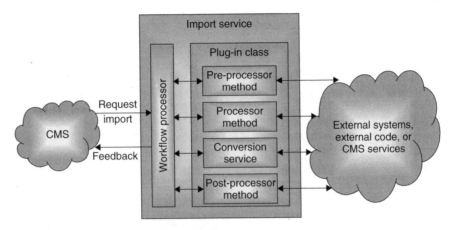

Figure 6.9 Import Service architecture

to accomplish certain tasks during the import process. A possible import workflow could be processed in a number of steps as follows:

- The workflow processor handles an incoming import request and forwards the request to the pre-processor.
- The pre-processor evaluates the incoming request and performs certain scenario-specific tasks to prepare the fulfillment of the overall import job. Part of these tasks is to check whether all preconditions for a successful import are fulfilled. Upon that the pre-processor decides whether the import can continue or must be aborted. If the import continues, the pre-processor hands the responsibility back to the workflow processor
- The workflow processor creates the import process list and calls the processor for each object to be imported.
- The processor performs scenario-specific tasks for each object to be imported. This can include creation or update of metadata, movement of files, etc. After completing these tasks, the processor hands back control to the workflow processor.
- The workflow processor may invoke the Conversion Service for format conversion if required. This is necessary when the business rules require that, upon import:
  - the object must be converted into a standard archive format, or
  - additional formats, such as browse or Internet copies, shall be created.
- The workflow processor now invokes the post-processor that can perform scenario-specific tasks to close the overall import job. After successful completion of the import job, the post-processor returns control back to the workflow processor.
- The workflow processor closes the task by providing feedback to the initial requester.

6.5.4 EXPORT

The Export Service is invoked whenever an application or a workflow processor requests full or partial export of assets from the CMS to an external system. The Export Service should allow export of single objects as well as batch export of multiple objects.

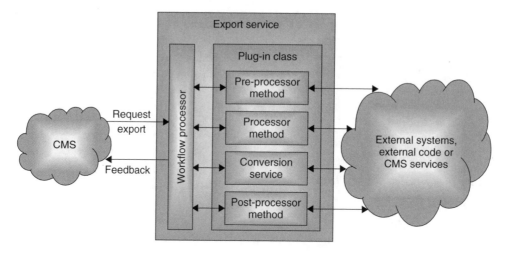

Figure 6.10 Export Service architecture

Comparable to the import process the export procedures strongly depend on workflow requirements, business rules, and the specific requirements of the various target systems.

Figure 6.10 shows that the architecture of an Export Service can be quite similar to that of an Import Service. Again, the use of a workflow processor, together with pre-processor, processor, Conversion Service, and post-processor should provide enough flexibility to handle various kinds of export scenarios. Each processor may again use external systems, external code or CMS services to accomplish certain tasks during the export process. A possible export workflow could be processed as follows:

- The workflow processor handles an incoming export request and forwards the request to the pre-processor.
- The pre-processor evaluates the incoming request and performs certain scenario-specific tasks to prepare the fulfillment of the overall export job. Part of these tasks is to check whether all pre-conditions for a successful export are fulfilled. Upon that the pre-processor decides whether the export can continue or must be aborted. If the export continues, the pre-processor hands the responsibility back to the workflow processor.
- The workflow processor creates the export process list, ordered by source material, and calls the processor for each object to be exported.
- The processor performs scenario-specific tasks for each object to be exported. This can include retrieval of metadata, etc. After completing these tasks, the processor hands back control to the workflow processor.
- The workflow processor may invoke the Conversion Service for format conversion if required. This is necessary when the business rules require that, upon export:
  - the object must be converted into a format supported by the target system, or
  - a partial export is requested that can be handled by the conversion engine.
- The workflow processor now invokes the post-processor to perform scenario-specific tasks to close the overall export job. After successful completion, the post-processor returns control to the workflow processor.
- The workflow processor closes the task by providing feedback to the initial requester.

6.5.5 EDITING

A CMS must be able to store content as assembled pieces, but also broken down into the elementary video and audio clips, stills, graphics, and text essences. Thus, a CMS should be able to segment content into smaller elements. The reverse operation (i.e. to assemble content from smaller elements) is also required. This functionality can be provided by a dedicated Editing Service.

For assembling essence from clips, the Editing Service can work in conjunction with a production system, a video server or a Streaming Server, selecting the relevant essence, transferring it to this system, and providing an assembly list to the respective system for auto-assembly or playout.

For disassembling, an Editing Service should be able to automatically perform simple editing functions like cutting, trimming, multiplexing, and de-multiplexing on certain well-known essence types. This functionality can be used, e.g. to segment a news feed into the single items. This is for instance required to eliminate irrelevant parts from a recording or split a continuous recording into units that are congruent with the logical content object.

These kinds of operations are not as straightforward as they might seem. The two major issues that have to be considered in this context are encoding formats and timecode accuracy for all essence copies and versions of a content object. If an encoding format uses inter-frame encoding (as for instance MPEG-1 and MPEG-2) it is not possible to segment the video at every frame. A new segment always has to start with an I-frame. Thus, the Editing Service has to select the closest preceding I-frame to a user-selected cut. Ideally the preceding set of frames up to the selected frame should remain hidden from the user. This functionality, however, depends not only on the Editing Service but also on the video player presenting the video to the user. Another solution would be a partial re-encoding of the edited segment where the first frame is encoded as an intracoded frame. All other frames up to the first regular I-frame would also have to be re-encoded. This is a very costly procedure but would allow playback of the video from any standard player.

In any case frame and timecode accuracy have to be ensured. It is crucial that the consistency and integrity of all time references is maintained across all copies of the essence even after editing has been performed.

6.5.6 CONVERSION

The *Conversion Service* provides the CMS with capabilities for transcoding between different media formats or file formats. There are a number of encoding formats that are used within content-rich organizations. They range from standard formats (e.g. MPEG- or DV-based formats) to proprietary products. In order to allow the conversion between these formats without going through a re-recording process every time the Conversion Service has been devised. However, it is not feasible (in the case of proprietary formats sometimes not even possible) to implement all the possible conversions within the CMS. Thus, it should provide a well-defined framework for the Conversion Service that allows easy integration of well-established third-party products into the CMS. This allows the system to leverage the expertise of specialists and to integrate other formats when the respective tools become available.

Important application areas for the Conversion Service are the automatic conversion between different production qualities. For instance in an organization that does not uniformly rely on one format the conversion MPEG-2 4:2:2P@ML and DV-based formats and vice versa might be required. Further transcoding between linear audio and MPEG-1/2 Layer 2 audio is another application area. Another area where transcoding might be frequently used is the automatic generation of browsing copies from high-quality archiving material. This can comprise the generation of an MPEG-1 video from production quality video or of MPEG-4 simple profile video at low bit rate for Internet distribution. For audio an equivalent procedure is the conversion from linear audio into a popular MPEG-1 Layer III version.

Apart from encoding format changes, file format changes are also supported by the Conversion Service. This includes the migration between OMFI, GXF, MXF, and AAF file formats and between WAVE and BWF file formats.

Other application areas for Conversion Services are easily conceivable, hence the requirement for easy inclusion of third-party tools.

6.5.7 ANALYSIS

The *Analysis Service* encapsulates advanced essence processing tools that analyze the media and produce metadata (including information about the structure of the content) and audiovisual abstracts. It accepts essence as input and uses this essence to automatically generate additional information. A multitude of analysis functionality can be imagined and an increasing number of interesting technologies are entering the market. A CMS should be able to integrate these technologies, as long as the service complies with certain requirements concerning API and output of results. Thus, comparable to the Conversion Service the Analysis Service also provides a framework for the integration of a number of relevant technologies. The following sections discuss some analysis technologies found in a number of today's CMS.

6.5.7.1 Video Analysis Service

In an environment where preliminarily video is processed (as is the case in television) a Video Analysis Service is valuable to assist the work of catalogers and provide additional visual information about a content object. It helps to cope with the growing amount of audiovisual material and make documentation and retrieval more efficient.

A basic Video Analysis Service supports shot detection as well as selection and extraction of keyframes. Based on this information a keyframe selection process can select a representative set of keyframes that gives full visual coverage of the video's image content with a minimum number of frames. These frames are a kind of reconstruction of the video's original storyboard, where each keyframe references a certain presentation timecode in the video stream.

A number of approaches regarding the selection of keyframes exist. The simplest tools select keyframes according to their temporal location (e.g. one keyframe every 2.5 seconds). However, this cannot be regarded as proper analysis since neither the syntax nor the semantics of the video are considered. In general a Video Analysis Service should only select a keyframe when there was a 'significant change' in the image content. There are several reasons why frames are be selected to be part of the keyframe set, such as being the first/last frame in the video and the first/last frame in a shot. Further, it can represent a camera operation or a transition. Another reason may be that a user selects a frame that s/he considers most suitable to represent a certain sequence.

A Video Analysis Service can support several granularity levels when selecting and extracting keyframes. Higher levels of granularity mean that more frames are extracted to cover smaller changes in image content. Using these granularity levels, applications can influence the overall number of frames presented to the user.

A more advanced Video Analysis Service may be able to extract information about the camera work, such as zooms, pans, and tilts, and may try to identify transitions. Supported classifications for camera work should be at least *pan, zoom*, and *tilt*.

The transitions can be further classified. Transitions are often used as artistic elements in the editing process. There are a number of transitions and not all of them can be detected automatically. For instance very long fades or chequerboard cross-fades are hard to detect. However, a video analysis should support the following classifications:

- cut
- fade-in
- fade-out
- cross-fade
- other.

As an advanced feature, a Video Analysis Service may provide means for shot clustering. The keyframes of each shot can be compared to the keyframes of the second but the next shot and similar looking shots can be selected to belong to the same cluster. This can be helpful to identify dialog scenes. It can also be used as input to a human-controlled scene clustering.

The functionality a Video Analysis Service can offer depends on the technology available. However, it is crucial that the technology works in real-time or taster on standard equipment. Any method that takes longer cannot be deployed in an operative system since it could not cope with the workload in today's content production. Thus, it is vital that the technology used for a Video Analysis Service has reached a mature and stable state and can operate within the boundaries given by the operative requirements.

6.5.7.2 Audio Analysis Service

An *Audio Analysis* Service analyses the audio track and adds additional metadata or can be used for indexing purposes. Audio analysis has been researched for a number of years and different features can be automatically extracted. The capabilities of available analysis tools range from simple audio classification to the creation of a transcript. In general, audio analysis tools with the following features can be identified:

- classification of segments containing music, speech or other sounds;
- speech recognition to derive transcripts or at least a text base for fulltext retrieval;
- speaker recognition;
- keyword spotting.

An Audio Analysis Service should have the capabilities to integrate all these tools depending on the application requirements. In this context it is important to manage the expectations of the user and only use specific tools when they are really appropriate. Depending on the environment speech recognition tools can reach a recognition accuracy of 95%. This is not sufficient for many applications such as news where especially names (which are very often not correctly recognized) are important. Thus, keyword spotting might be much more appropriate in this case. In general, the data provided by an Audio Analysis Service built on today's technology only delivers auxiliary data that has to be enhanced by human users to achieve a full documentation or is used together with other metadata to achieve full accuracy.

As is the case for video analysis, the technology used within the Audio Analysis Service has to work in real-time on standard computer equipment in order to be a feasible alternative or even enhancement to manual documentation.

6.5.7.3 Other Analysis Services

There are a number of other technologies that allow the automatic analysis and retrieval of certain characteristics of the media. This can for instance be Optical Character Recognition (OCR) from text on screen. Another interesting option for the application of automatic indexing technologies in images and video is face recognition, leading to person recognition.

More and more interesting technologies are arising, and a CMS must be prepared to integrate these technologies as they become available.

6.5.8 WATERMARKING

The *Watermarking Service* should handle issues involved in authentication of media objects and handling security issues involved in exchange of program material. At least two approaches to watermarking can be considered. One approach is to insert a visible or hearable watermark, that still allows a recipient to appreciate the content, but makes ownership clear and limits possibilities of reuse. The second approach is to insert invisible or unhearable watermarks that allow ownership to be established in case of unauthorized reuse of content. In both cases it is important that the watermark is preserved throughout several generations or production cycles.

At present several promising technologies in this area are being developed. Since these technologies will most probably be vital for future electronic commerce activities involving the exchange or delivery of content, a CMS must be open to include such a service when suitable solutions become available.

The term watermarking in the context of a Watermarking Service does not only refer to the watermarking technology as currently being developed. It includes all kinds of technologies that allow the authentication of content, i.e. the proof of ownership. Other technologies relevant in this context are for instance fingerprinting. This technology analyses unique characteristics of audiovisual material and stores them in a database. This data is compared with the data of material in question to find out if and in which form content was used. Thus, the Watermarking Service stands for all services that authenticate content.

6.5.9 ORDER MANAGEMENT

Behind an enterprise-wide CMS implementation is the vision of 'information at your fingertips', implying that everyone should be able to find and access 'anything, anywhere, anytime.' In reality, there often are several caveats:

- There are users who do not really care that there is a CMS where they can find assets themselves. They want to contact the archive, by phone or fax, and ask the archive personnel to do the research and provide them with material ('I need some images of airplane crashes for my feature show next Tuesday. Please find me something that has not been on air too often'). Ordering archive research is a standard workflow in many enterprises.
- Users may want to have the material delivered at a certain time to a certain area, such as 'please deliver this material to Edit Bay #7 on Wednesday next week, ready to work at 08:00 am.' Deferred delivery is also quite common in many enterprises.
- Often a user has not really the authority or the knowledge to decide the location the material may go to. There may be a facility manager who has the task of deciding which edit bay can be used for a certain project at a given time. Hence, this person needs to be involved in the delivery chain.
- It is quite common that material requested may have certain rights associated that do not allow use in a given context or at a given time, which may cause severe legal problems. It also may be that there is a considerable cost involved in using a certain asset that may go beyond the budget of a certain production.
- It may not be possible to deliver the material to the desired location. For instance older material may still be available only on conventional storage media and may require digitization or conventional delivery.

Obviously, there is a need for an additional service to be provided by the CMS. In the context of this book we call this service *Order Management*. The Order Management should at least provide support for requesting mediated research by archive personnel, i.e. the CMS user should have the possibility to ask the staff managing the archive and the CMS to take research requests and feed these requests into the system. Such a query might be more efficient since archive personnel know the structure and details of the documentation and system. The CMS is used for retrieval and delivery but also has to provide a feedback channel to the requesting user.

The Ordering Service also has to allow taking orders for delivery of material (i.e. essence copies) either conventionally or via CMS applications. These orders comprise assets that may be delivered as complete files, on conventional carriers, or as a list of clips (ordered by providing a rough-cut EDL). The person handling these requests has to be able to access each single element within the order, check rights and cost, and decide whether it may be delivered. If this is the case s/he triggers additional processes such as digitization or copying of conventional carriers. Further s/he has to be able to modify rough-cut EDL in case the material to be delivered stems from a different source that does not share the same timecode as the one requested (for example, there may be an asset that has been broadcast in a news show, and the user has requested a clip from this item. The archive, however, has access to the original item archived directly from the production system. Timecodes do not match, so the EDL needs to be adapted to these timecodes).

It should also be possible to change the target location (e.g. the user may have requested all material to be delivered as file to edit bay #7, but there is one item that s/he has to pick up on a conventional carrier at archive desk #3) and provide feedback to the requesting user, comprising the modified EDL as well as updates concerning delivery location and delivery times. Finally, support is required for scheduling and authorizing delivery. This implies that each delivery authorized by the archive is scheduled for a certain delivery time, but the facility manager has the option to change the schedule and modify final delivery locations.

Obviously, implementing an Order Management that matches the requirements of an organization can be a challenging task. It is important to build as much flexibility into the design of the Order Management as possible, so that customization to accommodate different workflows is possible.

6.5.10 WATCH FOLDER

Watch folders allow triggering of processes in the background upon appearance of a file. A *Watch Folder Service* reacts on appearance of files with a given suffix, and performs a configurable 'action.' This action can be a complex workflow, however. It may include operations performed by third-party systems as well as passing a job on to other watch folders for continuation of a workflow. It should be possible to implement the action performed by the Watch Folder Service as a plug-in, thus facilitating an easy way to enhance the possible actions that can be performed when a certain file appears.

Watch folders are very popular in import scenarios, where any kind of third-party ingest tool can create a new object and drop it into such a watch folder, thus triggering automatic import into the CMS. But watch folders can provide much more. To fully understand the implications of watch folders, let us consider a more elaborate example. The customer service of a broadcaster has received a request from a viewer who wants to have a DVD copy of a program that has been broadcast. The program is available in the archive but the archive holds it in a certain archive format (e.g. DVCPRO50) which is not equivalent to the one required on DVD (i.e. MPEG-2 long GoP). It could also be the case that the archive is located at the headquarter but the customer service is in a remote location 30 miles away. In all these cases the Watch Folder can keep track of the execution and status of the request.

Figure 6.11 shows a possible workflow involving watch folders that would allow the customer service to retrieve the material in the desired format:

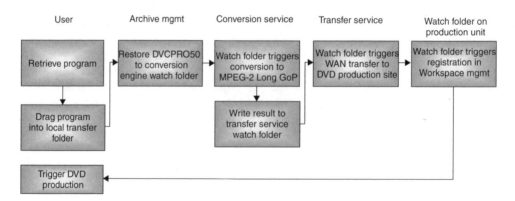

Figure 6.11   A workflow supported by Watch Folders

- The user retrieves the desired program from the archive using the standard query application, which may be provided by the Web Retrieval Service (see below).
- The user then drags the program from the hit list to a transfer folder provided by the Workspace Management (see below).
- The Workspace Management interprets this drop event and writes a transfer order file into a watch folder of the archive management.
- This watch folder initiates a transfer of the archived object to a watch folder of a suitable conversion engine.
- This watch folder triggers the conversion engine to transcoder the file to DVD-compliant MPEG-2 and write the file into a watch folder of a transfer service.
- This watch folder initiates a WAN transfer (e.g. via FTP) to a watch folder of the DVD production system.
- This watch folder then imports the file into the DVD production system and registers the object with the Workspace Management.
- Now the desired object is available to the user who has requested it and he or she can now proceed with creating the DVD.

Even though such a scenario would probably not be exactly implemented as described, the example should help to understand the possibilities that come with a flexible implementation of watch folders.

6.5.11 WEB RETRIEVAL

The *Web Retrieval Service* provides Web-based access to content. For instance, it may provide a Web-based retrieval interface, offering simple or advanced query interface and displaying retrieval results, like hit lists, metadata, and keyframes, via standard Web browsers on any desktop.

Depending on the network connection of the client machine requesting access, the Web Access Service may provide access to streaming media in various qualities. While an in-house client may be able to access an MPEG-1 preview video at 1.5 Mb/s, a client connecting via the Internet may only be able to access a low-resolution video proxy, a low-resolution audio stream, or even only keyframes. If only keyframes can be displayed, rough-cut functionality can be offered based on the keyframes set.

Another example for an application for the Web Retrieval Service is offering public access to archive content. Together with an e-commerce application, this can be a suitable way of advertising and marketing content to third parties, which opens up additional revenue potentials for archived content. For security reasons it may be sensible to set up the e-commerce presence as a separate entity into which selected content is replicated. This entity can be integrated with the core CMS via import and export filters.

In general, the Web Retrieval Service must provide a very high degree of flexibility and customizability with respect to the look-and-feel of the Web-based user interfaces. Many vendors or customers may want to use this service to integrate CMS functionality into their applications. Hence, the following requirements apply to the Web Retrieval Service:

- Each view provided by the Web Retrieval Service should be available via a dedicated URL, so that third-party applications can access each of the functionalities separately.

- The Web Retrieval Service should provide means to access asset status information via URL.
- The Web Retrieval Service must have user management support that allows enabling or disabling of each functionality for each user group separately.
- User interfaces should be easily configurable, ideally via modification of XSL stylesheets.
- The service should be scalable and distributable over multiple platforms in order to support large numbers of concurrent users.

6.5.12 WORKSPACE MANAGEMENT

The Workspace Management is a service component that provides a unique interface to the multiple facets of the user's daily work, such as organizing assets and queries, accessing databases, and browsing file systems. Other tasks that have to be supported are scanning devices and moving files between devices. This can also take place in conjunction with accessing system functionality such as import, export, analysis, etc. Further, automatically triggering actions upon appearance of an object or file in a folder should also be provided. An important aspect is also supporting interaction between peer users including sharing information, sending requests and receiving responses, and providing a generic communication platform that offers a context (e.g. project, or content object) for the communication.

The Workspace Management manages a folder structure. Each folder belongs to a certain folder class that defines the functionality provided by the folder. Each class of folder performs a certain kind of action when opened, such as reading the contents of a physical directory on a hard disk, performing a certain database query, or reading data sets from a device. Drag and drop operations initiate certain procedures, such as starting an export, import or file transfer, or triggering a process such as format conversion or video analysis. Certain folders (so-called watch folders) recognize when a file of a given file type appears in the folder and perform a predefined action as a response to this event. Part of such an action could be sending an email to a distribution list. Other folders may interface to a chat environment, thus providing peer-to-peer communication. Folders may also have properties (metadata) that can be used as default values where appropriate.

The following example highlights possible applications of the Workspace Management in a typical workflow. For a new project, an editor creates a project-related folder in his private folder space. In a specific subfolder, he stores queries that have created interesting hits by dragging and dropping these queries into this folder. In another folder, he collects references to media objects that may be used in later stages of his work, via drag and drop from the hit list. During the project he also creates scripts and schedules, storing them again in project subfolders. At a certain time he uses the messaging capabilities to submit a request for an archive query to the archive department. He receives a hit list in a dedicated folder, together with the query that has created the result. He consolidates these hits into his project folders, thus adding the archive query results to his own results. He also keeps the query in his query folder so that he can issue this query again by himself at a later time, to check if new interesting material has arrived in the archive.

Subsequently he accesses the material he has collected and creates a rough EDL, which he also stores in his project. He drags the EDL to a shared folder to grant other members of the team (e.g. a cutter) access to the work for further processing. In addition, by dragging the EDL to the inbox folder or the rights management department, he submits this EDL for clearance of the pieces he intends to use. The rights department answers with an element-by-element clearance of the material to be used. Since the item he is producing is to be reused in a versioned form in a later program, he shares the project folder with the persons who are responsible for creating this version. Finally, he drags the EDL to a folder representing a production system. By entering the necessary additional metadata, such as required time of delivery, he thus orders the transfer of material to the production system or a request to the archive to deliver the selected content on tape.

This scenario, even while covering only a small fraction of the possible daily work of a certain editor, highlights the possibilities that arise from the use of Workspace Management. When folders are implemented that show all objects present in the data management that have a certain status, it can even support workflows by letting these objects appear in a certain folder until a certain task is completed and the object obtains a new status.

Figure 6.12 depicts a possible architecture for the Workspace Management. It distinguishes between the management of the folder structure and the contents of the folders, provided by SOAP servers, and the presentation of the overall tree, the contents of folders, and the details of an entity present in a folder via a presentation server.

The SOAP servers provide the information to be displayed in the GUI as XML messages via SOAP, and they receive processing commands via the same mechanism. For generic

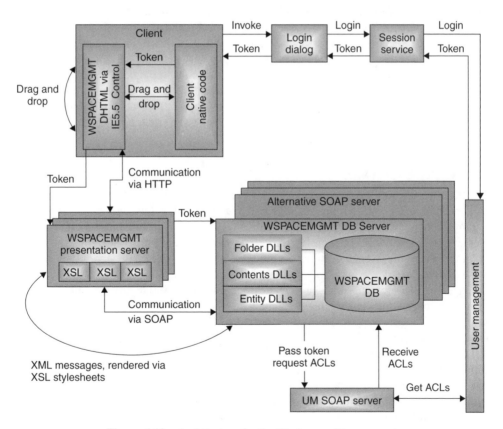

Figure 6.12 Architecture for the Workspace Management

collection folders, such a SOAP server could consist of a database that stores the folder hierarchy as well as references to the objects stored in the various folders, together with a set of plug-ins that, for instance, implement the functionality of the folder (open, close, create, delete, etc.) and allow access to the contents of a folder (creating the directory listing). They also implement the way in which details for the objects present in a folder can be accessed (i.e. opening the object).

It should be possible to add new folder types as plug-ins to the Workspace Management, for example by providing either the respective dynamic link libraries (as additional plug-ins to an existing DB Server) or an alternative SOAP server that delivers the necessary XML messages directly via SOAP to the Workspace Management Presentation Server.

In order to create the graphical user interface, the Presentation Server must be able to accept XSL stylesheets that describe how the respective XML message is to be rendered into a given layout. When dynamic HTML is used, available technology can be used to render the final GUI on each desktop, without the need for any installation of a local component.

Useful folder types comprise collections of arbitrary objects, folders that interface the file system interface of Device Servers (thus allowing browsing the content of a device and transferring assets from or to the device), and folders that interface services. The latter allow the service to be invoked for a given object by dragging it onto the folder, and the current list of jobs processed by the service to be monitored. There can also be folders that allow collecting certain queries to the data management and for the collection of compound objects such as edit decision lists. A further set of folders implement a predefined query to the data management such as the 'open folder' command, allowing folders to be provided such as 'What's New', which could implement a query that shows all objects added to the system within, say, the last 24 hours.

It is important that the Workspace Management supports user access rights associated to each folder, thus allowing definition of which user group may use a folder, remove or add a folder, or even see a folder.

6.6 APPLICATIONS

Applications are the interfaces that content producers, content distributors, content catalogers, and other user groups involved in the day-to-day business of a content-rich organization use to access the functionality a CMS provides. Examples of important elements of these workflows are planning and drafting, commissioning, ingest, import, logging, cataloging, retrieval, browsing, rough-cut, export, clearance, and administration. It is obvious that such a list of possible applications is always incomplete. New workflow and applications elements that have to be supported will be identified while a CMS is being introduced into an organization. In many cases the organization itself will even want to build its own applications on top of the CMS, or enrich existing applications with means to access the CMS.

Thus, it is reasonable that the CMS provides a kind of construction kit for such applications, building on reusable components and flexible layout mechanisms. Using this approach, applications can be easily modified to add or remove functionality as required, and new applications can be configured according to new customer requirements.

A classical application of this approach is the integration of the CMS into the user interfaces of popular newsroom systems (see Section 8.3.2.3). At present all major vendors of newsroom solutions support the integration of plug-in mechanisms for client GUI integration into their desktop application. Therefore a reasonable approach for a component-based approach to CMS-enabled applications is providing all relevant user interfaces as plug-ins for integration in the application GUI framework, integrating them into an XML/XSL-based framework where applicable.

Since applications play such an important role in CMS their design and the functionality of the major application components are discussed separately in Chapter 9. In this chapter the examples considered in the architecture, their features and characteristics are discussed in detail. This section concentrates on the architectural aspects of CMS application modules.

All components should run in a framework comparable to a Web environment. However, when there are no dedicated applications, there must be an entity that runs on the user's desktop that brings all these components consistently together since a simple set of client components has no application context. Thus, a *Session Manager* should be provided that defines the client status and context of all application components that run as part of a client-side user interface.

On a client workstation a considerable number of client applications or even client components such as ActiveX controls may be installed. Hence, there is a need for a single CMS entity that runs on the client computer that represents all these clients and components when managing sessions with the CMS. This single entity is required especially for authentication with the user management and authorization of CMS functionalities. The Session Manager manages the user session in this context and enables communication between clients and client controls.

To start a session, the Session Manager has to accept a login name and a password and to verify this login information with the user management. After successful login, the Session Manager holds the access token for the CMS and manages the required challenges to the user management. Clients use interfaces of the Session Manager when they check a user's rights to access certain functionality they provide.

The Session Management represents the CMS application framework that is able to host all relevant CMS application components. It allows customizing and configuring the application views according to roles, rights, and user preferences.

6.7 SUPPORTING SERVICES

There are a number of supporting services that are orthogonal to the other architectural planes. They may be used by the components in the Core, the Services, and the Application plane and hence need to be available throughout the whole system. In general they can be divided into job management, system management, and system administration. This section introduces the most important services in this context.

6.7.1 JOB MANAGEMENT

The *Job Management Services* are vertical services that support all components and workflows of the CMS whenever job processing is involved. Job Management Services comprise Workflow Management, Task Management, and Transaction Control.

6.7.1.1 Workflow Management

The *Workflow Management* should not impose certain workflows on users. Instead it should allow configuring how complex tasks could be accomplished by performing certain simple jobs sequentially or concurrently. Thus, a complex job like performing an ingest can be broken into more simple (primitive) jobs. Each of these jobs can be performed either by one of the CMS's services or by the CMS core itself. It can be useful to add a task management service (see below) to the architecture that provides frequently used workflows as additional simple jobs, thus providing means for optimization.

The workflow management can be compared to a state machine, where the states that a job can assume are handled by the workflow management, and the transitions between states are accomplished by invoking simple jobs. To do so, the Workflow Management needs to provide means to specify complex jobs from primitive jobs, e.g. a script language. When processing a complex job, the primitive jobs that need to be performed are submitted to the respective CMS component in the required sequence.

In addition, the Workflow Engine needs to provide means to specify user workflows, to monitor processes or workflows, and to allow assessing the status of an object in such a workflow. This can be done by providing status flags, together with possibilities to approve the result of a certain step.

Examples of monitoring workflows via flags are for instance tracing an essence to be ingested, using flags like *Ingest Completed, Logging Completed, Selected for Archiving, Formal Cataloging Completed*, and *Full Cataloging Completed*. A further example is tracing a production process, using flags like *Planned, Approved, Essence Available, Rough-Cut Completed, Video Edit Completed, Color Correction Completed, Audio Edit Completed, Formal Cataloging Completed, Full Cataloging Completed*.

For the administration of the workflow management an administration interface needs to be provided. This should support (besides configuration and standard maintenance issues) modification of jobs and job parameters, monitoring of job processing, and removal of jobs.

6.7.1.2 Task Management

The *Task Management Service* is a scheduling and job processing facility of the CMS that provides additional simple jobs. It basically is a workflow processor that allows hardwiring of frequently used workflows, thus allowing the use of code optimization for better performance.

The task management needs to support scheduling of one-time jobs and of periodically recurring jobs. Further, it queues jobs that are ready for processing and provides a report on the status and progress of a job under its control. One of the main tasks it is responsible for is the distribution of jobs to the CMS components that are processing them. These components are either core modules (such as the different Essence Manager components) or systems (e.g. video analysis service).

For the administration of the task management an administration interface needs to be provided, which should support (besides configuration and standard maintenance issues) the modification of jobs and job parameters, the monitoring of job processing, and the deletion of jobs.

6.7.1.3 Transaction Control

The *Transaction Control* ensures the security of distributed transactions that span over several servers and/or services. For many of the communication middleware platforms used for remote communication, a suitable transaction mechanism is already specified. An example for a standardized Distributed Transaction Processing (DTP) mechanism is the X/Open Standard (Microsoft Corporation (2003)), which forms the basis for transaction service implementations for several communication platforms. Other communication infrastructure proposals, such as, e.g. CORBA, suggest similar concepts.

This functionality is crucial since the components- and job-based approach relies highly on the full execution of all tasks. Otherwise the system or processed content objects might end up in an inconsistent state if any part of the processing fails. No component or module in the system would be able to discover that a series of related jobs have not been completed correctly and therefore left the system in an inconsistent state.

6.7.2 SYSTEM MANAGEMENT

The *System Management Services* are vertical services that support the CMS on a system management level. The system management typically includes a naming authority, a central event log facility, a distributed process monitor, and a resource management facility. These are management task internal to the system and involve only system processes and no user interaction. In normal circumstances System Management services are only known and visible to the system administrator. Since they depend very much on implementation details they can vary from system to system considerably. However, similar functionality will be required in most systems, and hence components that provide them should exist in some form in each CMS. In the following a set of basic system management services are introduced.

6.7.2.1 Naming

The *Naming Service* is the CMS's core naming authority, allowing identification of where CMS services, components, and objects are located in the system. This is the case for local as well as for remote services. Basically the naming service operates in conjunction with the communication middleware platform used in the CMS. Its functionality and features are comparable to the classic Internet Domain Name Service (DNS).

Each component of the CMS needs to be able to register with the Naming Service. In this process the registering entity provides information that allows other components to locate and contact it. Once registered, a component has to be able to check the validity of its registration at any point in time, update or renew the information, and un-register from the Naming Service when it ceases to operate.

6.7.2.2 Event Logging

The *Event Log Service* is the CMS's facility for internal system messages, like notifications, warnings or errors messages. It should support multiple levels of severity for these messages. This service can be distributed and co-located with different core and services modules. It is also possible to use one Event Log Service per physical server or to have multiple Event Log Services distributed throughout the system that are acting as servers taking messages from other services and core components. The structure and design of the Event Log Service also depends very much on the implementation of the system. Different ways to organize it (i.e. ranging from a hierarchical client–server setup to a peer-to-peer like infrastructure) are possible.

However, each component of the CMS needs to be able to register with an instance of an Event Log Service, to check the validity of its registration at any point in time, and to update or renew its registration information. The main purpose of the Event Log Service is to record messages from other CMS components. Thus, other services or core components have to be able to open a log channel that enables these components to write messages to the Event Log Service, to actually write messages to this channel, and to close open log channels again. Further, when a component ceases to operate it has to un-register from the Event Log Service.

6.7.2.3 Process Monitoring

The *Process Monitor* is the CMS's internal monitor. It constantly controls the status of all system processes and restarts systems that have failed. The Process Monitor also provides an interface to start, stop, and restart services manually. This is used by system administrators or content managers for maintenance work. A possible implementation of the process monitor is running an instance of the service on each server that hosts CMS services or core components. In this case the process monitor should run as a background task using only processor idle time. Using this approach the Process Monitor does not interfere with the productive processes of the CMS.

Each back-end component (i.e. core and service modules) of the CMS needs to be able to register with the process monitor, check the validity of its registration at any point in time, and update or renew its registration information if necessary. A registered process has to provide the Process Monitor with information about how to start, pause, continue, shutdown, and restart itself. Further, it has to provide the Process Monitor with status information about its health state, which is used by the Process Monitor or system administrator to react to failures. Further, when a component ceases to operate it has to be able to un-register from the Process Monitor.

6.7.2.4 Resource Management

The *Resource Management* provides resource reservation, allocation, and monitoring for all resources provided or controlled by the CMS. A CMS service that supports resource reservation needs to interface with the resource management where it registers the resources it can provide. A service reports on the current status of resource allocation whenever a change in resource usage or availability occurs.

A client (being an application or a service) can now query the resource management for availability of resources. Reservations can be made by providing a time span when a resource (or a number of resources) is (are) needed, a priority for actually accessing the resource, and a confidence level indicating how sure it is the reservation option is actually used. Note, the latter is required since there is a level of uncertainty in the occurrence of some events.

The confidence level offers a possibility to overbook resources in order to optimize system usage, while the priority decides on who will finally get access to a contested resource in case of overbooking. In the case of overbooking with identical priorities, the situation is resolved on a first come, first served basis.

In order to support mission-critical applications, the resource management should allow services to assign a minimum priority to each resource registered. The resource management then can only assign a prioritized resource to a client that requests this resource with a priority that is at least as high as the minimum priority registered.

6.7.3 SYSTEM ADMINISTRATION

The *System Administration Services* are vertical services that support the CMS on an administrative level. This includes the administration of users as well as the support of software maintenance processes and the management of licences. Thus, they comprise a range of services that are concerned with the administration of the system that are not directly related to the status management of the CMS. System Administration Services often include *User Management, Accounting and Licencing, Messaging, Configuration Service, and Remote Installation.*

6.7.3.1 User Management Service

The *User Management Service* is an essential orthogonal service in a CMS. All components that restrict access according to user rights or use a role classification scheme for different users to configure applications or the presentation of content have to have access to the User Management Service. Theoretically each component could implement its own access control scheme. However, this would be very inefficient and system-wide rights could not be considered. Thus, a service that distinguishes users and groups on a system-wide level is required.

Users are the core elements in the User Management Service. Different properties relevant to the work with content in a system process pertain to a user. This information is part of the User Management Service. Further, a user may be a member of several groups. Thus, apart from the notion of individual users the concept of groups is fundamental to the User Management Service. In this context it should be possible to configure an arbitrary number of groups and assign an arbitrary number of groups to each user. Further, it is valuable when User Management Service supports group hierarchies.

Membership of certain groups defines which part of the overall functionality offered by the CMS a user can access. Further, it specifies in which applications this functionality is presented. Hence, for each group it should be possible to define all applications that are accessible, and for each application it should be possible to specify the accessible functionality.

Besides access to overall functionality, membership of certain groups also grants or denies access rights to objects managed by the CMS.

In order to facilitate a unified enterprise-wide management of users and groups that is shared by all systems in use at the company, the user management system employed in the CMS should be able to contact a central (i.e. company-wide) repository for authentication and authorization. Examples of such repositories are directory services based on the Lightweight Directory Access Protocol (LDAP), as well as proprietary services such as Microsoft Windows Primary Domain Controller (PDC).

Each application and each content object that the CMS manages should have an associated Access Control List (ACL). The ACL comprises information related to its properties and possible usage restrictions such as the ID of the group that owns the object, the ID of a number of groups that have group-specific rights to the object, and the ID of a default group that describes the general access rights for all users. The rights pertaining to all groups (including the default group) could for instance include simple access rights on a content object such as to see, read, access, and replay an object. More advanced rights allow users to write, edit, and modify content objects, and create, copy, and delete them. Further, there are administrative rights linked to a group, for instance to change owner and the ACL. Rights can be granted or revoked by authorized entities or users.

There are some important limitations that should be taken into account. While members of the owner group should always be able to change the owner and all user rights for all groups, a user who is not the owner of the object, but a member of a group that has the right to change user rights, can only change the user rights for this specific group.

Besides the configurable groups there should be a possibility to configure a number of system-specific groups that can handle the access rights to all objects within the CMS. An example would be a super-user group that has full access to all objects.

The following functionality should at least be offered by the user management and should be accessible via the user management administration interface:

- add, modify, delete, enable, disable user;
- add, modify, delete, enable, disable group;
- assign/remove group to/from user;
- assign/remove group to/from group;
- assign/remove access to application functionality to/from group;
- assign/remove access to asset to/from group.

6.7.3.2 Accounting and Licencing

The *Accounting and Licencing Service* is responsible for registering access to the various parts of the system as well as to the content objects stored in the system. It ensures, together with other facilities, compliance with software licencing conditions. The data gathered by this service is the basis for status reports and statistics.

A CMS component that supports registering activities with the Accounting and Licencing Service needs to interface with this service, to register its unique identification and location, and to report each access to its functionalities, together with user ID, date, and time of the access. In addition, any access to a content object present in the CMS needs to be reported to the Accounting and Licencing Service, by providing content ID, user ID, date and time of the access, and the kind of access that occurred (i.e. read, write, modify, delete, etc.).

Based on the data provided to the Accounting and Licencing Service, usage statistics can be generated that support billing the usage of CMS services to the accessing parties of the organization. Further, it can help to optimize the system by identifying bottlenecks or by optimizing caching strategies.

In addition, the Accounting and Licencing Service is required to resolve software licencing issues that will arise in a system that comprises multiple vendor solutions.

6.7.3.3 Messaging

The *Messaging Service* provides client-to-client communication facilities within the CMS. The functionality of the service comprises sending and receiving email messages, direct peer-to-peer communication as in chats, and instant notification messages. This service operates within the context of a CMS. Hence, information exchanges via the Messaging Service can be directly related to content objects, projects or any other component managed in the system.

The Messaging Service enables information interchange and collaborative work within and beyond work groups, using instant notification to alert a user that another user has provided some information or even content that the user should take notice of, and allows for asynchronous branching of workflows, such as requesting an archive search as an offline process, results to be delivered at an unspecified time, requesting a clip level rights clearance for a new production, again to be answered at an unspecified time, or submitting proposals for commissioning and receiving answers some time later.

A messaging service can also be used to provide guaranteed delivery of messages between system components.

6.7.3.4 Configuration

The *Configuration Service* is the system registry, holding all configuration parameters for all system processes as well as for all user-dependent client configuration data. A service that is part of the CMS should only keep the absolute minimum configuration information locally. This is the information required to boot the service successfully, and to identify the Naming Service in order to register with it and retrieve the registration information.

In most cases this means that the only configuration information that a service needs to keep locally is the network identifier of the Naming Service and a system identification in case a Naming Service manages more than one CMS instance. Any other configuration information should be read from the Configuration Service. Thus, the Configuration Service facilitates system maintenance since changes have to be performed only centrally within the Configuration Service. The structure of the Configuration Service can be distributed, and even a hierarchical structure (e.g. similar to the broker management concept) is possible.

6.7.3.5 Remote Installation

The *Remote Installation Service* provides automatic installation and update facilities, minimizing the administrative overhead involved in installation and update of services and core components of a CMS that is distributed over multiple platforms. Such a service should be able to download components from an installation repository, install the components, and perform necessary configuration steps automatically. Like the process monitor, such a service should run on each server system, using processor idle time only.

7

Content Management System Infrastructure

The system architecture introduced in Chapter 6 is concerned with the overall architecture (i.e. the system framework), system functionality, and the different modules that provide this functionality. In this chapter the CMS infrastructure is introduced, i.e. hardware and software components, and the communication subsystem required to build a CMS in an enterprise-wide environment. The system infrastructure is the physical system level that comprises servers, encoders, storage systems, and the various communication networks that are operated in a media-rich environment. These are not pure hardware systems but in some cases also run their own software that has to be integrated by the CMS. Essence Management Archive Manager and Archive Transfer Servers integrate with the software controlling systems near-online storage. Standard IT servers host the different CMS components and can be considered as hardware components.[1] Thus, the system infrastructure provides the platform for the CMS.

The demand for shared access to content, faster-than-real-time transfer, first-generation production, rapid transcoding, and support for multiple heterogeneous delivery channels calls for a file-based production, delivery, and archiving environment. Such an environment must also provide means to administer rapidly increasing program volumes, must be able to manage increasing content fragmentation, and must support the distribution of content via traditional channels and non-traditional networks to individually addressable interest groups. This has to be considered in the system architecture as well as in the system infrastructure.

Traditional tape-based media and broadcast tools are not flexible enough to accommodate these requirements. The only conceivable solution is to apply IT-based technology with a high degree of automation, managed by a CMS. IT-based solutions have the flexibility required, and provide a rapidly improving cost-to-performance ratio. In order to make the best use of the rapid development cycle of IT-based solutions, it is important to

---

[1] Note, the operating system running on a server is also software but conventionally it is not regarded as separate element that provides specific functionality for a CMS.

Professional Content Management Systems: Handling Digital Media Assets   A. Mauthe, P. Thomas
© 2004 John Wiley & Sons, Ltd   ISBN: 0-470-85542-8

design the system infrastructure around standard IT products. Interfaces to domain-specific systems increase complexity and cost and hence should be avoided by design wherever possible. Where such interfaces are required, they must be clearly specified. Also, maintenance and re-invest strategies must be reconsidered and revised to make the best use of the IT development cycles.

This chapter introduces the most important components of an IT-based system, proposes a hardware architecture for a distributed CMS, and discusses important aspects such as migration options and cost drivers. In conjunction with the system architecture introduced in the previous chapter, this chapter provides the full view on the structure and components of a CMS. Both chapters form the core of the book concerning the functionality and design of a CMS for content-rich organizations.

7.1 INFRASTRUCTURE: BASIC PRINCIPLES

When designing the infrastructure for an enterprise-wide CMS two general points have to be considered, namely the system components and certain design principles. The former provide the platform on which the CMS is built. These components and sub-components become an integral part of the system. However, there is no blueprint for implementing such an infrastructure since a number of factors and parameters have to be taken into account. For instance the structure of the operation and organization are important factors to consider. However, there are some basic principles that can be used to help design an infrastructure that supports the management of content and the operation of a CMS optimally.

7.1.1 INFRASTRUCTURE COMPONENTS AND SUB-COMPONENTS

The proposed CMS infrastructure is largely based on IT technology. This comprises the entire range of hardware equipment from the low-end PC platform to the high-end disk-based storage system. In an IT-based system typically the standard hardware components of the infrastructure providing a service fall into the following categories:

- *Servers*: computation platforms that run parts of the system software. Typically, standard servers run popular operating systems such as Microsoft® Windows™ or one of several UNIX flavors.
- *Storage*: refers to disk- or tape-based mass storage systems accessible by servers. Optical systems are less popular in large-scale CMS.
- *Network*: refers to communication links between servers (Local Area Network, Wide Area Network), or between servers and storage (Storage Area Network).

In addition, a system may comprise other solution-specific components, such as encoders, decoders, crossbars (i.e. matrix switches), etc. Client computers (typically PC) are also part of the system infrastructure and are standard off-the-shelf products. Depending on the component, sub-components for any IT-based system could be, for instance, power supplies and fans, network adapters, internal disks, random access memory (RAM), etc.

Since storage is of the utmost importance for an enterprise-wide CMS (and since storage systems are available in very different technology flavors) it is discussed in more detail in the following sections. In general, storage is available as *Server Attached Storage* (SAS), *Network Attached Storage* (NAS), and *Storage Area Network* (SAN). Separate from these concepts are mass-storage systems (also called near-online storage) such as tape libraries and automatic jukeboxes. These are low-cost alternatives to online storage. The trade-off of these systems is that it requires some time to make material available (the so-called staging process). Whereas with online storage material is always online, near-online systems have to seek the desire object, load the storage medium (e.g. tape or DVD) that contains the object into the drive, and put it onto a disk-based staging area. Therefore the different storage systems are discussed separately. However, in the case of SAN, near-online mass-storage systems can be part of the SAN infrastructure.

Different network types have to be considered at the design of the infrastructure. This is due to the various IT and broadcast components that have to be interconnected. Besides data communication networks there are broadcast connections (such as SDI and SDTI) or machine connections such as Fibre Channel and SCSI. Further, content-rich organizations are often spread across a number of locations. Thus, LAN and WAN technologies have to be supported.

7.1.2 INCREASING SYSTEM RELIABILITY VIA HARDWARE REDUNDANCY

CMS will become a vital component of a content-rich organization. It is crucial that content is available and the system is operable round the clock (i.e. 24 hours a day, seven days a week). Especially in broadcast systems downtimes can result in serious financial losses. Thus, provisions have to be made to prevent system failures or to compensate a failure in such a way that it does not do any harm. Hardware redundancy is a major principle in this context.

Hardware redundancy means adding additional hardware to systems or system components, thus improving the availability of these systems. Redundancy can be applied to system components by adding redundant sub-components, such as redundant power supplies, redundant network interfaces, mirrored system disks, and so on. This level of redundancy reduces the risk of a complete failure of a single hardware system. An example is increasing the availability of disk-based storage systems by grouping single disks into a *redundant array of independent disks* (RAID).

A well-established way to apply hardware redundancy to servers is clustering. Clustering is used to support software that requires a very high level of availability, such as databases. In the most basic configuration a cluster consisting of a couple of servers connected to a shared storage device is equipped with special cluster software. In certain configurations this is supported by special hardware connections. In case one of the servers fails the second automatically takes over, providing the same resources and connectivity as the failed server. Hence, as seen from the outside, the server continues to be available. For more advanced applications, there are cluster software solutions available that allow platform-independent clustering over multiple servers, even without the need for shared storage.

To increase redundancy, certain software applications are capable of being run in service groups or distributed over multiple servers.[2] In this case, each server adds resources to the service the application provides. In case a server fails, the service is still in operation, but it lacks the resources provided by the failed server. By over-engineering such a group (adding more servers than required to run the service at a given quality level), the service can accommodate failure of one or more servers without loss of quality of service.

---

[2] This kind of group concept is discussed in detail in Section 6.1.2.

Another way of adding redundancy is adding hot standby servers. Provided the application supports hot standby, an instance of this application running on a hot standby server detects the failure of the instance on the primary server and immediately takes over.

Finally, redundancy can be added by adding cold standby servers. In practice this means that a server is added to the system that has an instance of the application that needs to be protected installed, but not running. In case of failure of the server that actively runs the application, this passive instance is manually launched on the standby server, thus allowing continuing operation after a minimum downtime.

Which of these redundancy concepts should be used depends on the accepted failure rate, structure and capabilities of the system components, and cost considerations. In general the introduced concepts can provide a high-level of system reliability. Whether hardware redundancy is required, and if so, which form, has to be assessed on a case-by-case basis.

7.1.3 HARDWARE DESIGN AND CONFIGURATION PRINCIPLES

In order to gain the most from an IT-based infrastructure, it is important to apply certain principles when designing and configuring the infrastructure for an integrated enterprise-wide CMS. This section explores some of the principles relevant in this context. They can be used in the system design process as well as when evaluating a system proposal. The rules introduced in this section should assist the system design process and help to assess alternative design proposals.

However, these rules should not be regarded as tenets. They are meant as recommendations and should always be considered in the context of a specific project. Other parameters such as cost, operational and user requirements, etc. are also part of such considerations.

7.1.3.1 Deciding on Server Configurations

A CMS typically comprises a considerable number of heterogeneous applications that interface to each other. Each application provides a subset of the overall functionality of the CMS and has certain requirements with respect to the configuration of the server that hosts the application. If each server is tailored to the requirements of the respective application the overall system finally has a good chance to be a pandemonium of different servers with different hardware and software configurations, and even different operating systems. Such a system is very difficult to maintain and operate.

It is one of the advantages of IT-based systems that to a certain extent servers and storage are a commodity that can serve different applications. Hence, it is advisable to consider the various applications that should run on the servers carefully and try to identify if there are common denominators in their system requirements. The different servers should be classified into categories such as 'Application Server', 'SAN Server', 'Database Server', etc. Subsequently applications can be assigned to the categories and the final configuration of the servers is then defined accordingly.

The following rule summarizes this:

> Try to deploy a minimum number of different server configurations. In order to do so, try to classify the servers according to common requirements, assign applications to the categories, and derive the required server configuration from this.

7.1.3.2 Deciding on Redundancy

When deploying an infrastructure for a CMS it is important to carefully balance availability requirements against cost. In many cases the availability of system components can be increased (as discussed above) by adding redundancy to the most important sub-components, such as power supplies, cooling fans, system disks, network and host bus adapters. It is sensible to apply the following design principle:

> Add redundancy to the sub-components of the various system components, especially in the area of power supplies, cooling fans, system disks, network and host bus adapters.

Please note that adding redundancy to network and Fibre Channel adapters also requires increasing the number of ports in the respective backbone, which may significantly increase the total hardware cost.

By deploying all system components on server clusters, a very high level of availability can be reached. However, not all applications are cluster enabled, and the cost of deploying clusters is very high. Thus, the following design principle is proposed:

> Use server clusters only where necessary to support the availability of a very sensitive application. Ensure that the application is cluster enabled.

Databases are a good example of applications that benefit from clustering. Others might be more suited towards a service group concept (as is the case for a video analysis process).

A very cost effective way of adding redundancy to the server environment is using hot or cold standby servers (see above). Often, adding a single server to the system as the standby server for a large number of other servers suffices to ensure a high level of overall system availability and quality of service at low cost. When doing so, it is necessary to consider if applications can work in hot standby or cold standby mode.

> Add a limited number of standby servers into the overall architecture. Choose the number and configuration of these standby servers by considering the requirements of the applications that should be installed on these servers.

7.1.3.3 Deciding on Storage

Considering the performance requirements of an enterprise-wide CMS in television and film production the storage systems should be deployed in a SAN-like manner. Ideally, the systems are deployed on top of a switched Fibre Channel fabric.

> Use Fibre Channel storage devices wherever possible. Integrate these devices into a SAN deployed as switched Fibre Channel fabric.

Having deployed hard disk storage devices as a SAN, they can be shared between the various servers via a SAN file system. Tape based devices are integrated by introducing suitable archive management solutions.

While for each kind of content, independent of the resolution and bit rate a certain amount of online storage needs to be provided as a cache for fast access, an important decision to be taken is whether the archive should also be on disk or should use data tape library systems. Production quality video and film have average bit rates that still prohibit using disk as an archive storage medium simply due to the enormous costs involved. However, disk-based storage systems are rapidly getting more affordable, and for browsing or Internet quality video as well as for production quality audio, it is already economically feasible to archive on large hard disk systems. The advantages are clear: very fast access to content combined with short retrieval times. Disk-based systems are also very easy to maintain. Thus, consider the following principle:

> Use hard-disk-based mass storage as archive systems wherever feasible. Use data tape libraries only for content qualities that prohibit archiving on disk due to economic reasons, and for backup.

7.1.3.4 Designing the Networks

When SAN are deployed at least three different networks have to be put in place:

- the Fibre Channel network for the mass data transport within the SAN;
- a SAN communications network for the metadata communication between servers attached to a SAN file system and the respective file system server;
- a CMS communications network that interconnects the components within the CMS and connects the CMS to the enterprise network.

The Fibre Channel network, as stated above, should be deployed as a switched fabric. When larger distances have to be covered, optical fiber is a good choice.

The SAN communications network should be a non-routed private network that is only accessible by the servers connected to the SAN file system. Please consider this principle:

> Interconnect all servers that share a SAN file system with a non-routed private network. The SAN file system server should be connected to this network via 1000baseT.

When designing the CMS communications network, it is important to understand the application that will be hosted by a server. Some applications will mostly require processing power; the server hosting such an application will act as computational node and will not transfer mass data via the network. For those servers, no specific provisions need to be made with respect to the network interface. A standard 100baseT connection typically is sufficient. Other applications, such as video servers, keyframe servers, Web services, or transfer services, deliver mass data to clients or move mass data between storage systems. Here, it should be considered to equip the hosting server with a 1000baseT network interface.

In a well-designed system, servers that move mass data are also connected to the SAN, since the SAN is where the mass data is stored. Thus, the following basic rule can be applied:

> Servers that do not move mass data do not necessarily need to be connected to the SAN. It is typically also sufficient to connect them to the CMS communications network via Fast Ethernet. Servers that move mass data should be connected to the SAN. They should be connected to the CMS communications network via Gigabit Ethernet.

Besides servers, other hardware components requiring a certain quality of service (QoS) may have to be connected to the network. Examples are real-time encoders, playout systems or systems that require a guaranteed minimum transfer rate. These kinds of systems require a minimum guaranteed bandwidth on the network. It often is sensible to connect such systems via a private point-to-point connection or a V-LAN.

> When a component must be connected to the CMS that requires a minimum guaranteed bandwidth, consider using private connections when feasible.

Apart from these IT connections a CMS might also include studio connections such as *serial digital interface* (SDI) (Society of Motion Picture and Television Engineers, 1997b) and *serial data transport interface* (SDTI) (Society of Motion Picture and Television Engineers, 2000) links. These are point-to-point connections linking studio devices such as NLE and playout servers. Many of these devices are now equipped with both IT and broadcast communications links. Depending on the set-up the CMS might have to coordinate the communication via these different channels.

7.1.3.5 Isolating Interfaces

Broadcast and media production systems often have specific requirements with respect to real-time capabilities or domain-specific interfaces. In addition, integrating proprietary systems such as legacy databases may require the implementation of custom interfaces.

In order to gain the most from an IT-based infrastructure it is important to avoid using domain-specific technology or supporting domain-specific interfaces wherever possible. However, in many cases the CMS is the bridge between the broadcast and the IT world. Thus, it has to provide certain functionality to unify both. An option is to isolate the required interfaces and to allocate this bridging functionality on dedicated system components. They are then treated separately in the CMS infrastructure by systems called *Interface Servers*.

A good example of an Interface Server is a disk recorder or video server. A professional disk recorder is a broadcast device in the sense that it typically has SDI and AES/EBU in and out ports, supports LTC and VITC timecodes, allows external control via RS-422 connectors and standard device control protocols, and is frame-accurate and operates in real-time. However, a disk recorder also has a number of features that make it an IT device. For instance it typically records to and plays from internal hard disks and stores essence on these disks as files. Further, it has standard IT connections that allow file transfer (often via Gigabit Ethernet or Fibre Channel). Thus, such a disk recorder can be used as an Interface Server to interconnect the IT-based CMS with broadcast technology dominated production and delivery environments. In a similar fashion, interfaces to other systems can be isolated on specific Interface Servers that form a gateway between the CMS and those other systems.

> Isolate domain-specific interfaces by employing Interface Servers as gateways between the CMS and other systems wherever possible.

7.2 STORAGE SYSTEMS

In recent years storage systems have seen considerable advances. Various ways to structure, organize, and access physical storage (i.e. disk space and associated storage media) have been devised. Initially these systems have been mainly deployed in areas where a large amount of structured data has to be stored reliably and safely (such as for insurance and finance data, scientific applications, and security systems). Therefore reliability and system security are of prime concern to these systems. Their use in the Web environment and for the management of multimedia content has only happened in recent years. Here the requirements are slightly different. Apart from reliability and security, timing, band-width, and throughput requirements are vital. This is due to the nature of data and the operation. In order to retrieve multiple production quality video files (storing MPEG-2- or DV-based formats with up to 50 Mb/s) a high load is placed on the I/O interfaces of a storage system. Ideally, files should also be retrieved faster than real-time. Thus, there is no real upper limit to the I/O throughput of the system.

These requirements have to be taken into account when considering the different storage options for a CMS. Apart from their operational features easy integration into the system architecture and infrastructure of a CMS are also important criteria at the assessment of the most suitable storage subsystem for a specific project.

7.2.1 SERVER-ATTACHED STORAGE (SAS)

In the past a server-centric architecture has been the dominant IT infrastructure. One of the great disadvantages of this approach is that the storage system depends on the operating system of the server that is also responsible for the handling of the files. In this set-up the server acts as a *master*, while the attached storage system has a *slave* status (as shown in Figure 7.1). The general-purpose server has to fulfill a number of tasks simultane-ously, e.g. serving applications, performing database read/write operations, providing file and print services, performing communication tasks and data integrity checks, etc. Client

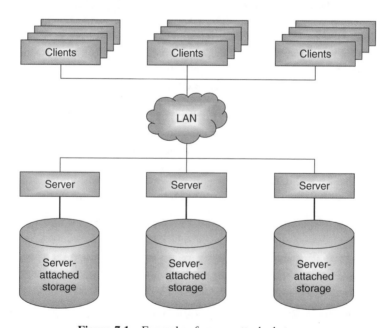

Figure 7.1   Example of server-attached storage

requests for data compete with these other services the server provides. As the number of connected clients increases, server response time decreases proportionally. In addition, the direct interconnection of server and storage device suffers from limited interconnection lengths and low bandwidth. LANs (such as Token Ring, FDDI or Ethernet) keep the operating system busy with communication tasks. Today's technical and economic requirements have made SAS systems mostly obsolete. Today, SAS is mostly used for installation of the operating system or for locally installed applications. Their relevance as storage platforms for content is in the area of separate smaller units (such as a content server at a departmental level). In this case SAS can provide a low-cost alternative to large-scale solutions since they can be easily integrated into the wider CMS infrastructure.

7.2.2 NETWORK ATTACHED STORAGE (NAS)

With an increasing number of application servers and a dramatic increase of LAN-connected workstations, network services have been becoming increasingly important. For that purpose, special servers for LAN-attached disk arrays have been developed. They are called network attached storage (NAS) and are exclusively optimized for tasks such as file serving and data management. Thus, storage and file management are separated from other server tasks, which are still performed by 'traditional' servers. Figure 7.2 demonstrates the principle of NAS using an example of a possible configuration.

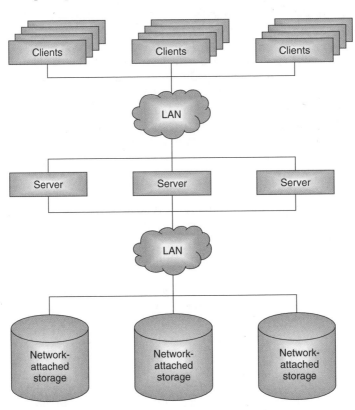

Figure 7.2   Networked-attached storage

The basic idea behind NAS is the *file-centric* IT paradigm. It defines a dedicated file server as a NAS appliance that owns files and data in the LAN and is operating on the basis of file-level I/O commands. NAS devices can be installed as sharable Plug-and-Play devices within a LAN without the requirement of a shutdown and reboot of an application server. To bring a NAS into operation it is only required to allocate an IP address and to make the NAS known to the LAN by setting the right control and access parameters. Conventional file sharing protocols such as Network File System (NFS) or Common Internet File System (CIFS) allow file sharing between heterogeneous NT and UNIX clients. Thus, the NAS can be directly accessed from clients on the file level and not on the communication level. This allows the use of heterogeneous client systems in the CMS. Note, these clients may be machines running services such as the Video Analysis Service or Essence Manager modules such as a Cache Server or Archive Transfer Server. The most suitable operating system can be chosen for each of these services regardless of the NAS storage subsystem.

NAS is based on a multi-layer network that preferably uses Internet protocols. Clients transmit file level I/O commands that are transported to the appliance server via the LAN using TCP/IP or IXP protocols. TCP/IP is a communications protocol based on acknowledging the reception of data and therefore provides guaranteed delivery of data. IP is responsible for routing data through the network, while TCP provides the control mechanisms for data security.

In the context of the CMS this approach also has some drawbacks. With the requirement for sustained high data rates for video streaming, LAN bandwidth may soon become a bottleneck. Further, NAS shows restrictions in the area of QoS such as transport priorities, routing reliability, security level, and performance. This is due to the fact that TCP/IP has originally not been optimized taking into account the specific requirements of storage applications, the transport of large data files or continuous media requirements. The partitioning of data into small IP packets and the high traffic volume within the LAN (originating from a large number of clients) can lead to delays and reduced performance. For applications where NAS reaches its performance limits, SANs can provide a more scalable and flexible solution.

7.2.3 STORAGE AREA NETWORKS (SAN)

A SAN is a dedicated high-speed storage network over which a high data volume can be transported between heterogeneous server and storage systems at high speed. Figure 7.3 shows an example configuration of a SAN. SAN are the most recent storage technology. They are now reaching a level of maturity that allows deploying them for industrial and business applications.

The SAN infrastructure is based on the standardized *Fibre Channel Multi-Layer Network Architecture*, which supports multiple protocols, a switched Fibre Channel network and service classes optimized for various interconnection distances. Since a SAN uses a separate network any conflict between clients and application server (are typical for a LAN) are avoided. A SAN based on Fibre Channel combines the advantage of a high speed I/O channel with general network connectivity over great distances. First generation Fibre Channel offered a bandwidth of 100 Mbyte/s. This is equivalent to about three SDI streams (with about 270 Mb/s each). Recently, next-generation Fibre Channel devices have even increased the available bandwidth to 200 Mbyte/s.

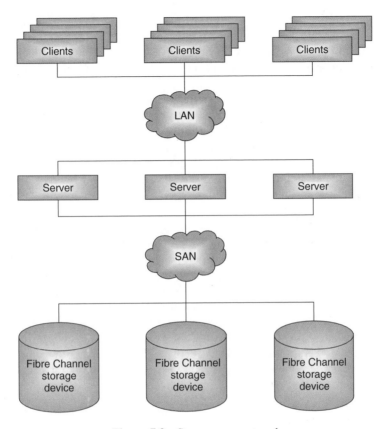

Figure 7.3 Storage area network

Comparing SAN to SAS it offers substantial saving potentials via pooling and resource sharing. For instance the SAN concept allows the exploitation of few, powerful storage systems by many servers or applications. In contrast to a NAS architecture (which centers on LAN protocols and file level I/O) a SAN architecture is *data centric* and relies on channel protocols with block I/O for data access.

A SAN hardware structure comprises three levels:

- *First level*: server and client systems;
- *Second level*: a SAN fabric composed of gateways, routers, hubs, and switches;
- *Third level*: storage devices such as hard disks, tapes, and tape management systems.

All SAN components are interconnected via fiber optic cabling, providing flexible communication paths between the different components, i.e. for server-to-server, server-to-storage-device, and storage-device-to-storage-device communication.

Each device is identified via a unique ID (the so-called World Wide Port Number; WWPN); storage devices may also offer to subdivide the available storage and assign Logical Unit Numbers (LUNs) to these partitions. Access can be restricted to give distinct rights to different accessing devices (and hence to the respective users). Which device can see which other device can be configured by granting or denying access to WWPNs and LUNs.

The any-to-any communication within a SAN offers new options. Depending on the type of the fiber optic cable employed, distances of about 80 km can be bridged (campus or metropolitan area coverage).

Many modern storage devices are equipped with Fibre Channel connectors and therefore are easily integrated into a SAN architecture. Conventional SCSI-based storage systems can be integrated into the SAN using existing SCSI–SAN fabric converter products.

The interconnection of servers and storage systems via SAN has a number of advantages. For instance SAN provide high bandwidth with guaranteed QoS and non-blocking access. Further, secure operation and high availability ensured by fault tolerance and multi-pathing are core features of a SAN. Also, SAN are inherently scalable and flexible with respect to interconnection distances.

The heart of an enterprise-level SAN is based on Fibre Channel switches, interconnected in a fabric-like structure. Thus, a SAN architecture is specially suited for online transactions, e.g. large databases, the transport of high data volumes as required in video applications as well as for streaming. In this context it has to be noted that the performance parameters and individual infrastructure of NAS and SAN architectures are optimized for specific requirements. They are in fact congenial and do not compete with one another.

Based on such a Fibre Channel hardware infrastructure, quite different applications can be implemented, supporting, for example, the pooling and resource sharing of tape and disk systems. Others are applications for LAN-free/server-free copying for backup and archival (including remote mirroring), clustering with fail-over for high availability, and NT/UNIX data sharing within SAN.

7.2.3.1 SAN Management

SAN management provides central control and administration of all geographically distributed components and interconnections on a single graphical user interface (GUI). Status information in a SAN should be provided via standard protocols, such as the simple network management protocol (SNMP). It may be distributed either *inband* via the SAN interconnections or *outband* via existing external LAN interconnections. Further, it is important that it should be graphically visualized and displayed as *SAN topologies*.

Error messages, warnings or the violation of user-defined tolerances within the SAN should be shown so that appropriate action can be taken. SAN management solutions must allow monitoring and reconfiguration of SAN components provided by different vendors. Disk management options should included administration of LUNs (which represent the partitioned disk capacity) across heterogeneous storage systems of different vendors. In addition, policy-based supervision of file systems should be supported, such as automatic assignment of additional LUNS in case a certain threshold is exceeded.

7.2.3.2 SAN-Based Backup

Backup in a client–server environment is traditionally carried out in the four following steps:

- exchange of metadata between the target server that performs the backup and the source server that is to be backed up;

- reading of user data from disk and subsequent transfer to the source server;
- data transfer from the source server to the target server via a LAN;
- securing of user data within a library system connected to the server.

A SAN implementation offers a much more efficient solution as shown in Figure 7.4. The first two steps are carried out as described above, but data is subsequently directly transferred from the server to the storage device via the high-capacity, high-speed SAN interconnection. Thus step 3 is eliminated and with it the transport of user data between individual systems. Here, two options may be considered, namely LAN-free backup and server-less backup. In the former case the data is no longer transported over a network. It is rather transferred from a source device via the SAN fabric, and from there again over the SAN fabric to the backup device. With server-free backup the data is directly transferred from the source device to the backup device. In both cases, the management of the backup process may reside on any server in the network.

7.2.3.3 SAN Requirements in Content Production

A SAN allows resolving or reducing many problems and restrictions in file-based production, archival, and distribution in content production, in particular within television. These environments require time-critical, high-speed delivery of data and compliance with defined quality criteria. To accomplish this a SAN employed in content production and television applications has not only to comply with, but also exceed, several requirements for distributed video production. For instance it has to be able to transfer huge data volumes in real-time or faster than real-time. Further, QoS guarantees are required to ensure the timely delivery of the data while ensuring high system availability. A SAN

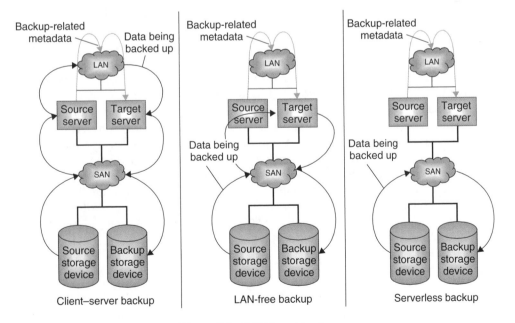

Figure 7.4 SAN-based backup

deployed in this context also has to allow shared and concurrent access to content. A crucial requirement is the provisions of absolute data security and integrity. Ideally it should also provide the possibility to prioritize processes of specific applications within the production process. Last but not least, SAN should allow integrating storage products of different vendors into the total system.

The major applications within digital content production that rely on SAN are:

- archival and access of audio files, video files, and metadata within a distributed archive (local or remote);
- file sharing for NLE applications;
- local or remote control of playout or transmission in real-time and live contribution of video streams generated at a remote location;
- program exchange;
- video-on-demand or e-commerce solutions for video distribution.

In order to use SANs in a content production environment they have to be designed in a way that allows the integration of production, editing, playout, and archival. They may further be used in conjunction with rights management and licence management, which might be carried out locally or in geographically dispersed areas. This could virtualize the total production process by making content available at different locations at a defined time.

7.3 MASS STORAGE SOLUTIONS

Despite the advances in chip development and the decrease in costs for online storage it is still not feasible to keep all the content in a content-rich organization online. Take the example of a medium size video archive with 100 000 hours of content. In order to store a broadcast format of around 4 Mb/s 180 terabytes are required. If the material is also stored at a higher bit rate (for instance at 25 Mb/s) additional 1.125 petabytes are necessary. Another 68 terabytes for MPEG-1 browse copies will be required. Thus, alongside the online storage system there should be a cost-efficient mass storage system that allows accessing the content stored on it without undue time delay.

Tape-based mass storage systems are still the most economic solution to cope with the enormous capacity required in video archiving. Taking primarily video into account there are two main options, i.e. technology supporting standard videotape formats and standard IT storage devices. Videotapes as such are still used throughout the content production chain, and there are a variety of tape libraries that are dedicated to support these tape formats and matching recorders. Relevant IT storage devices are, for instance, data tapes, optical storage devices or hard disks. There is also a relatively large number of robotics that support these devices. Both approaches have their specific advantages and disadvantages, as will be shown in the following sections.

7.3.1 VIDEOTAPE-BASED SYSTEMS

Video library management systems (LMS) are considered to be a broadcast equipment component. They are typically offered by traditional broadcast suppliers. Some vendors have proposed systems that allow storage and handling of 200 000 program hours at

50 Mb/s, resulting in a gross storage capacity of 4.5 petabytes. In such a system the number of videotape recorders integrated with the tape robotics determines the number of simultaneous playouts and recordings. The overall performance of such LMS is comparable to data-tape-based mass storage systems. However, in a broadcast environment they have additional benefits. For instance a real-time LMS output stream is directly usable in current production. Further, timecode-controlled partial retrieval of video, audio, and metadata can be easily accomplished. If video compression is used, even faster than real-time transfer can be achieved.

Nevertheless, there are also a number of significant limitations to the LMS approach. Since standard VCR are used within the library the solution is intrinsically linked to formats supported by the recorders installed in the library. Thus, the approach lacks flexibility. Further, the solution is essentially video specific and thus is incompatible with audio and data archiving. A separate solution will be required for them. Since only VCR technology is used only streaming but no file transfer is supported. With the growing amount of content automatic support for checking the data integrity migrating content from one format to another (or from one storage media to another) is required. This is not provided by currently available LMS. Thus, LMS offer a huge storage pool but not enough operational support. Another disadvantage is also that broadcast media formats are not supported by mainstream IT industry products.

In recent years data tape libraries have been established alongside the traditional LMS. The migration from video-based carriers (tapes) to data tape carriers has been recognized as a necessary step to fully exploit the potential of a modern automated archive. All the same, it is clear that such a migration process may take several years. Therefore, a parallel management of traditional videotapes and newer data tapes is compulsory for each CMS established in this environment. This is at least true throughout the migration period but may also be required in an established system where videotape continues to be a storage alternative. For these scenarios the LMS approach is a valid intermediate step on the way to the fully integrated digital file-based solution or an accompanying technology receding alongside the more efficient IT-based mass storage systems.

7.3.2 IT-BASED MASS STORAGE SYSTEMS

IT-based mass data storage systems (MDS) are already industry standard in traditional IT. The main benefits of MDS are their flexible architecture and the widespread availability of management tools. These tools for instance allow achieving automatic migration from old to new storage formats as a background operation. Further, media can be logged to ensure data integrity and subsequent preventive action can be triggered at a higher level. This approach also allows timely and automatic adaptation to new technology as a means to control the physical size of the archive and to utilize improvements to achieve a competitive advantage. MDS also provide backwards read compatibility over at least two or three technology generations.

Since these storage technologies enable a more integrated approach towards content, the integrity of vulnerable metadata can be maintained. IT-based storage systems are also truly format agnostic, i.e. they can accept content in various formats and with different data rates. It is also possible to administer different storage media within a range of storage architectures.

In contrast to LMS that only support streaming, the file transfer mode is an inherent feature of MDS. They can also adapt to different network bandwidths and transfer requirements by means of buffers. Thus, streaming and file transfer are supported. However, streaming via SDI is not yet part of the standard feature set.

This latter point already indicates one of the main issues regarding the usage of MDS technology in a broadcast context. A drawback is that most of the above listed benefits can only be fully exploited if MDS systems are customized to interface with broadcast production environments, and standard IT systems do not provide these interfaces. Exploitation of this niche market has led to the introduction of special archive management solutions, which are software systems that add the relevant functionality. Examples of such functionality are, for instance, partial file restore based on timecodes, support for file formats relevant for the broadcast industry, or interfaces to proprietary production systems, to the mass data storage system offerings.

The major advantage of IT-based mass data storage solutions is the flexible storage architecture that allows matching user requirements with the specific properties of certain storage media (data tape, disk drives, CD-R, DVD, etc.), in terms of cost, capacity, throughput, access, functionality, longevity, etc. This allows tailoring such storage systems to a specific application, such as online archive, near-online archive, deep archive etc.

7.3.3 STORAGE DEVICES FOR LONG-TERM STORAGE

When selecting the technology for a mass storage system it is important, however, to take the rapid development cycle of storage technology into account. Improvements in hard disk storage density and the significant capability improvements expected for recordable optical media will offer alternatives to data tapes as mass data storage in the long term. Although hard disk and optical disk are currently not as cost effective as magnetic tape, they do offer one very important advantage over magnetic data tape, i.e. the substantially shorter access time.

The following sections explore the various options for storage devices for long-term mass storage. This includes a discussion of videotape recorders (VTR) that are part of the LMS infrastructure.

7.3.3.1 Videotape Recorders

A VTR drive records an incoming video signal and writes it to a videotape. This procedure is not necessarily data transparent. That is, when video compression is involved the data rate is reduced, which leads to an irrevocable degradation of the signal. The amount of degradation strongly depends on the compression algorithm and the resulting bit rate (see Section 3.2).

For playback a program segment is selected with reference to the timecodes that are associated with the material on the tape. The timecodes are used as start and stop values. At the start of the playback process the tape is shuttled to the beginning of the clip, and frame-accurate start and stop of playback of a selected video, audio, and/or metadata segment takes place. This can happen in real time (or faster than real time) depending on the properties of the VTR. Digital video and audio is streamed via an SDI or SDTI interface to its destination at a constant data rate where it is directly usable in the production process.

Errors can be handled to a certain extent by a VTR. The residual bit error rate is in the area of 10^{-11}. In case of an error correction overload, concealment mechanisms in video and audio are used to mitigate the subjective effects of bit errors. These concealment mechanisms result in a degradation of image or sound quality, but do not lead to an interruption in the signal transfer. Since streaming video or audio via SDI does not allow the retransmission of data all frames need to remain in sequence so that the viewing or listening experience is not impaired. Thus, concealment mechanisms have been designed to keep the stream flowing, even if this means accepting signal degradation.

7.3.3.2 Optical Disk Drives

Optical disk formats suitable for the archival of content currently comprise CD formats, the DVD product family, and small disk magneto-optical formats. Most relevant in the context of an enterprise-wide CMS are DVD-based formats. Capacity and throughput limitations of DVD currently restrict its use to a near-online storage library device for streaming of browsing quality video and audio at about 1.5 Mb/s. This makes it currently only suitable for EDL browse quality video or uncompressed PCM-encoded audio. Thus, it can only be used for a specific application class. Further, experience shows that the mechanics of conventional jukeboxes have not been designed with the heavy access load of enterprise-wide CMS in mind. This is especially an issue when considering that in general data has to be available in a 24 × 7 environment. Despite these drawbacks CD and DVD might be useful in a specific context at a department level (for instance within radio broadcasting) as a near-online storage alternative.

7.3.3.3 Data Tape Drives

Data tape drives can be used to store digital media in the same way any other data is stored on them, i.e. in the form of files. The major difference between VTR-based near-online storage such as LMS is that data tape drives do not offer broadcast streaming interfaces such as SDI or SDTI. All storage operations are file based. When content is delivered in a container file format containing metadata as well as the different essence components, the complete file including all metadata and housekeeping information is treated as one unit and stored as such. To the system the data stored on it is entirely transparent. It is basically interpreted as a sequence of bits.

Data tapes can record and play at very high data rates but they work best when they receive and deliver the data stream at a determined constant sustained data rate. Differences between sustained data rate and the data rate delivered at the input and output should be smoothed out using a fast buffer. An example of this is the high performance disk system available in a SAN. The data transfer from such a buffer to and from tape then may be accomplished by a specific transfer protocol to guarantee delivery, i.e. to ensure that all bits are received and written correctly to tape.

For data retrieval the control system will first locate the cartridge with the content and find a drive to playback the tape. It then shuttles to the exact position of the start of the selected file on tape. Retrieval of the complete payload contained in the file will be at the native sustained data rate of the drive. IT-based tape libraries without modifications always operate on a file and bit level. There are certain upper and lower bounds on the retrieval operations, but no guarantees on the retrieval time can be given. In most cases data will not be read continuously off tape but rather in bursts, depending on the relationship between the sustained data rate of the tape, the buffer size, and data rate limitations of the network. If read-errors have to be corrected using retransmission, further delays may be incurred. Thus, a buffer or cache must adjust the outgoing file transfer speed to the requirements of the connected network (e.g. Gigabit Ethernet or Fibre Channel) for file delivery. Residual bit errors are quoted to be in the area of 10^{-17}. However, there are no concealment mechanisms.

Standard IT solutions do not provide timecode accurate access to the contents of a broadcast file. Where partial file retrieval is possible it is based on a method that operates on byte offsets. In contrast, partial file retrieval of audiovisual content requires access to parts delimited by timecodes. Thus, in order to provide this kind of partial file retrieval within a data tape library where the content is stored in files, timecode information or index tables contained within the file that correlate the position of the segments in the logical file structure are required. This means, even if only certain elements (such as video or a selected audio track) have to be recovered, access to the whole data structure must be possible. A special software module is required that allows partial file retrieval based on timecodes in a data library environment. This component has to be able to interpret content files (such as the ones introduced in Chapter 5) and use the information about file structure and essence encoding to map timecodes to the respective byte offsets. This kind of functionality is often provided by a broadcast-specific archive management solution.

Practically all modern IT-based storage formats offer a memory in cassette (MIC) as a means to improve load/unload time and access to files, and for providing data for performance monitoring.

7.3.3.4 Hard-Disk-Based Online Storage

Hard disks are not used as storage devices in robotics but are deployed in SAS, NAS, or SAN systems. They are discussed here for two reasons. First, they are part of larger storage infrastructures. As can be seen in the case of SAN, these infrastructures are becoming more and more transparent to the CMS that uses them. Second, for low-bandwidth media (such as browse audio and video) disks are a cost-efficient long-term storage alternative.

Data security in the case of storage infrastructures based on hard disks is achieved by installing hard disks in a RAID (i.e. *redundant array of independent disks*) mode. Different RAID modes have been developed that range from full disk mirroring (RAID level 1) to the duplication of a subset. The development of hard disk technology is still occurring at a breathtaking pace, and is starting to be competitive to other storage technologies, even up to storage volumes of several terabytes.

7.3.3.5 Combining Technology

The various types of hard disk and robotic tape-based storage options can be combined to achieve optimal functional and cost efficiency. When combining storage options it is important to select components and networks and design the combined infrastructure such that no control, bandwidth or other bottlenecks are introduced. It is particularly important that the sustained data rates of the various storage options are compatible. If this is not the case provisions have to be made to ensure that no data is lost due to overload.

Combining different storage technologies can actually provide a good way to bring together all the different requirements placed onto the storage subsystem and solve them in a unified system context. However, it has to be ensured that the combined systems can be integrated without major changes and adaptation to the system architecture. This might lead to proprietary solutions that are difficult to support and maintain. Any such solution should be avoided at any rate.

7.4 SUPPORTING WORKFLOWS BY A BASELINE SYSTEM INFRASTRUCTURE

Large-scale CMS are not off-the-shelf products since they have to be adapted to the specific requirements of the organizations that want to deploy them. Thus, there is no infrastructure blueprint or standard architecture according to which all systems could be designed. However, before designing a particular technical infrastructure for a large-scale CMS, it helps to consider a generic baseline hardware infrastructure. This infrastructure contains all relevant elements of a CMS infrastructure and also considers important design principles such as expandability to meet the expected increased demand for capacity, availability, and performance. This baseline infrastructure allows the components required in a specific CMS deployment to be identified and the scalability potential as well as the costs involved when scaling the system to be estimated. This section proposes such a generic baseline hardware infrastructure that puts the components into an architectural context. Further, it gives some insight into the workflow aspects covered by the architecture.

7.4.1 GENERIC SYSTEM LAYOUT

The infrastructure within an organization together with the way essence and metadata are processed through this infrastructure are the basis for enabling workflows and executing processes. Figure 7.5 outlines a possible implementation of a generic infrastructure for a CMS designed to meet the requirements in terms of functionality and scalability. This infrastructure can also be used to discuss the interaction of system components and the flow of essence and metadata through the various networks.

The proposed generic architecture supports the following steps in standard workflows:

- ingest, storage, editing, and playout of high-resolution material in various formats;
- creation and storage of low-resolution proxies such as browse copies and keyframes;
- generation, storage, and modification of metadata;
- automatic indexing;
- advanced retrieval capabilities;
- access to metadata and essence in suitable quality from distributed work places for improved overall workflow.

In accordance with the design rules given above, three classes of servers are used to build the CMS infrastructure, namely clustered database servers, application servers, and SAN servers. In the example the application servers have Fast Ethernet interfaces only, whereas the SAN servers, have Fast Ethernet, Gigabit Ethernet, and Fibre Channel interfaces. In addition, interface servers link the IT-based system to the broadcast environment.

The following subsections briefly highlight the various modules of this proposed infrastructure. To achieve maximum scalability, certain services may be distributed over multiple server systems. For simplicity reasons Figure 7.5 shows only one server per service. How to scale the system and the different components is described below.

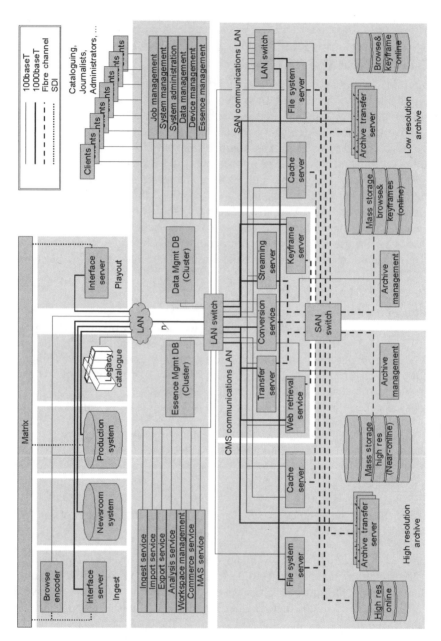

Figure 7.5 Baseline hardware architecture

7.4.2 CONTRIBUTION AND INGEST

In an organization dealing with content there are many ways to acquire material. There are a number of possible sources such as tapes in various formats, film in different gauges, cable transmission, satellite downlinks, and networks using file transfer.

When essence is to be ingested into the system via signal (as in the case of ingest from tape, film, cable, and satellite) the signal is routed to selected recording devices via a matrix switch (or crossbar), using SDI or SDTI. The crossbar (or matrix switch), the switch controller, and all connected recording devices are controlled via standard protocols, such as RS-422. The recording devices typically are disk recorders or video servers that act as interface servers as described in Section 7.1.3.5.

In the case of file transfers the essence is not routed via a matrix switch but via standard IT networks. The target system can be any suitable IT device in the CMS; a natural choice again are the interface servers since they will be able to immediately provide the essence received as files, as signal to the matrix switch, and hence also to the standard broadcast infrastructure.

7.4.3 INGEST CONTROL AND RECORDING

As default the incoming essence is recorded onto an interface server as described above. If a signal is recorded the interface server encodes the essence into the standard production format. In case of an SDTI transfer or a file transfer the essence is directly written to the interface server's disk. File transfer via IT networks then is used to bring the recorded essence from the interface server into the domain of the CMS. Note, in most cases the broadcast equipment is not under the exclusive control of the CMS since it is also used by other components such as the studio automation system.

The choice of the encoding facility (the interface server) strongly depends on the choice of production and archiving format. When using disk recorders as encoding devices it can be sensible to let several or all disk recorders share a single SAN. SAN-enabled disk recorders can address the SAN as one local disk and can therefore avoid unnecessary material movement. Such a configuration, however, puts high demands on the performance of the SAN and introduces a single point of failure for the ingest process. This can be difficult for live recordings since such recording cannot be repeated.

It is important to define a common interchange file format for the recording of essence. Good examples for such a standard interchange file format are the Media eXchange Format (MXF) or the Advanced Authoring Format (AAF) (see Chapter 5). If an interface server does not support such a standard interchange file format, file format converters must be provided for conversion between incompatible file formats. This is required when various production devices use different formats while still using the same essence encoding format. File format converters may directly access the interface server's disk or SAN, or they may access files in the high-resolution online storage area of the CMS.

Ingest and parallel recording into multiple target formats are typically managed by issuing commands to an automation or broadcast control system. The automation then takes control over all devices involved in the recording.

In case of ingests from signal, low-resolution copies for browsing purposes should be recorded simultaneously by feeding the incoming stream to a low-resolution (browse) encoder (e.g. an MPEG-1 encoder that supports frame-accurate recording and SMPTE timecode) and writing the encoded stream onto the low-resolution online storage area (browse and keyframe online SAN) of the CMS.

In case of file or SDTI transfers, the low-resolution proxy can either be generated via software transcoding of by replaying the received files via interface servers to baseband and recording these signals in low-resolution only.

In the transition period, when conventional video-centric operations still dominate, it is important to provide tape copies for in-house production and playout of essence to professional tape formats accepted for program exchange.

7.4.4 TRANSPORT OF PRODUCTION QUALITY ESSENCE

Many organizations have an existing SDI/SDTI infrastructure that can be used for the transmission of high-resolution quality content during the transition from traditional production to file-based production. Since SDI is a point-to-point protocol these networks require signal routing via matrix switches (i.e. crossbars or routers). In such a scenario interface servers (disk recorders) integrate the SDI infrastructure and the IT infrastructure as depicted in Figure 7.6. File transfer can be used to copy essence from interface servers to file-based production systems (such as nonlinear editors) while the recording is still in progress.

Ultimately, the aim must be to migrate all production and delivery processes to file access and file exchange. Here, high-performance network connections (e.g. Fibre Channel or Gigabit Ethernet) are the preferred technology.

7.4.5 ACCESS TO BROWSE COPIES AND KEYFRAMES

In content-rich organizations several hundred to several thousand desktop workstations may access and replay browse-quality content via the standard corporate network. For scalability reasons and to minimize the replication of content (which would increase storage costs) the architecture proposed in this chapter connects multiple browse streaming servers to a single SAN (browse and keyframe online SAN as shown in Figure 7.7). This allows scaling the number of concurrently available browse streams by introducing a suitable number of streaming servers. All servers share the same disk space, thus avoiding the replication of content in the online storage.

Keyframes are sometime an even more important way to give a user a fast overview of the visual content than browse video. Thus, fast access to keyframes is an important feature of a CMS and it is therefore sensible to keep all keyframes permanently online on the browse and keyframe SAN. This avoids latencies introduced by the low performance during the staging process of near-online storage systems. To ensure scalability and reduce costs when dealing with thousands of concurrent user queries, the strategy is again to use multiple keyframe servers, sharing the same SAN, for load balancing.

Figure 7.7 shows the configuration where streaming servers and keyframe servers (being connected to the browse and keyframe online SAN) are based on servers of the class 'SAN server'.

7.4.6 REAL-TIME ANALYSIS AND ANNOTATION

During the ingest process the CMS should provide options for real-time generation of metadata. Metadata generation in this context can be an automatic or manual process. The

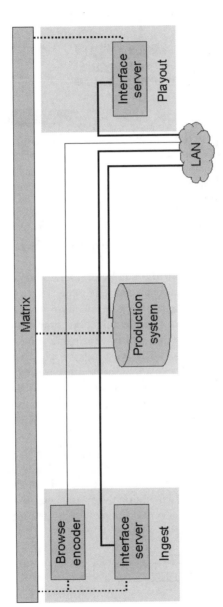

Figure 7.6 Integrating the CMS with an SDI infrastructure via interface servers

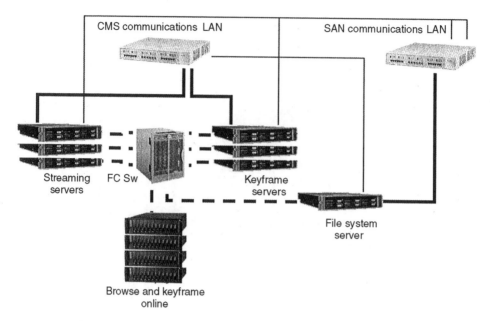

Figure 7.7 Multiple streaming and keyframe servers sharing a SAN

former case for instance occurs with the automatic extraction of ancillary data (closed captions, ANC data, etc.), automatic video analysis (shot detection, keyframe extraction, etc.), automatic image analysis (face recognition, OCR from screen, etc.), and automatic audio analysis (keyword spotting, speaker identification, simple audio classification, etc.). The manual annotation is called logging and is a real-time process.

Automatic analysis processes tend to be very demanding with respect to processing capacity and computing time. Powerful processors are required to complete these processes faster than real time. Thus, it must be possible to distribute video and audio analysis services over multiple servers in order to scale the number of concurrent analysis processes, as shown in Figure 7.8. This kind of set-up also adds redundancy and increases availability.

If the Analysis Services use Streaming Servers for accessing content they can reside on servers of the class application servers. If they have to access the essence at file level they have to be connected to the SAN. In this case they are of the class SAN server.

In addition to automatic metadata generation the system should provide client interfaces for close-to-real-time manual annotation (logging). During the logging process the users

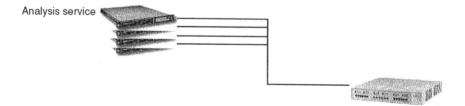

Figure 7.8 A group of analysis servers

need access to browse copies while they are being recorded. Ideally there is only a very short time delay (i.e. in the range of seconds) between the actual recording and the viewing of the material.

The result of the essence analysis typically is timecoded text, but may also be new essence (e.g. keyframes). Manual annotation creates alphanumeric metadata. While keyframes are stored on keyframe servers, metadata may be distributed over various databases, such as the Data Management System's databases or existing (legacy) cataloging systems. Essence-related metadata (for example locations of files, encoding formats, file sizes, etc.) are kept in the Essence Management database.

7.4.7 ARCHIVE

Keyframes, and high-resolution and low-resolution copies of the essence are permanently archived in mass storage systems. Figure 7.9 shows the relevant part of the baseline hardware infrastructure that supports the archive. Depending on the amount of storage required, the core of the mass storage system can be either an automated data tape library system (in case of production-quality video and film) or a hard-disk-based storage system (applicable for production-quality audio, video and audio browse copies, and keyframes). While video and audio essence in production quality as well as video browse copies are stored as near-online copies and staged to the respective online storage SAN file systems on demand, audio browse copies, Internet quality essence, and keyframes should always be kept online. For these essence formats the mass-storage system serves as a backup device.

The transfer of data from the SAN file system to data tape is done via high-performance data tape drives. The throughput of modern tape drives is considerable; transfer rates of more that 30 Mbyte/s (i.e. more than 240 Mb/s) are currently state-of-the-art. To cope with these transfer rates the system design again employs distribution. Each tape drive is connected to the Fibre Channel switch and thus is available as a resource on the SAN. A sufficient number of archive transfer servers are connected to the SAN and thus see on the one side the data tape drives, and on the other side the SAN file system. Ideally, each archive transfer server should primarily access one dedicated tape drive. Depending on the performance of tape drives and servers, it may be possible in certain configurations to use two tape drives or more via a single archive transfer server without performance penalties.

The fact that the SAN architecture allows a tape drive to be assigned to more than one server can also be used to increase the availability of the solution; in this case the loss of one Archive Transfer Server no longer means also losing the attached tape drive. When more Archive Transfer Servers are added to the solution than are required to operate the tape drives the loss of an archive transfer server does not even lead to any loss in quality of service.

The Archive Transfer Servers directly transfer content from SAN file system to tape and back. Since all units are connected to the SAN, resources can be reallocated when a server or a tape drive fails. Archive Transfer Servers are also used for transferring files between the online SAN and disk-based mass storage systems. Depending on the software solution employed they may also be able to copy files to and from interface servers. Archive Transfer Servers are of the server class SAN server.

The mass storage systems are the physically long-term archive. Online SAN are only caches that grant optimized fast and direct access to archive content, i.e. content stored

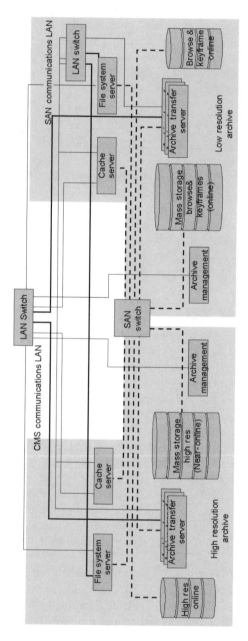

Figure 7.9 Archive infrastructure

on these components is volatile. Files are migrated between the online SAN and the mass storage systems via archive transfer servers (see Section 6.4.1.3).

Some additional services are required to manage such an infrastructure. First, there is a need for a library index that knows what is archived and where it can be found. Further, this index accepts requests for archiving, retrieval, and deletion of content (the so-called archive management server). Second, when employing a number of Archive Transfer Servers there has to be a coordinating entity that decides which Archive Transfer Server processes which incoming request. This involves queuing these requests and distributing them to the available Archive Transfer Servers according to priority and the availability of resources. This component is the Archive Transfer Manager. Finally, the available storage capacity of the online SAN as well as the capacity of the managed interface servers needs to be kept between configurable high and low watermarks. This is done by the cache servers. With respect to the hardware architecture, these services may be located on a single server or be distributed over more than one server (depending on the system load). While Archive Management Server and Archive Transfer Manager can reside on an application server, the cache server needs to run on a SAN server platform, since it needs to access the SAN to be able to monitor available storage space and delete files.

7.4.8 QUERY AND RETRIEVAL

The standard query interface provided by a CMS should be Web-based. The Web retrieval service that provides this interface is hosted by one or more server systems. This approach is again taken to distribute the load and make the system more scalable.

The Web retrieval service may run on an application server class of platform with a high-bandwidth network interface. This is especially required in the case where larger sets of keyframes are to be presented to the user. In installations where embedded video or audio players are used to play video or audio files residing on the SAN, the platform should be a SAN server class system.

The CMS uses internal databases to persistently store data such as descriptive or essence-related metadata, or processing information. The servers hosting these vital databases should be designed as server clusters in order to assure maximum failover capability and performance. Figure 7.10 shows two database server clusters for the databases of Data Management and Essence Management.

Queries are processed by the various components of the Data Management. In real-world applications these Data Management components will reside on separate servers for load distribution. Typically, these servers interface to databases in which a specific subset of the overall metadata is kept, or they embed or interface to specific search engines. The Data Management federates searches over multiple databases and merges resulting hit lists via the data broker. Since these services typically have no requirement to access essence, in most cases their hosts can be of the class application server.

It is also possible to apply the concept of distribution to single functional components of the Data Management. An example would be using multiple fulltext search engines (each running on a dedicated server) in order to distribute the query load. These fulltext search engines then either directly index attributes from database tables or they index documents residing on a shared storage medium. In the latter case these fulltext search engines should reside on SAN servers.

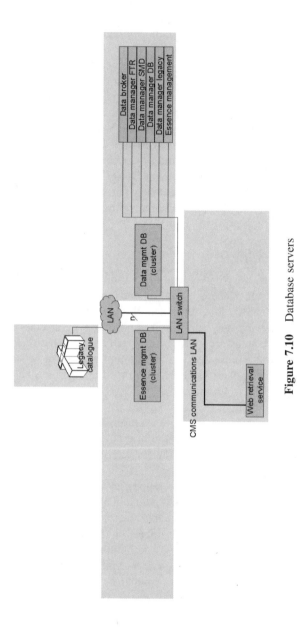

Figure 7.10 Database servers

In addition to traditional search capabilities a CMS may offer advanced retrieval functionalities (such as image similarity search). These search engines often have considerable performance requirements and hence should also be supported via distributed server systems. The following simple example highlights the problem:

> Considering 1000 keyframes per hour of video, 100 000 h of video yield 100 000 000 keyframes. For user acceptance, an image similarity query, comparing one image to all keyframes stored, has to be answered in less than 10 s.

This represents a remarkable challenge for the performance of such a search engine. Since the usefulness of this kind of advanced retrieval system still has to be established in a real-world deployment (especially for large broadcasting organizations) it is important to evaluate whether the benefits justify the expenses.

The Essence Management processes requests for essence-related metadata (such as formats, timecodes, durations, file sizes, locations, etc.). It accesses the Essence Management databases but, like the Data Management, it handles data and no essence. Thus, it can run on one or more servers of the type application server.

7.4.9 INTERFACE TO PRODUCTION AND PLAYOUT

Based on the decisions made during retrieval and rough-cut, production-quality essence is transferred from the high-resolution archive to one of the production systems. Production systems can be stand-alone editing systems or server-based editing solutions supporting multiple editing clients sharing the same content, as depicted in Figure 7.11.

The transfer of essence from the archive to the edit suite can be carried out in three different ways:

- *Real-time streaming of baseband video via standard studio SDI connections.* Depending on how the control is shared between a CMS and a broadcast control automation, the procedure can be implemented as follows:
  - The CMS transfers the files to an interface server (playout server). The transfer to the edit suite is then done upon manual request, under control of the edit suite.
  - The CMS transfers the files to an interface server (playout server). The transfer to the edit suite is controlled by a broadcast control automation.
  - The broadcast control automation transfers the files to an interface server and controls the playout to the edit suite.
- *Streaming of native compressed video in real-time or faster than real-time via SDTI connections.* This kind of transfer can be managed in the same way as an SDI transfer. The major difference is signal quality. SDI transfers typically implies decoding to baseband and re-encoding to the compressed domain on the target system, which results in degradation of signal quality. SDTI transfers allow exchanging data in the compressed domain using SDI connections. The signal quality is not impaired.
- *Copying via an IT network (preferably Fibre Channel or 1000baseT).* This transfer is directly managed by the CMS. It may be necessary to change the file wrapper format via a file format converter to accommodate proprietary file formats. When production systems are interfaced that use incompatible encoding formats, software transcoding to this encoding format is required.

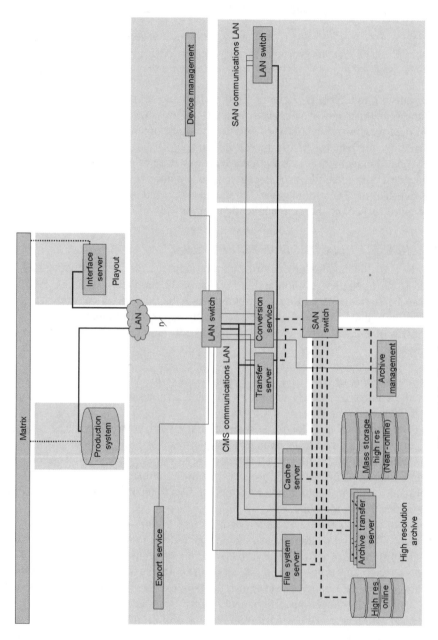

Figure 7.11 Interface to production and playout

From the infrastructure point of view the CMS handles the physical transfer of file between archive and interface servers or production systems via Transfer Servers. Since these servers need to interface the SAN, they need to be of the class SAN server. The control of the process involves interfaces to the Interface Servers and the broadcast automation system (components of the device management) and managing the transfer processes (via the export service). As controlling components, these system components can reside on application server class hosts. If format conversions are required, Conversion Servers have to be invoked. Ideally, Conversion Servers are installed on SAN servers.

On the production system, material is edited and composed to form a newly finished program. After final conformation of the program on the editing workstation the content may be transferred to a Playout system. This transfer may be handled in the same fashion as the transfer from the archive to the production environment, or it may use transfer capabilities provided by the production system itself. The playout itself is fully managed by the broadcast control system.

A browse copy of the finished material can be generated by moving finished copies onto an interface server and replaying via baseband to real-time browse encoders, or via software transcoding using conversion servers.

Finished pieces may be transferred from the production system to the archive for permanent storage, using the same transfer functionality.

7.4.10 CATALOGING AND ARCHIVING

Material selected for permanent storage and archival is subject to in-depth cataloging and indexing by professional archivists. The cataloging staff access and manipulate browse copies, keyframes, and metadata from their desktop computers. The major task during cataloging is refining the quality of metadata. The result is accessible throughout the organization via established IT-based access protocols. With respect to the hardware infrastructure, all systems involved to support these tasks have already been described.

7.4.11 ADMINISTRATION

System administrators use their desktop computers to monitor and administer the CMS. Since the administrators need access to full system functionality in order to maintain and supervise smooth performance, all system functionalities must be accessible from their desktop.

7.5 OPERATIONAL CONSIDERATIONS

After the introduction of a CMS the infrastructure as well as the software has to be operated, managed, and maintained. For IT-based components these are standard maintenance processes that are carried out in this context. Most computer equipment is depreciated over a time period of three to five years. After it reaches the end of its lifetime it needs to be replaced or be updated. Software maintenance is a more continuous process. Software updates are usually part of the maintenance contract. Thus, common IT regulations and agreements can be applied for CMS as well.

However, there is one area where operation, maintenance, and replacement are of special importance, i.e. the maintenance of the mass storage system and its data carriers.

It is crucial to inspect the integrity of the data carriers on a regular basis and take action when there is a danger of losing data. A special migration strategy has to be devised for this case that ensures that no data is lost. Ideally, this kind of maintenance should be an automatic background process that works without human intervention or even without the users noticing it.

Another aspect in this context is the cost of data management and the migration process. Considering the amount of data that has to be managed in a content-rich organization the sums that are required for these processes are substantial. Thus, the data maintenance and migration strategy not only depends on the technical parameters but also on cost factors. In the following both, the technical issues connected with carrier maintenance and migration as well as the cost issues are introduced.

7.5.1 MIGRATION

One of the key reasons for digital, file-based archival is creating an *eternal data set* that retains its integrity over time, independent of changes in storage technology and media. This promise obviously cannot be easily fulfilled. IT hardware technology continues to rapidly evolve. Examples are storage density and computing power, as well as the quality of media-related software technologies such as compression. Further, new encoding formats are also being developed continuously. While in traditional media production these formats were tied to the physical carrier (i.e. tape), new digital formats are now independent of the physical media.

However, to keep the digital archive technologically up-to-date this technological progress and advances in encoding developments require that migration is an inherent design paradigm of a CMS. Migration in this context may even require changing the original data set. Since the technical progress and developments are constant migrations will also not be a rare incident but are going to be an integral part of the operation of the digital archive system. This is in contrast to the initial migration from the videotape-based production discussed in Section 8.2.2. Thus, continuous automatic migration has to be an in-built functionality of a CMS.

7.5.1.1 Using Automatic Tape Libraries

A digital data archive must be able to guarantee the integrity of the data for an unlimited time span. This implies that physical deterioration of the storage medium must not affect the integrity of the stored data. Tape formats (regardless of whether they are videotapes or data tapes) are prone to physical deterioration. Thus, a digital data tape-based archive must provide means to automatically detect endangered storage media and perform automatic recovery procedures. Automatic tape libraries are an adequate means to achieve this goal. Automated tape libraries are systems that host tapes and tape drives, move tapes automatically between slots and the drives, and control mounting and un-mounting of tapes in drives.

When such a tape library is used as the mass storage system for permanent storage of high-resolution media, maintaining data integrity can be software controlled and performed automatically, without human intervention. Therefore automatic tape libraries are not only suitable as near-online mass storage but also facilitate the preservation of content by automatic migration processes.

7.5.1.2 Monitoring Carrier Integrity

Monitoring carrier integrity is the key element in detecting whether the data set has been transferred correctly to a storage medium or carrier. Further, it is also important to discover whether the 'health condition' of the carrier has deteriorated beyond a predefined limit. If this is the case, preventive measures initiate an automatic transfer of the data to a new carrier.

A suitable indicator of the data integrity of a data tape is the bit error rate (or an equivalent indicator provided by the vendor of the tape drive) when reading data from tape.

It is necessary to understand where an error in a data set occurs, and what influence an error will have on a block of data. A block of data typically consists of:

- an address
- a synch word or synch block
- a data packet
- a check word or checksum, which delivers the error protection for the block.

Since an error in the sync block could lead to the loss of the whole block, additional error protection mechanisms are introduced which use information across multiple data blocks in order to generate so-called product codes. Using the product code, errors in sync blocks can be accommodated. Some vendors also use shuffling techniques to distribute error protection over multiple data blocks and their respective product codes.

In order to really understand what is happening with the data on a certain data tape a CMS would have to fully understand the way the error protection has been implemented in each single data tape drive solution it supports. This is challenging, especially when considering the fast product release cycles and the fact that vendors want to protect their competitive advantage.

The CMS does, however, need an early indication if a certain tape or tape drive deteriorates and is no longer reliable. This implies that sufficient information must be made available to at least establish the class of errors, the error frequency, and the increase of errors over time. For example, when burst errors occur on specific tapes, this may indicate a defect in the tape itself, while an increase of single errors on multiple tapes may indicate a deterioration of the writing head of a specific drive.

In order to protect the competitive advantage of vendors and to minimize the knowledge that the CMS needs to have about the specific implementation of an error protection algorithm, *a traffic light* system could be used. The interpretation of the different lights could be such that the first traffic light could indicate the maximum error rate over a given time span and the second traffic light could indicate the integral of all error rates over a given time span.

The time span is a system specific constant value. In case of helical scan recording, for instance, a reasonable time span would be the time required to write one track. The traffic lights could be defined as:

- green - well within design parameters;
- amber - close to design parameters, still recoverable;
- red - unrecoverable errors occurred.

It would be up to the vendor to relate the internal error statistics to a vendor-specific normative value in order to deliver the result in this neutral traffic light way. Also, it would be sensible to have an independent organization test and qualify the results delivered in order to ensure compliance.

The CMS then is responsible for recording this information, storing it, performing the necessary statistical evaluations, and initiating appropriate action to protect the overall data integrity of the data set.

Ongoing statistics of error rates or error indicators give an indication of the current state of health of a carrier and tape drive as well as the development of its health over time. Additional information can be helpful to derive a possible end-of-life date for a carrier via empirical models or by applying information provided by the vendor. Examples are a maximum tolerable number of mounts for a certain carrier, a maximum tolerable number of read passes for a certain carrier, and a maximum tolerable number of write passes for a certain carrier. Further, there are also certain parameters provided by the vendors or based on experience such as the maximum theoretical lifetime of the carrier when kept in a controlled environment. In this context the time the carrier spent outside of the controlled environment also has to be considered. The remaining lifetime of the carrier can then be derived from maximum theoretical lifetime and recognizing that a carrier ages much more rapidly when it spends time outside of a controlled environment. Further, an arbitrary end-of-life date that is specified according to a company or organization policy should be specified and implemented.

If archiving on data tape within a robotic mass storage solution is the preferred choice, the responsible management system must keep track, for each tape, of at least the following information:

- maximum tolerable error rates;
- maximum tolerable number of mounts;
- maximum tolerable number of read accesses;
- maximum tolerable number of write accesses;
- maximum tolerable time between servicing procedures;
- maximum tolerable lifetime of the tape;
- user selected end-of-life date and time;
- date and time of first recording, in order to derive the age of the tape;
- an ID of the recording device, in order to follow up on potential systematic failures due to technical problems of a certain recording device;
- traffic light statistics of the last read accesses;
- current number of mounts;
- current number of read accesses;
- current number of write accesses;
- scheduled date for next servicing;
- status.

Based on this information the integrity of a carrier can be derived and if necessary the data stored on the carrier can be migrated to a new carrier. There are a number of tasks that need to be performed automatically in order to ensure the integrity of a data-tape-based carrier. These tasks for instance include mounting the tapes that have just been written by a tape drive in a second, different tape drive and re-reading the data in order to eliminate the possibility of write errors due to misalignment of heads or other technical defects in the write assembly of the drive. Further, rewinding and re-tensioning tapes that have not been accessed for a certain amount of time is required to avoid sticking of tapes. This should be done on a regular basis. Reading tapes at periodic intervals in order to check the bit error rate (or an equivalent indicator of the health state of the tape) is also required. If problems have been detected (i.e. any of the specified parameters have been exceeded) action has to be taken. This includes copying the tapes that feature a bit error rate exceeding a given safe bit error rate margin to new tapes, and copying tapes that have reached end-of-life to new tapes. A tape has reached end-of-life when its overall age due to its time spent inside a controlled environment is larger than a given safe age.

It is important that these processes are considered in the design of the CMS infrastructure. Although the inspection of the storage media and the migration process itself can be run in the background capacity has to be reserved to ensure that this is possible on a regular basis.

7.5.1.3 Carrier Replacement and Decommissioning Policies

There are two major reasons why carriers may be decommissioned and discarded from the robotic library, namely the carrier has reached end-of-life and the data is migrated to a replacement carrier, or a new technology is introduced that makes the carrier obsolete. In the latter case the material needs to be migrated to a new carrier based on the new storage technology. In both cases, however, the CMS is responsible for migrating the content stored on the carriers to be replaced to the new carriers. After completion of the migration, the old carrier can be decommissioned.

Decommissioning is an automatic process. The management system:

- selects a spare carrier from a collection of spares that is kept in the robotic library;
- mounts the carrier to be decommissioned in a playback device;
- mounts the selected spare carrier in a recording device;
- reads the data from the carrier to be decommissioned and writes it to an online system;
- reads the data from the online system and writes it to the new carrier;
- un-mounts both carriers;
- assigns a slot in the robotic library to the new carrier;
- transfers the new carrier to this slot;
- transfers the carrier to be decommissioned to a ejection port of the robotic library, thus removing the carrier from the robot.

The only elements of human intervention in this process are necessary to provide a sufficient number of spare carriers in the robotic library, and to remove decommissioned carriers from the library's ejection port.

7.5.2 COSTS

Costs are one of the main arguments in the discussion surrounding the introduction of an enterprise-wide CMS. This book does not deal with the economic aspects of content management. There are a number of tangible but also intangible benefits in the introduction of CMS. These are offset by the cost of implementing and introducing a CMS into a content-rich organization. So far neither the benefits nor the costs have been fully established. This is an ongoing process since a plethora of parameters and aspects have to be considered. However, the main cost drivers have been identified. These cost drivers are the mass storage system, the internal costs for realigning business processes, and system migration during operation.

Initially the cost drivers in a CMS project are the initial costs for the investment in mass storage systems. During the deployment the internal costs of re-aligning business processes and gaining system acceptance by the users are among the largest cost factors. During operation the migration costs due to scaling of the system in order to meet increasing demand, and due to replacing system components that reach end-of-life with new, state-of-the-art components, are the greatest cost factors.

Using a baseline hardware infrastructure such as the one presented in this chapter it is possible to derive models to estimate hardware-related costs. In the case of internal costs the real cost drivers can only be identified by doing a thorough business process analysis within the organization. Some of the most common internal cost drivers, however, are discussed below, and approaches to minimizing these are suggested.

7.5.2.1 Initial Cost

Typically, the initial costs for a CMS project distribute as follows:

- Near-online storage (tape library and tape drives)   20%–35%
- Online storage (storage area network, network attached storage)   20%–35%
- Servers   10%–20%
- Miscellaneous (network switches, racks, cables, etc.)   1%–2%
- Software licences   10%–20%
- Services (project management, design, installation, support, training, etc.) 10%–30%

In addition, there are investment costs for desktop PC if the units already installed do not meet the technical requirements for certain CMS applications.

Since storage cost typically ranges between 40% and 70% of the overall initial project costs, probably the most important leverage for reducing initial project costs is the mass storage system. This means that these costs offer the most powerful way of controlling the initial expenditure by reconsidering the requirements that impact its capacity and performance. This includes the storage of essence (especially video) in the compressed domain versus the storage of uncompressed material. For instance storing video essence in a 50 Mb/s format compared to full resolution ITU 601 at 270 Mb/s reduces the costs of the near-online part of the mass storage system to approximately one fifth. A further cost factor in this context are the requirements concerning access times and availability. Longer acceptable time-to-data and lower requirements concerning availability and planned downtimes allow the deployment of lower performing and hence less expensive hardware components.

Next, the envisioned scenario for ingesting content into the new system strongly influences the system cost. These costs are again mostly related to the mass storage system. There are options that allow reducing the costs of this process. For instance this can be done by defining how much of the incoming and existing material really needs to be permanently archived. It is possible to derive realistic storage volumes and thus to minimize costs. Further, it is sensible to design the initial size of the mass storage system in such a way that it can store approximately the content that can be ingested in two years. Making use of Moore's law, it is possible to double the density of the near-online storage system approximately every two years by introducing a new generation of, for example, tape drives. Hence, considering that migration is a standard CMS operation, upgrading the hardware with a new generation of storage systems and performing migration of data to this new technology helps to increase the storage density while keeping the footprint of the archive constant. The most cost-effective approach therefore is to design the initial archive capacity along the storage needs within one technology cycle. Last but not least, it is important to remember that the system is initially empty. A major design criterion hence is the maximum amount of content that *can* be ingested per year.

These considerations help to define the required initial capacity of the mass storage system and thus have an immediate impact on the initial cost by helping to avoid over-engineering of the system.

Another important issue is the deployment scenario. A phased approach with a clear view of the volumes needed for different applications will help to reduce cost. Consider the following guidelines:

> Making use of decreasing storage cost: the later in time a system is upgraded to hold more content, the better the price.
>
> Introducing in phases: consider which functionality is really needed where and when, and who really needs access at what point in time.

Adhering to these principles distributes the initial cost over a certain time, making use of decreasing hardware cost. Although a phased approach obviously increases costs for professional services, compared to the gain in decreased hardware cost, in most cases there will be a considerable net gain.

7.5.2.2 Internal Cost

The cost for re-aligning business processes and bringing the users to accept the new system can make up a substantial part of the overall project cost. This is a fact that is often underestimated.

First of all, introducing a CMS *will* change workflows and job profiles. It is vital for the organization planning to introduce such a system that it puts initial efforts into designing these new workflows and preparing the users to work in the new environment. If the end users are not involved in the project, there is a huge risk that the system is initially rejected as unusable, just because the opinion of the users has not been heard and addressed properly. This often leads to unnecessary redesign steps, postpones the time when the system goes into productive use, and increases internal friction in the organization.

The following guidelines should be considered in this context:

- Involve key users from all departments and work groups that are touched by the system. Use their expertise to derive requirements.

- Take a clear look at the current workflows and processes, and determine where improvements can save time and money.
- Define improved workflows and processes that incorporate the possibilities that the planned CMS offers. Remember that the new workflows should recognize the capabilities of the CMS. Projects that require substantial modifications to existing products tend to be quite expensive and result in proprietary solutions.
- Involve key users in the definition of the re-aligned processes.
- Use the enthusiasm of the key users to carry the flag into the groups they are working with. This minimizes friction and helps to achieve a maximum level of acceptance by the users.
- Find an innovative user group within the organization that is highly motivated to act as a first target user group for the new system. Deploy the system to this innovative group first as part of a pilot or extended pilot installation, acquire feedback, and introduce the feedback into the system design or revision of processes where applicable.
- Devise an extensive training program. Consider the need for starting with training basic computer skills, like the general interaction with graphical user interfaces using the mouse, before going into details concerning the use of the new system

The reasoning behind these recommendations is simple: get people involved, listen to their needs, make them understand that the new system is *their* new system, and try to minimize internal costs this way.

7.5.2.3 Cost of Operation

The two major cost drivers of the operation of a CMS are support and maintenance costs and migration costs. Support and maintenance cost are typically in the order of magnitude of 15% to 20% of the initial system price charged on a yearly basis. Conventionally this includes software upgrades as well as hardware maintenance. However, support and maintenance costs do not include the costs for replacement of units that reach end-of-life or units that have to be deployed in order to meet increasing demands for availability or performance. These costs are addressed as *migration costs* and are one of the major cost drivers that have to be recognized.

Migration in this specific context refers to replacing outdated technology with state-of-the-art solutions, or simply adding additional components when scaling the system. Since IT technology has much faster development cycles than standard broadcast technology, organizations will have to get accustomed to innovation cycles of two to three years. This means not only that there is a considerable improvement in performance for each cycle, but also that maintenance for older products becomes expensive after nominal end-of-life. Thus, migrating to new technology has the benefits of increasing performance while keeping costs for maintenance and support in reasonable ranges.

Migration can happen gradually and is a step-by-step process. It involves migrating to new generations of data carriers and data recorders such as digital data tapes and tape drives, which offer more capacity and faster transfer rates. This is necessary to increase the storage density of data-tape-based mass storage systems and to keep up with technological innovation. Further, disk-based storage systems also have to be upgraded or replaced to keep up with technological innovation of disk drive systems. Updating servers with models that provide higher performance to either meet increasing demand or add functionality that requires more performance is also part of the migration process.

An intelligent migration scenario tries to combine the introduction of new technology with the necessity of introducing more resources to meet increasing demand on the system. The following sections discuss a possible strategy regarding the migration of an enterprise-wide CMS. This strategy also addresses system scalability for some representative components of the overall system architecture.

The proposal is based on the assumption that there is a linear increase in demand for resources provided by the CMS (more users, more functionality). Further, it is assumed that Moore's law is valid, i.e. that advances in technology happen within regular intervals.

Based on that, a straightforward approach is taken to derive a reasonable indication of the cost involved in such a migration scenario.

7.5.2.4 Case study: Near-Online Systems

A digital data-tape-based near-online system (responsible for approximately half of the costs of the total mass storage system) mostly scales with the number of tape drives installed and the number of tapes used. Each tape has a certain storage capacity.

In order to model costs for the near-online system over time, the following assumptions are made:

- each quarter around 2500 h of material are ingested into the near-online system;
- a 50 Mb/s archive format is used;
- the first-generation tape cartridge has a net capacity of 50 GB;
- every two years a new generation of tape drives offers doubled net tape cartridge capacity;
- migration is started whenever a new generation of tape drives is available;
- a new tape drive technology is introduced every two years;
- on average, for migration purposes two tape drives are fully dedicated for read access, one for write access;
- the model begins in the year 2000, when 50 GB tapes were state-of-the-art.

Matching the availability of the new tape drives, the first migration step starts when the second-generation tape drive is available. Keeping up with the development cycles of two years, every two years a new migration step is taken. For migration two drives are dedicated to reading, one to writing data.

This approach yields an estimated total number of tapes kept in the tape library versus time as indicated in Figure 7.12. Without migration, the number of tapes to be managed increases linearly, matching the linear increase of 2500 h of material per quarter. Using the migration strategy indicated above, the maximum number of tapes to be kept concurrently in the library is reached after the first two years and steadily decreases after four years.

This means that it is sufficient to design the near-online system in a way that is able to keep the tapes needed to store all material that is ingested in the first two years, plus a certain level of redundancy as required. Thus, one of the cost drivers for the near-online system (i.e. the constant increase in the number of tapes) is effectively eliminated. An additional benefit is that the time a data tape stays in the library is clearly defined, i.e. a minimum of two years and a maximum of four years. This minimizes data losses due to aging of the carriers. Therefore, additional possible costs for ensuring data integrity are kept minimal.

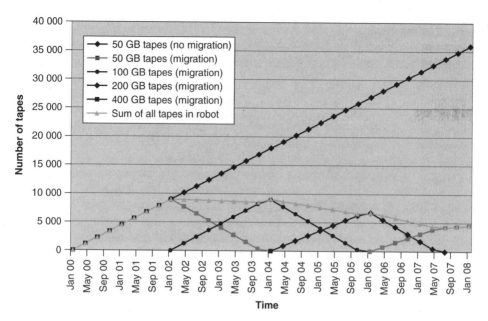

Figure 7.12   Number of tapes in tape library versus time

Since this approach implies that tapes are frequently exchanged and discarded (decommissioned), the question arises if this migration scenario introduces additional costs for buying all these tapes. In order to investigate this, we make an additional assumption:

- A tape cartridge compatible to a new-generation tape drive initially costs approximately the same as the last-generation tape did at its point of introduction.

While the tape cartridge price for an older tape generation may be less than the price of the next-generation tape cartridge at its point of introduction, in the long run it will get more expensive, since the overall market demand for these tapes cartridges will decrease. Thus, we assume that, on average, a tape cartridge basically always has the same price, independent of its generation-related storage capacity.

Figure 7.13 shows the number of tapes that have to be bought over time. If we continue to increase the archive storage capacity by using the first-generation tape drives and the matching tapes, the number of tapes to be bought increases linearly over time. Interestingly enough, using the migration approach keeps the number of tapes to be bought always at or under the same level as in the linear case. In the long run there is even an advantage for the migration solution in that the number of tapes to be purchased decreases compared to the non-migration case.

Applying the constant tape cartridge price assumption the overall cost for tapes is not higher when migrating. Moreover, there is even a net gain in the long term. Thus, when using a migration scenario as proposed, we can summarize as follows:

- The overall footprint of the near-online system remains constant, i.e. the maximum number of tape cartridges to be stored in this system is reached after approximately two years. This can be used as an initial design criterion for the size of the tape library.

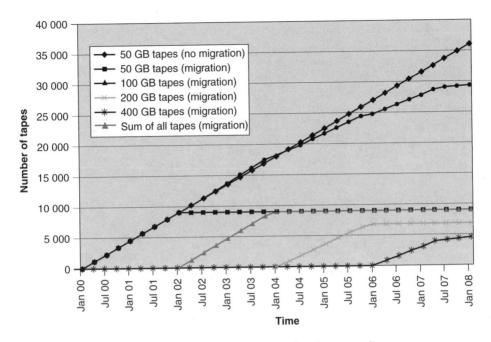

Figure 7.13   Number of tapes to be bought versus time

- The migration scenario does not impose additional costs for carriers (data tape cartridges), i.e. in the long run, the number of tape cartridges to be purchased even decreases.
- No data tape cartridge in the library is older than four years. Therefore it is even possible to keep the effective time of use for a cartridge down to two years. This is very helpful with respect to long-term data integrity.

In order to take advantage of this, a set of guidelines for designing near-online systems has been devised. First, a suitable data tape drive family has to be selected that meets the initial needs for capacity, mount time, seek time, and transfer rate, and that offers a perspective for keeping up with the migration scenario. This latter point means that it should have a credible development roadmap for increase of capacity and transfer rate. Further, a data tape library should be chosen that supports the selected tape drive family, and that is sufficiently expandable with respect to number of slots and number of installable tape drives. The size of the library should be selected so that initially two years worth of input are covered. Provisions should be made for unexpected events. This includes additional capacity that may be required for internal redundancy on unexpected increase in input. Further, a set of spare cartridges is required to be used for carrier integrity management.

The final point to be considered in the costs of the near-online system is the cost of the tape drives. In order to estimate costs involved with the drives, a number of assumptions are made. Initially the system shall have four tape drives installed. Further, as it becomes increasingly popular, the demand for system resources increases over time. To meet this demand, it is necessary to add another four tape drives per year. A tape drive shall reach its end-of-life after 3 years. Finally, the average cost of a tape drive is estimated to be always 1 'Accounting Unit'.

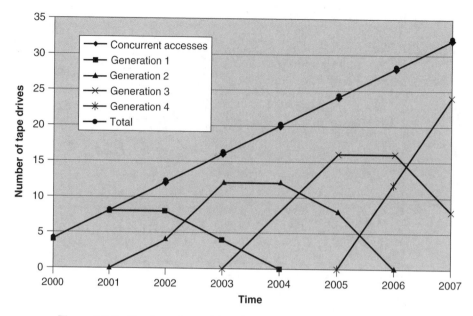

Figure 7.14   Number of tape drives installed in tape library versus time

Figure 7.14 shows the total number of concurrent accesses possible versus time. The number of concurrent accesses directly matches the total number of tape drives installed in the tape library. In addition, the figure indicates how many tape drives of which generation are installed at a certain point in time. The model as outlined above is based on the assumption that tape drives of the latest generation are always installed to meet increasing demand, and that the units which reach their end-of-life are replaced with units of the latest generation.

Following this approach, the system is growing with the increasing demand while keeping the technology always state-of-the-art. This minimizes costs for maintenance and support, since typically support for older systems tends to get increasingly expensive over time.

In order to estimate the cost involved in this permanent re-investment scenario, the average cost for tape drives as indicated in the assumptions given above are applied.

Figure 7.15 shows the result of this approach. Basically, the migration scenario, together with a constant upgrade of the system to support increasing demand, results in an almost constant annual investment. Depending on the scenario derived for a real-world project, using this approach allows estimation of the annual cost of re-investing and using the result for planning of financial resources.

7.5.2.5 Case study: Online Systems

Online systems account for approximately half of the costs of a CMS mass storage system (i.e. the storage costs are equally divided between online and near-online storage). It scales almost directly with the amount of storage capacity required at a certain point in time. In order to model costs for an online system over time, the assumption is made that the online storage system is deployed as a SAN solution.

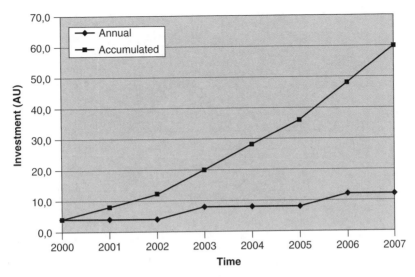

Figure 7.15   Annual and accumulated cost for tape drives installed in the tape library versus time

For simplification it is assumed that the SAN solution can be divided into certain subsystems that are called SAN units. A SAN unit may be a set of disks together with the necessary periphery. In order to simplify the model it is assumed that only complete units are added or replaced (in contrast to an upgrade, which may be handled by only exchanging or adding disks).

In addition, the following assumptions are made:

- The first-generation SAN unit shall be capable of storing 1 TByte.
- Initially, the system shall use one of these SAN units; thus, it is able to store 1 TByte of data online.
- The demand on the system increases over time as it is deployed and gets more and more popular. To meet this demand, it is necessary to add 1 TByte of storage capacity per year.
- The storage technology advances, which means that every two years there is a new generation of SAN units that provides double capacity compared to the previous generation.
- A SAN unit reaches end-of-life after three years.
- A SAN unit, on average, always costs 1 'Accounting Unit,' independent of the generation.

Figure 7.16 shows an overview of the number of SAN units installed concurrently over time, as well as an indication of how many units are deployed in which generation. It is obvious that (even when responding to increasing demand) over time the system reduces in physical size, i.e. in the long term the number of units is reduced to one. This implies that in the long run the overall system complexity will decrease. This has of course a positive influence on the operation cost (total cost of ownership).

Figure 7.17 shows the cost involved in this approach. Basically, the migration scenario (together with a constant upgrade of the system to support increasing demand) results in an almost constant annual investment. In the long term, the investment may even

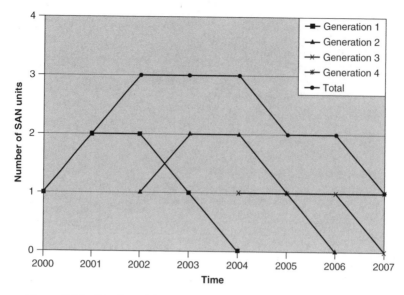

Figure 7.16   Number of SAN units deployed in the system versus time

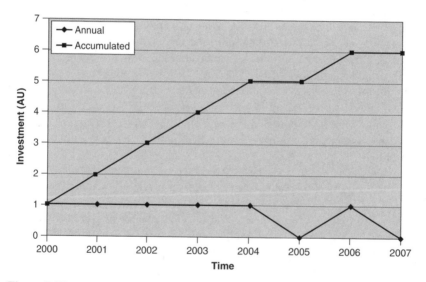

Figure 7.17   Annual and accumulated investment for the SAN system versus time

decrease. Depending on the scenario derived for a real-world project, using this approach allows estimation of the annual re-investment cost and using the result for planning of financial resources.

7.5.2.6 Case Study: Browse Server

The baseline hardware architecture as presented in this chapter makes use of distributed servers in order to deliver the resources required for a certain service to the system. As an example of such a service this section gives a cost case study on Browse Servers.

The service offered by the Browse Servers is streaming browse-quality video to a given number of users. Depending on its capabilities each server can handle a certain number of concurrent streams. If more concurrent streams are requested than one server can handle (or if there is a request for a certain level of redundancy) additional servers have to be installed.

In order to model the server cost versus time for such a service, the following assumptions are made:

- A server of the first generation can handle up to 50 concurrent streams.
- Initially, the system shall support 50 concurrent streams
- No redundancy options are considered.
- The demand on the system increases over time as it gets more and more popular. To meet this demand, it is necessary to add the capability to support 50 additional concurrent streams per year.
- The server and network technology improves over time, which means that every two years there is a new generation of servers that is capable of supporting twice the number of streams than the previous generation.
- A server reaches end-of-life after three years.
- A server, on the average, always costs 1 'Accounting Unit,' independent of the generation.

Figure 7.18 indicates the total number of servers actually installed versus time, together with the number of servers from various generations present at that point in time. The example scenario presented shows an almost linear increase in units to be deployed concurrently over time in order to meet increasing demand and replacing units which approach their end-of-life cycle. In the long run this will level out when the performance of even more advanced server generations will cover an increased number of concurrent

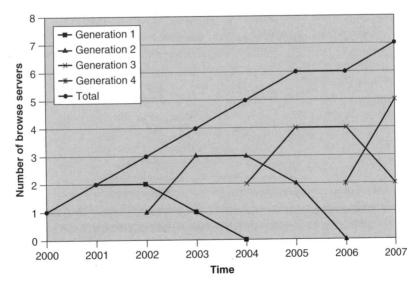

Figure 7.18   Number of browse servers deployed in the system versus time

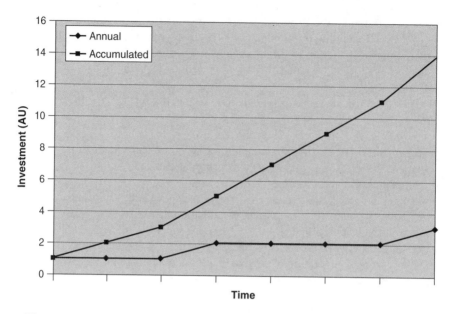

Figure 7.19   Annual and accumulated investment for browse servers versus time

streams. If a linear increase in demand is assumed (50 streams per year) and the service quality level of a server doubles every two years (50, 100, 200, 400, etc. concurrent streams), at some point in time the technological advancement will grow faster than the increases in demand. In that case, the diagram will look similar to the one shown in Figure 7.15 for the SAN units.

Figure 7.19 shows an example for the annual and accumulated investment required to follow this migration scenario for the Browse Server resulting in an almost constant annual investment. In the long term the investment may even decrease. Depending on the scenario derived for a real-world project, using this approach allows estimation of the annual re-investment cost and using the result for planning of financial resources.

The model presented for the Browse Server in principle is valid for all servers deployed as server groups that host a certain distributed service. It is thus possible to derive an estimated annual investment necessary to keep the complete system state-of-the-art at an increased service level to match increasing demand. This contributes to the definition of the annual budget required to operate the complete system.

8

System and Data Integration in CMS

A CMS is regarded as a fundamental platform or hub for content. Essence and metadata are ingested, handled, accessed, exchanged, and played out. Hence it is crucial to provide a number of basic interfaces and integration options for the applications and systems that access content. Further, a CMS very often has to incorporate a number of existing system components and manage legacy content. Thus, being able to integrate or interface to other systems, system components, and existing content is crucial for an open, versatile CMS. As discussed in Chapter 6, provisions for this have been made in the system architecture within the Device Management and also as part of the Data Management. However, on an inter-system level this might have to be enhanced by other forms of integration.

All content-related processes and applications within a content-rich organization have to be supported. There is a plethora of systems, applications, and application components that interact with a CMS. Figure 8.1 shows an example of different CMS and third-party components within a broadcast architecture.

The bottom layer contains the management components of the different systems. The generic CMS modules are marked in light grey whereas the third-party systems and general infrastructure components are dark grey. The third-party systems in this example range from control systems (such as studio automation systems and on-air scheduling) over production support and planning tools (e.g. newsroom systems, on-air promotion system, airtime sales, program planning) to information systems (rights management system and cataloging system). They are interfaced via a communication middleware layer and application control sets. This example shows the wide variety of systems present in a standard broadcast environment that have to be integrated and interfaced to. Some of these systems are existing databases and information systems. The CMS in this context has to provide the integration platform by providing hooks and interfaces.

The integration with (and of) other systems that are part of workflows and the content life cycle is a fundamental feature of a CMS. Relevant systems in this context are,

Professional Content Management Systems: Handling Digital Media Assets   A. Mauthe, P. Thomas
© 2004 John Wiley & Sons, Ltd   ISBN: 0-470-85542-8

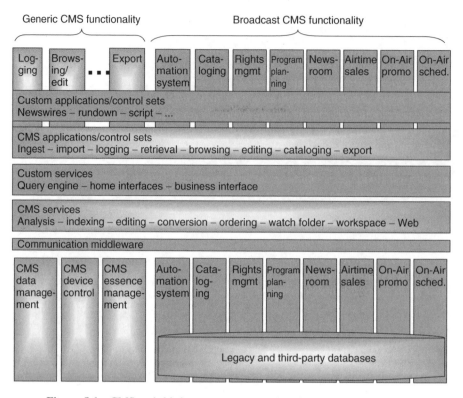

Figure 8.1   CMS and third-party components in a broadcast infrastructure

for instance, newsroom systems, automation or media management systems, production systems, nonlinear editing (NLE) systems, and also rights management systems, enterprise resource planning (ERP) systems, etc. The presentation of content on the Web and e-commerce systems is also becoming more and more relevant. A CMS may serve as a central repository providing interfaces for all these systems or may provide interfaces and workflow support that enables cross-system interaction on a peer-to-peer level. Depending on the kind of integration, the interaction can range from a simple exchange of metadata and essence copies to direct access of the relevant information out of (or within) tightly integrated third-party systems.

The legacy content and the information contained in legacy systems are particularly relevant for well-established media-rich organizations (such as for instance public archives and broadcasters). The existing content is an important part of their operating capital. Thus, it is vital that it can be incorporated or managed by a CMS. The migration of essence to new formats and the incorporation of all existing information are essential prerequisites for the successful introduction of CMS in such organizations.

System integration is a parallel task to the actual system design. Thus, the integration process should be carried out in conjunction with the actual system design and implementation. A later integration of system components is possible but preparations should be made in advance and already be considered in the original architecture and (long-term) implementation plane.

8.1 INTEGRATION PRINCIPLES

Integration is basically the cooperation of different (possibly independent) systems or system components to achieve a more comprehensive functionality. Further, it is also used to refer to the incorporation of information units in a system context. Through integration the overall functionality and capabilities of a system can be enhanced. *System integration* can be defined as (Institute for Telecommunications Sciences, 1996):

> The progressive linking and testing of system components to merge their functional and technical characteristics into a comprehensive, interoperable system. Integration of data systems allows data existing on disparate systems to be shared or accessed across functional or system boundaries.

This definition stresses the process of integrating the system components as well as focusing on the outcome. This is an important aspect very often neglected by just specifying interfaces without defining the different integration steps. Thus, two aspects are important in this context, namely the overall enhanced functionality, characteristics or features that should be achieved by integrating different system components, *and* the process of integration. In contrast to the above definition, however, it is important to stress that individual system components in the case of the integration of third-party systems in CMS still can maintain their autonomy. They will usually work in a stand-alone manner but provide a more comprehensive feature set for seamless operation in the case of integration. The nature of the integration is also determined by the physical location of the systems and the (communication) link between them. In case communication failures occur it is important that each system can still function independently.

Within CMS, data integration is also important since it is often the case that data has to be shared across system boundaries. This is particularly the case when part of the information is relevant in another system context. For instance IPR-related information is kept in a rights management system but also has to be accessed from the CMS. In the case of data integration it has to be considered how important timing aspects and up-to-date information are. If information is changing frequently and it is essential that the latest information is always available, direct access to the data is required. On the other hand, if the nature of the data is more static and temporal inconsistency can be accepted the data might also be exchanged (i.e. copied) between the different systems and kept redundantly in each system.

8.1.1 TYPES OF INTEGRATION

The kind and depth of the integration between systems varies depending on the functions, features, and shared data set of the different integrated systems. In this context the characteristics and appearance of the overall system are also important. Further, the role of the specific component within the overall system influences the kind of integration. For instance a hardware server or storage system is much more tightly coupled with a CMS than an autonomous software tool such as a newsroom system. In order to choose the right type of integration for a specific problem it is also important to consider aspects such as platform dependencies, and costs and limitations of certain solutions. Thus, there is a wide range of possible approaches towards integrating different system components. However, three principal types of integration can be identified, namely integration via data exchange, integration via application programming interfaces, and integration of application components. Sometimes the different forms of integration are mixed within one integrated system. For instance a message can be used to notify another system component that data is now available over a certain interface.

8.1.1.1 Integration via Data Exchange

Integration using messages and data exchange is a loose form of system corporation. A special set of exchange protocols and data encoding standards such as KLV, MOS, and SOAP (see Section 4.5) are used in this context. At a lower level protocols such as FTP, NFS, NDCP, VACP, etc. can be used. Messages encoded in the respective format are passed between the different systems. Further, plain file transfer also belongs to this group. Either a request–response scheme can be used or information can be pushed to the other side.

Essence and metadata may be exchanged in this manner. Communication at a system- rather than device-level frequently uses data exchange for metadata. The exchange of essence between systems depends very much on the kind and format of the essence. Continuous media can for instance be streamed via SDI/SDTI or data networks using appropriate streaming servers and protocols. File transmissions with standard communication protocols can also be used to exchange both continuous and discrete media. In this case standard file formats (as described in Chapter 5) play an important role.

8.1.1.2 Integration via Application Programming Interfaces

The integration between systems is generally higher when using application programming interfaces (API). Via an API a program offers a set of functions that can be used by other programs or subroutines. They include database access client/server, and peer-to-peer interaction, transaction processing, etc. In general two main types of API interaction are distinguished, namely Remote Procedure Call (RPC) and Standard Query Language (SQL). RPC and the use of a SQL provide a very tight integration. In the case of RPC, programs communicate on procedures or tasks that act on shared buffers. For data sharing between applications, access to a common database using SQL can be employed. This is a non-procedural data access. Other ways to access databases directly via filters are also possible. For instance XML-based interfaces can be used that keep the actual database structure transparent but still provide direct and detailed access to the stored data.

In order to be less platform dependent and also operate in a distributed environment standards such as the Common Object Request Broker Architecture (CORBA) or Web- Services can be used.

8.1.1.3 Integration of Application Components

Component-based application development technologies (such as OCX, ActiveX, and JavaBeans) allow integrating various application components from different systems within one application framework. To the user this appears as one application without any visible boundaries. Parameters and information can be passed directly between the different components by using messages. Within the application framework each system is represented by its own application module.

This kind of integration is very attractive when a seamless application interface should be presented to the user while leaving it to the different systems how to present the information. However, there are limitations when information from different systems has to be pre-processed. For instance in the case of intellectual property rights linked to a specific content object one could use ActiveX to present the rights situation, though in this case a query asking for an object that has certain content properties but no rights restrictions would not be possible.

8.1.2 INTEGRATION PROCESS

As already indicated by the above definition, system integration is a step-by-step process in which the different system components are linked and tested to finally form a comprehensive interoperable system. Usually the systems integrated in the context of content production and management are not designed from scratch but are modules that have been existing for some time. For a CMS system integration is an essential feature and therefore hooks and interfaces for possible integration should exist in the original design. However, the overall system architecture and infrastructure as well as the capabilities of the system components to be integrated have to be considered in the integration process.

The integration process within a CMS can be divided into three main steps:

1. requirements study and component analysis
2. system and interface design
3. implementation, testing, and installation.

For each of these steps deliverables and milestones should be defined to monitor the progress. Depending on the project different approaches might be taken. The following discussion only provides an outline of the tasks and possible deliverables of each of the integration steps.

8.1.2.1 Requirements Study and Component Analysis

In the requirements study the user requirements are derived. In this process existing work steps are analyzed. This includes the identification of the workflow but also the documentation of systems, tools, and components that are used to support them. The whole analysis is documented in a description of the relevant workflow and involved systems.

A parallel task is the evaluation of all components and (sub)systems that can potentially become part of the overall system. This includes an analysis of their functionality and features as well as their interfaces and integration capabilities. A feature and interface list should document the capabilities of each alternative. At this stage a ranking of the different alternative components may take place to identify those that are most suitable for a specific task.

8.1.2.2 System and Interface Design

In the system design phase the overall system architecture is determined. Inputs are user and functional requirements, technical and environmental parameters and restrictions, financial and budget limitations, etc. The list of alternative system components that was drawn up in the previous step also has to be considered at this stage. Those components providing the best match to the user and system requirements and which also allow easy integration should be selected for the overall architecture. The result of this step is the overall system architecture specifying the different system components and information flow between them.

Subsequently the actual integration task starts with the specification of the component functionality and interfaces. Which types of integration should be used between the different components is determined at this stage. The capabilities of the individual components as well as the overall system have to be taken into account in this context. In order to keep the system manageable it is important that there are not too many different protocols, encoding schemes, and API standards. For each integration type there should be one, maximum two, principal standards, which should be used for the implementation of the component interfaces. This is also a selection criterion for the different component alternatives.

The capabilities of each component have to be clearly specified in a Functional Design Specification (FDS). This is the reference document with respect to the functionality of a specific system component in an overall system context. It is accompanied by the Interface Design Specification (IDS), which specifies the types of interfaces offered by a component.[1] These documents together with the information flow charts of the overall system architecture specify in detail the kind of interfacing and types of information to be exchanged. This has to be supplemented by sequence diagrams that state exactly the order of events and interaction that can occur between the different system components.

8.1.2.3 Implementation, Testing, and Installation

Each component is implemented (or adapted) separately in accordance with the overall system architecture, FDS, IDS, and sequence diagrams. Initial test are carried out separately to verify the error-free operation of each component. After successful completion of this step directly interacting components are tested against each other. In the case of failures or errors their cause has to be established and they have to be fixed in the respective component. This can be an iterative process but in a well-engineered system not too many steps will be necessary until the different components are stable.

Finally all the different components are installed in the overall system context. Only then can the overall system be tested and evaluated. Ideally there should be a phase where the entire system can be tested thoroughly offline (i.e. without being linked to any productive part of the infrastructure). Especially in established operations this might not be possible. In this case a migration strategy is necessary where different components are introduced and linked up in a stepwise manner. In each migration step new functionality is added and the interaction between a clearly identified set of components is changed. It is important that in case of failure the old system state can be reconstructed in order not to disrupt the operation unnecessarily.

8.2 CMS AND LEGACY SYSTEMS

Content has been managed for centuries and even multimedia content has been around for decades. There are a number of existing systems in content-rich organizations that are

---

[1] Compare Section 6.3, where FDS and IDS are first introduced.

already dealing with certain aspects of managing content. In particular there are existing databases, picture and media libraries, production tools, and automation systems. These systems support the existing workflow but also keep a significant amount of information related to content. In order to utilize this information and the knowledge enshrined in the existing infrastructure and the staff currently handling content, a CMS has to integrate existing system components. There are various factors that have to be taken into account when introducing a CMS in a well-established operation.

One particular problem that has to be tackled in this context is the migration of legacy essence. The preservation of material is one of the foremost tasks of the archive. The introduction of a CMS that manages digital content in a fully automated environment also changes the preservation task. For instance automatic processes can be established that take over the preservation of essence and the migration to new formats. However, format-related issues still remain and decisions about how and when to migrate essence are not only technical.

A second important factor is the wealth of information stored in existing databases and information systems. The technology they are based on as well as the data modeling might not be ideal for a modern CMS. However, to replace them with new technology might sometimes not be feasible. Thus, a way has to be found to integrate these systems or data into a CMS.

8.2.1 INTRODUCING CMS IN EXISTING INFRASTRUCTURES

At present media and television production is moving away from analog technology towards digital processes that involve computer- and disk-based operations. It is already foreseeable that future production processes and archives will no longer be based on conventional videotape recording formats. Server-based TV production platforms will have to be linked with data storage, in particular mass data storage. A policy for migrating existing systems into future CMS replacing traditional archives will have to take into account that a number of system changes might occur concurrently. For instance there is the change from tape-based to server-based production, the migration of traditional host-based databases to advanced information systems accessible to various user groups, and the integration from production, CMS, studio automation, and newsroom systems, inter alia.

The following goals should be achieved by introducing a CMS within an existing infrastructure:

- introduction of a file-based transfer of the program material between the components involved, e.g. production platform and archive;
- automated storage and retrieval of program material as part of the overall system;
- generation and handling of metadata from acquisition, through the production and archiving process.

If legacy material has to be transferred to new formats the (possibly pre-degraded) initial technical quality of the program material stored on the archived media should not be further impaired through the migration process. The migration process should be supported through automatic or semi-automatic processes.

Any information kept in existing databases and information systems has to be fully accessible through the CMS. This has to be reflected in the data model(s) developed during the design phase. It also has to be ensured that all departments and users that have to access data will be able to do so in the introduction phase and when the CMS is fully established. How to improve the information gathering and flow within the new system will have to be considered in the early stages of the CMS design. Ideally, any kind of metadata created during the lifetime of a content object should be stored in the system. Thus, there are three factors that have to be considered when deciding about the integration or replacement of legacy information systems:

- existing databases and information systems and how to integrate them in a CMS;
- current usage and access rights and how to handle them in the future;
- improved metadata and information handling by the CMS.

Another important issue is existing workflows and work processes. They have to be analyzed and should be reflected in the CMS design. In this context it is important to recognize that there might be a conflict between the possibilities new technologies offer and the well-established processes. It is therefore crucial to involve representatives of all affected user groups during the system design process.

8.2.2 MIGRATION OF LEGACY ESSENCE

Essence is the physical copy of a content object and represents the actual information or message. In media-rich organizations such as broadcasters it is closely associated with the storage media (i.e. the video- or audiotape). Almost all processes center on this physical representation of content. Up to now the adoption of a new tape recording format in production was mainly justified on technical and economic grounds. The introduction of a new tape format in acquisition and production also has implications for the archiving format. It usually implies that this new format also has to be supported in the archive since a separate format specifically tailored for the archive cannot be afforded. The task of the archive in this context is the long-term storage and preservation according to certain guidelines. It has to ensure that the archived videotapes remain usable for an almost indefinite time. Actions have to be taken as soon as the physical carriers start to deteriorate.

With the introduction of CMS and server-based production a general decoupling of content and storage media is taking place. In an existing environment this decoupling process has to be supported by the CMS. It is represented by the migration of essence from traditional tape-based storage media and formats, to digital formats independent of the storage media. The migration process should coincide with a change of the production and archive infrastructure:

- The initial step is the introduction of an automated tape library system that can contain conventional videotapes. The essence is streamed via conventional video interfaces. It can be recorded on servers that are also able to connect to other devices or transfer data on communications networks.
- Subsequently the implementation of a mass data storage system using data tapes should take place. This system should be integrated into an emerging server-based production

environment. An automatic transfer from the content in the tape library to the new storage media is then possible. The exchange of program material takes place using file transfer via network interconnections.

With this kind of infrastructure the existing essence can be gradually migrated onto the new platform. However, during the migration period different cases have to be managed simultaneously. The essence that is still archived on tape-based formats must be provided for new and tapeless production areas. Thus, there needs to be a strategy about which essence is migrated when and how to deal with requests for specific material that has not yet been migrated. Further, the essence already archived on new storage devices must be provided for productions that are still using tapes during the transmission period. This has to be part of the migration strategy developed for a CMS.

The decision about the kind of digital formats that will be used in the CMS depends on the kind of legacy essence, planned production workflow, and preference within an individual organization. For the CMS it is only important that the respective hardware and software systems can be integrated into the CMS infrastructure and allow automatic control.

8.2.3 INTEGRATION OF EXISTING DATABASES AND INFORMATION

For the integration of information already existing within a media-rich organization two basic cases can be distinguished, namely the migration of data into a new database (or information system), and the integration of the existing database into the CMS architecture and infrastructure. Which integration approach is used depends on the legacy systems (i.e. their size, relevance, and technology), the planned extent of the CMS, and the implementation strategy. In both cases it has to be carefully evaluated what the advantages and disadvantages are and which is the right approach.

8.2.3.1 Migration of Legacy Databases and Information Systems

When existing databases and information systems should be replaced by a database that is part of the CMS all entries of the legacy database have to be transferred to the new database. This is not a straightforward process since the data model as well as the data elements and attributes might be different. Although the existing data set will have to be considered in the design of the CMS data model, it will not be a one-to-one mapping from the original data set. The view on content can be considerably different. Whereas new CMS are usually content centric older systems are very often carrier centric. In the former case the content object is the principal entity whereas in the latter it is a carrier that can hold a certain number of content objects. In this case a mapping between elements and attributes of the old and new data model is of secondary consideration. More important in this context is to find a way to represent the information from the old data model in the new one. This might involve an aggregation of data and splitting of data elements in order to use certain attributes within the new model or provide information for new elements.

It has to be carefully analyzed if the new data model covers all existing data and queries. Further, it has to be ensured that all applications and users operating on the database can also operate on top of the new CMS (or can be served by it).

In general migration is done with small databases that only contain aspects of specific topics that are represented within the overall CMS already. For instance a set of departmental databases might be subsumed into the larger CMS database. A general migration project might have to be necessary if the principal database of an organization is based on old and outdated technology. In this case the migration project is a separate task within the CMS design and implementation. The general design of a system should allow the integration of this database as part of the Data Manager/Data Broker infrastructure as introduced in Section 6.1.3. This allows decoupling the migration-specific aspects from the CMS implementation-specific tasks.

8.2.3.2 Integration of Legacy Databases and Information Systems

Existing and legacy databases and information systems can be integrated in the CMS exploiting the Data Manager/Data Broker concept. This allows integrating multiple systems in parallel and presenting the information to the user in a uniform manner. The basic idea is to access each system via a separate Data Manager that provides a uniform interface to the other system components. The different basic query concepts a Data Manager supports are:

- fulltext query
- query for labels
- native query.

The fulltext query searches for given expressions or strings within all elements and attributes of the underlying database or information system. In the case of query for labels, a set of labels representing some generic concepts (such as place, person or date) are defined. These labels are valid for the entire CMS. The Data Manager of a specific legacy system carries out the mapping onto specific elements and attributes. Native queries represent direct access to the legacy database. Queries for segments are only relevant for systems that also include the concept of segments. In most legacy systems this is not the case.

Figure 8.2 shows how the Data Broker/Data Manager handles requests and how different legacy databases are integrated. The Data Broker accepts an incoming request and distributes it to the relevant Data Managers. A federated search can only be performed for fulltext queries and queries for labels. Federated search means that a query is passed to multiple systems. The replies are collected by a central instance that removes duplicates and organizes them in a uniform hit list. In the case of a federated search a Data Broker passes the request to the Data Managers and subsequently collects the replies, processes them, and presents them in a unified hit list as a response.

Each Data Manager consists of a generic part and a legacy-system-specific part. The task of the generic part is to parse incoming requests and generate the response according to the specified interface. The system-specific part translates the requests into legacy-system-specific queries and handles the return results. This includes for instance the translation of queries for labels into native database queries for all the elements and attributes within the database that are represented by a specific label. The return values are processed by the generic Data Manager part handed back to the Data Broker in the standard format.

Native queries are handled differently. A native query directly targets a specific legacy system and is therefore passed straight to the system-specific part of the Data Manager. The Data Broker only selects the respective Data Manager for such as query. Only the

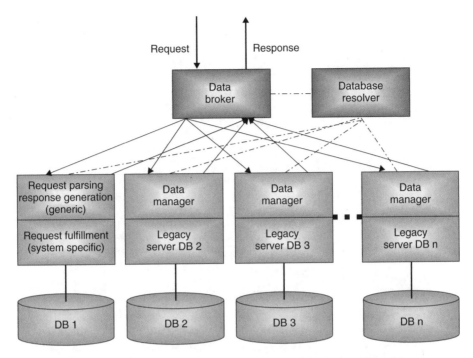

Figure 8.2   Integration of legacy systems using Data Broker and Data Managers

system-specific part of the Data Manger has to handle such a request. In this case there is no federated search since other databases and information systems cannot be queried using the same request.

8.3 INTEGRATION OF THIRD-PARTY SYSTEMS

A CMS can only realize its full potential as part of a fully integrated infrastructure. In such a system it provides a platform together with other cross-departmental platforms such as an enterprise resource planning (ERP) system or a rights management system. On top of these platforms there are the different output channels ranging from traditional radio and TV broadcast to Internet-based output channels (such as Webcasting, Webpublishing, and e-commerce). It is also conceivable that in future this will be enhanced by electronic services based on WAP, SMS or UMTS technologies. Figure 8.3 shows this vision, including the different components relevant in this context. Apart from the platform components there are channel-specific tools and systems that also have to be integrated. Some of them have a similar function within each output channel (such as program planning) but there have to be separate instances of these systems since the application context is considerably different.

In addition to cross-platform integration with other enterprise-wide systems there is also the integration with components that fulfill a certain task in a specific application context. In broadcast systems these are for instance studio automation systems, newsroom systems or NLE.

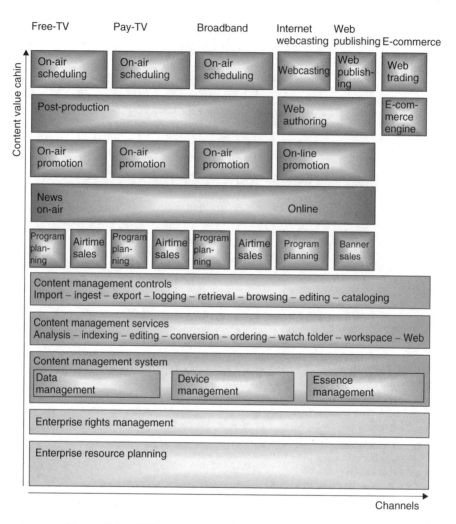

Figure 8.3   CMS in a multi-channel content-rich organization

8.3.1 INTEGRATION WITH ENTERPRISE-WIDE THIRD-PARTY SYSTEMS

The two major systems that are important on an enterprise-wide level within a media-rich organization are ERP systems and digital rights management (DRM) systems. The CMS and these systems are usually at the same level, i.e. there is no lead system but the integration is among equals. Each of these systems continues to provide the same service as it used to but there is additional information accessible from these systems.

8.3.1.1 Integration with Enterprise Resource Planning Systems

ERP systems are mainly based on standard commercial software that is adapted according to the needs of a specific organization. Within such a system all business-relevant data is managed. Regarding content this includes accounting data and all commercial transactions related to content, content production, and exploitation.

The systems are mainly integrated at the level of data exchange, i.e. information about objects or processes is exchanged between the systems. This includes budget information, invoices and purchasing information, and statistical information about content usage. The latter is required to calculate expenditures and financial obligations towards rights holders.

Initially the integration will not be very strong since in the current work process this kind of data is kept separate. However, with tighter financial budgets it is valuable to associate financial information directly with projects and make it accessible within the CMS for the responsible production staff. Depending on the organization this can be either through asynchronous updates of the respective information duplicate held in the CMS or by sending a direct query to the ERP system when this data is required. In the case where the information has to be up-to-date, inconsistencies between systems have to be avoided. Thus, the latter option would be preferred.

Integration via API or application components is at present not relevant. In the former case the two systems would operate on a common data set, which is not possible because of security reasons. The latter case is prevented due to the fact that most commercially available ERP systems do not provide the necessary application components that could be easily integrated within a CMS application framework.

8.3.1.2 Integration with Rights Management Systems

While a CMS can focus on the management of essence together with formal and descriptive metadata a true asset management system needs to manage rights as well. However, the management of rights is a complex issue that requires considerable domain knowledge. Most content-rich organizations have established a rights management system within their legal department that is based on proprietary software. Depending on the organization this information may be strictly protected from any user outside the department. In this case integration is neither desirable nor required. However, more and more of these organizations are recognizing that it may be beneficial to provide some information about the rights situation of a content object to the users of a CMS. In this case it is necessary to select the kind of data that should be made available and the way it should be presented. In general, if this kind of data is provided it is only to be used as an indication. Further clarification by the rights department is necessary to ensure that the rights situation is fully cleared. In future full access by other departments to the rights database might also be relevant. A very tight integration is necessary in this case.

Taking this into account it is not necessarily required that a CMS provides a native solution to the management of rights. It may be more sensible to integrate solutions that are tailored to handle specific rights situations that may occur in certain domains of media production and distribution. The CMS needs to provide a possibility to integrate rights management solutions into the overall framework.

Standardized DRM applications are slowly emerging. They are built around modern database technology (often using relational databases) and provide easy to integrate application components. Depending on the kind of integration envisaged three approaches could be taken, namely message exchange, integration via a Data Manager (i.e. API), and the integration of application components. In the case of message exchange certain general information about the rights situation of a content object is included in the presentation within a CMS application view. Therefore the system needs to query the DRM to find out if the usage of a specific content object is restricted. Important in this context is the representation of the content object within the different systems. If the entities are different the CMS has to know the structure of the rights database to find the appropriate data. Ideally this kind of query should be synchronous to provide the most up-to-date information about the rights situation. However, in order to prevent system overload due to frequent queries the relevant rights data might also be kept within the CMS where it is updated periodically (for instance once a day as a batch process). The users have to be made aware of the fact that there might be temporal inconsistencies. In general only basic information such as 'restricted,' 'not-restricted' should be presented within the CMS in this context. This information might have to be visualized in the CMS applications. For instance if a shot in a hit list has restrictions this shot might have to be visually marked.

A direct integration via a Data Manager provides the tightest coupling of the different systems. In this case rights-related information is just another set of metadata that has to be displayed in an appropriate manner by the CMS applications. The generic part of the integration component provides a standard interface to the CMS modules equivalent to other Data Manager interfaces. The rights-management-specific part translates the CMS system query into the rights management system query, recognizing different representations of content if necessary. The query results are translated into a notation recognized by the CMS.

Integration via application components provides the possibility to view the rights-related metadata in a DRM-specific view within the application context of the CMS. This view is provided through the integration of the respective DRM application components into the CMS presentation of content object information. The DRM view in this example is accessible through a tab folder in the detailed view for a content object. Whenever the user selects this view, the application view of the DRM is evoked, using the identifier of the respective content object as call parameter. The presentation of the data and the interaction with this application view is then entirely in the control of the DRM.

For this kind of integration it is necessary that both systems use the same representation of content objects. For instance if a hierarchy between the objects exists (such as between programs and program items), this hierarchy has to be identical in both systems in order to provide a consistent view.

8.3.2 INTEGRATION OF FUNCTION-SPECIFIC THIRD-PARTY SYSTEMS

Function-specific third-party systems are systems that perform a certain function in the context of content management, production, and/or transmission. They usually focus on a specific functional aspect regarding specific use cases or workflows. Important within a broadcast environment are for instance automation (or media management) systems, newsroom systems, and nonlinear editors (NLE).

8.3.2.1 CMS and Automation System/Media Management System

Automation systems are an integral part of the broadcast infrastructure. They control and coordinate the material transfer between the different studio devices (e.g. servers, LMS tape libraries, NLE, playout systems, etc.), control the central matrix switch (or cross-bar), and provide applications to record incoming material and to schedule playouts. Their functionality is orthogonal to, but to a certain extent also overlapping, that of a CMS. However, the main operational area of an automation system is in post-production and transmission within a broadcaster whereas a CMS covers the entire organization and workflow, including production planning, production, post-production, archive, etc. Thus, these systems have to be integrated to provide a seamless service to the users in areas where they are both operating.

The degree of integration between the two systems depends on the functionality provided by both systems. Basic automation systems control the recording of incoming material onto servers and tape recorders. They move the essence between the different server systems and are responsible for a scheduled playout of the programs. All the main broadcast devices are under its control to ensure timely playouts using a proper resource management scheme. A major part of the automation system is the database that holds metadata about the content objects under its control, and data about the scheduled material movements and program playouts. The application an automation system provides is basically a list of scheduled processes for recordings, material transfers, and playouts. Within this list the processes are ordered, resources are reserved, priorities are given, etc.

The integration between the CMS and the automation system starts at the time of recording. As soon as a new content object arrives the CMS has to be notified. Ideally the automation system also generates a browse copy of the incoming material that is then placed under the control of the CMS. The exchanged metadata can range from a simple identifier the essence carries in the automation system to a full set of all available descriptive and technical metadata elements. Another stage of interaction between the two systems takes place when program material is prepared by editors or journalists using rough-cuts. The CMS has to hand over the generated EDL and coordinate with the automation system the high-resolution material movement to an NLE. Since program making is often an iterative process, versions of the produced programs might have to be managed by the CMS. In this case the system automation has to send updates of the work in progress to the CMS. The final interaction between the two systems takes place after a program has been played out. The program itself but also raw material and rushes that should be preserved are handed over to the CMS. Further, the metadata collected during the production process should also be handed to the CMS for further use by catalogers and archivists.

Figure 8.4 shows the example of the integration between a CMS and a media management system (MMS) in a newsroom environment. The recording onto the high-res and browse servers is controlled by the MMS. All information about the object being recorded is placed into the MMS database. The CMS is notified about the incoming material via MOS message. It can access the MMS database to retrieve all the metadata that is associated with the content object. A copy of the browse material is placed into the CMS. From this stage onward all the usual workflow processes as described in Section 2.2 are carried out. Whereas the high-res material remains entirely under the control of the MMS, the browse videos are also available within the CMS. In the MMS this copy might be deleted since only current and in-progress material is kept there. If it is removed the CMS imports all the information and metadata from the MMS database and puts it into its own data storage. This happens via the MMS database API. If material is identified within the CMS that is required for production it triggers its movement from deep storage to the production and broadcast area with a MOS message sent to the MMS. It might also put a copy of the metadata and browse material back under the control of the MMS. The actual production then takes place within the MMS domain.

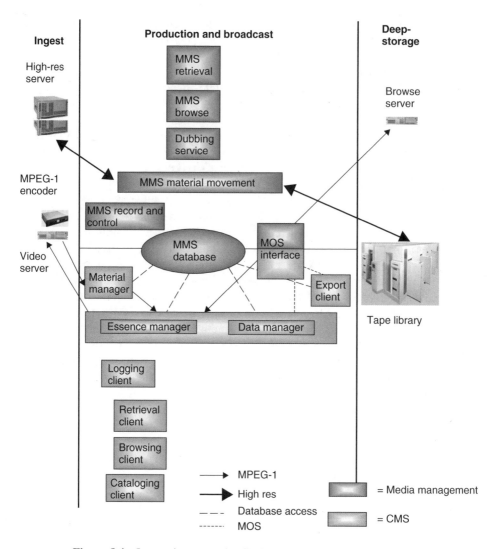

Figure 8.4   Integration example: CMS and media management system

In this example the integration uses message exchange and API integration. Depending on the systems and workflows other possibilities also exist. They have to be evaluated in the context of a specific project to find the most optimal solution. However, in general the interaction between these systems is related to the exchange of material, metadata, and the coordination of production processes and material movements.

8.3.2.2 CMS and Nonlinear Editing Systems

NLE are systems for the production of video and audio content. These are specialized hardware and software devices mostly based on computer technology. Digitized audiovisual raw footage, rushes, and original archive material are used to produce a new program item according to a program idea or draft EDL (rough-cut) from an editor or journalist. The craft edit is produced by trained picture or sound editors who have the necessary technical as well as artistic skills. The editing process also includes the creation of visual and sound effects such as edit effects (e.g. wipes, fades, cross-fades, etc.).

In an integrated infrastructure interaction between the CMS and NLE is based on material selections and rough-cuts. The users produce these selections in the CMS application environment. In order to ensure seamless integration the formats of the rough-cuts and material selections have to be compatible to the EDL format used by a NLE. This also implies that the reference timecodes within the rough-cut have to correspond to the timecodes in the high-res material used for production. The NLE is used to automatically load the material from servers or record it from tape to make it available to the NLE. System-wide material identifiers for the high-res material are required to unambiguously identify the essence copy referred to.

If an automation system or MMS is part of the integrated infrastructure it is responsible for the material transfer between the server systems, video recorders, and the NLE. In this case the EDL is used as coordination and information instrument between the different systems. If this is not the case the CMS in collaboration with the NLE has to ensure that the right material is made available according to the rough-cut or material selection. This interaction depends on the capabilities of the NLE in question. For older systems it might be necessary to produce a continuous set of raw material that they can import as if it would be stored on videotape. The CMS in this context has to prepare all the material located on a server and simulate the VCR environment. With upcoming server-based editing the NLE itself will be in charge of the material transfer. The CMS in this context only has to ensure that the right copy of the essence is available on the server in time for the edit.

The integration types employed in this context depend on the capabilities and interfaces provided by the NLE. With older systems it is based on message exchange. This can sometimes also take the form of file exchange if the system has no communication interface. In this case a file containing the necessary information in the form of an EDL is placed in a directory that can be accessed by the craft editor. In the same way, essence may be transferred to an exchange area, such as a shared file system or a mutually accessible video server. With more advanced systems API integration is possible. This kind of integration has to be part of a project since the interaction types have to be carefully designed and validated to avoid unwanted side effects. A better option might be to use the integration of application components in conjunction with message exchange. CMS application components (e.g. project folders or search and hit list views) can be integrated in the NLE application. The information between the systems is exchanged using standard messages (e.g. MOS message exchange). The user can initiate this using drag-and-drop from the CMS application view onto the appropriate NLE application component.

A particular problem in the context of CMS/NLE integration is work in progress and versioning. The editing work is in general carried out outside the control of the CMS. At a certain stage in the workflow it is re-imported. This version is then considered the latest valid representation of a project. The re-import process also has to include the generation of a browse copy and possibly an automatic essence analysis. It is desirable that the progress of the editing process can be observed from the CMS applications. However, the problem in this context is the availability of a browse copy within the CMS. Since the craft editing uses high-res material, which is not accessible within the CMS standard network infrastructure, a browse copy has to be available. At present this copy has to be actively generated.

8.3.2.3 Integration with Newsroom Systems

A newsroom system records and holds information on news items coming from news agencies. This includes textual excerpts on news stories and increasingly also audiovisual representations of the video, image, and audio material provided by the news agencies. Further, newsroom systems also provide an environment for the creation of news items and the planning of news programs.

The integration of newsroom systems and a CMS is mainly linked to the management of audiovisual material and the support of the program production process. When news clips from agencies arrive they are recorded and a browse version of the material is produced. This, together with the manual annotation and automatically extracted information, has to be made available to the newsroom system users. Ideally this material is searchable and can be associated with news stories and ongoing projects. Thus, links between the two systems have to exist in which the audiovisual material is associated with certain newsroom projects or items. Further, the content managed within the CMS should be searchable from the newsroom system and any information (textual, visual, audio or audiovisual information) found there should be presented to the user in the context of the newsroom application. The production process using browse editors and integration with other third-party tools (such as NLE) should also be supported. Therefore the CMS tools providing this kind of functionality have to be accessible from the newsroom system as well.

Another integration aspect is the archiving of raw material and program items. After a news item has been broadcast all relevant information (including final program, rushes, scripts, agency texts, etc.) should be handed over to the CMS for preservation. From this point onwards it is not under the control of the newsroom system anymore.

Depending on the capabilities of the newsroom system the degree of integration between the newsroom system and CMS can be very high. CMS application components can be seamlessly integrated within the newsroom application. The interaction can be coordinated using message exchange. Associated Press has specified the MOS protocol for this kind of message exchange (see Section 4.5.3). In most modern newsroom systems application views from other systems can be integrated. To the users these are part of the newsroom environment. Interaction is often supported using drag-and-drop.

However, in most cases API integration will also be required. For instance if a federated search on items stored within the newsroom system and the CMS should be performed, the newsroom system has to be able to query the CMS database. Another example is the association of agency feeds and content objects. Shared database tables or lists storing this information are often used in this context. This shows that with such a high degree of integration all three types of integration might be used alongside each other.

8.4 CMS AND WEB INTEGRATION

With Web outlets becoming more and more popular, integration with these systems is also becoming more important. Further, a well-organized CMS should also facilitate electronic content sales. Both the CMS and the e-commerce front-end have to be integrated to provide an e-commerce service that optimally exploits the characteristic of content as an electronic good.

In contrast to the already discussed systems with which a CMS has to integrate, the Web and e-commerce systems offer their service to the wider public and are therefore accessible from the outside. This also implies that there are security considerations that have to be taken into account. Application components based on ActiveX or Java technology could for instance be easily integrated into Web pages or e-commerce applications. However, it might be sensible to clearly separate the publicly accessible parts of the system from the part that stores and manages the assets. Hence, in contrast to Web content management systems that are part of the presentation application, an enterprise-wide CMS should provide the content and relevant information in a secure way. One possibility is, for instance, using message exchange between the Web application and the CMS. Once relevant content has been prepared and approved for publication on the Web it can either be actively sent to the Web application or placed in a shared storage area from which it is accessed. Depending on the workflow the content in this context might be already encoded for publication on the Web or is just raw material that is included in Web pages. In this context the format of the messages and files in which the content is provided is crucial. The integration can be facilitated if the content is already encoded in a way that is directly usable in a Web context.

In the case of e-commerce integration the CMS also has to be protected from public access. Further, the selection of content for sales is an active process and not all the objects within a CMS might be offered at the same time. Although a direct integration via application components and API could be easily realized it is more appropriate to separate the two systems and use message or file exchange as the integration type. According to the e-commerce workflow processes discussed in Section 2.3, there are three major interaction points between a CMS and an e-commerce system:

1. *Delivery of content information:* the CMS provides metadata and a proxy representation of content objects that have been selected for sales. This information should be provided in an exchange format that can be easily interpreted, processed, and included in the e-commerce front-end. Either message exchange or file exchange over a shared storage area can be used in this context.
2. *Fulfillment request:* a fulfillment request is a message from the e-commerce system to the CMS that includes a content object ID (that has to be unique in the context of both systems); further information about the intended usage, and a delivery address for the customer are included in the information. The information about the intended usage is required to clarify the rights and also for the calculation of the royalties. In a content-rich organization this information is kept in the enterprise-wide rights management system that is part of the back-end infrastructure. A sales-oriented organization might make this part of the e-commerce front-end. In this case the usage information would not have to be exchanged between the two systems.
3. *Rights clearance:* the CMS sends back a message stating the rights situation. This is the input for the further processing of the sales request.

Another interface relevant in this context is the interface to the ERP system, to which a message about the usage of a content object (i.e. the completion of a fulfillment request) is posted. Further, the system requires an interface to a delivery system. In the case of electronic delivery the acquired content is delivered via high-speed networks, satellite, etc. This is equivalent to an exchange of content between two organizations, i.e. standard file formats and a subset of metadata should be delivered alongside the actual essence encoded in the required format.

9

Applications

The set of applications users employ to interact with the CMS are the most visible part of the system. Although almost the entire functionality is provided by back-end components, it is this part that makes the system accessible to the users. Thus, it is crucial that the set of applications a CMS provides supports the user requirements and workflow steps related to content management optimally. All the different processes have to be considered to build a comprehensive set of user applications.

Ideally user applications are developed in close cooperation with the target user group. They will provide crucial input about the functional but also operational requirements that ought to guide the application design. The user input also includes aspects such as application ergonomics that might easily be neglected when focusing on technical issues. Theoretical requirement studies may not be sufficient to get a good feeling of the real needs of the users. They are very often too abstract and do not provide any visual or physical experience of the system. Thus, in order to involve users it might be necessary to produce various prototypes, which can be used to extract the users' needs. The initial prototype should be based on observations of the current workflow and studies of the tools that are used in a specific work context.

The applications presented in this chapter have all been developed taking this approach into account. They are the results of a number of research and commercial projects with major organizations in media production and broadcast. However, one of the most crucial results of these collaborations was that flexibility and configurability are key success factors. Almost all organizational units have very specific requirements regarding the application layout, operation, and partially even regarding the functionality. Apart from the user requirements there is also the overall infrastructure to consider. A number of systems might have to be integrated and CMS application views might become part of other systems. Therefore, it is crucial to provide not a monolithic set of basic applications but a selection of application modules that can be integrated and flexibly configured according to the user and overall system requirements. A component-based approach is ideal to satisfy these requirements. Examples of such application components are discussed in this chapter.

Professional Content Management Systems: Handling Digital Media Assets   A. Mauthe, P. Thomas
© 2004 John Wiley & Sons, Ltd   ISBN: 0-470-85542-8

However, discussing only application controls is too abstract and does not provide enough detail on how the workflows discussed in Chapter 2 can be captured by applications. Thus, in this chapter examples of applications that support specific workflows are introduced. They are derived from application use cases commonly found in a broadcast and media production environment. Many different workflows, ranging from deep archiving, over news processing to retrieval, have been considered. Other application domains may require different clients than those described here. However, the basic functionality is the same. How the set of application modules has to be configured in such a context depends on the specific requirements that have to be evaluated on a project-by-project basis.

9.1 APPLICATION DESIGN PRINCIPLES

The design and development of applications have been changing over the years. In the past a system had a user interface via which the users interacted with the system. This interface was entwined with the functional part of the system and the entire structure was referred to as an application. Within the last decade this outlook has changed. Initially the advent of client–server systems decoupled the user interfaces from the actual execution environment. Instead of monolithic applications running on hosts, distributed systems became the norm. This in conjunction with the development of windows-based applications led to a more interactive and user-friendly application environment. The next major development that changed the way users interact with applications was the emergence of the Web. The user interaction takes place within a Web browser. Information and messages are exchange via standardized scripting languages. There is no need to install any specific applications apart from a Web browser. The appearance of the user interfaces is determined by the Web designers. Web applications have become more and more sophisticated and are nowadays also able to host downloadable application components that are executed within the Web browser environment. Many users are accustomed to the kind of interaction taking place within a Web environment.

These developments have to be considered at the design of CMS applications. Other aspects are the technical and functional requirements alongside the actual user requirements. However, there is no blueprint on how to design and implement applications. Merely certain principles can be established that facilitate application design and their integration into the system and with other system components.

9.1.1 APPLICATION AND USER REQUIREMENTS

The developments concerning application design and interfaces always had two main drivers, namely the technology that allowed new applications and interaction modes to be realized, and the user requirements of those who have to work with the applications. A third factor that is also important in the context of large-scale application environments is user and technical support and maintenance. For a CMS it is crucial that applications can be easily installed, configured, and updated. Since there are various views on the content and a potentially very large number of users with different roles and rights, CMS applications also have to support individual application views taking these aspects into account.

Interoperability and integratability of (and with) other application components are also important factors. This implies that components should be developed in a way that allows running them within other application contexts. The CMS should also offer such a hosting environment. In order to integrate these components transparently to the user their appearance should be easily adaptable to different application layouts.

When designing application interfaces the user requirements have to be taken into account. These requirements concern all operational features, including functional and non-functional aspects. Workflow analysis is one way to establish these requirements. Further, the handling of the application and ergonomic aspects are design parameters that have to be established. It is crucial that these requirements are captured accurately to ensure that users can operate and interact with the system efficiently and on a continuing basis.

The various requirements concerning the application development can be distinguished according to their source and how they are determined. There are:

- *Functional requirements* dealing with the feature set and characteristics of applications. They are mainly derived from the examination of existing and potential workflows. Functional requirements only specify the kind of functionality that should be available in a specific application component. They do not state how they should be presented.
- *Operational requirements* describing the handling and use of applications. In contrast to functional requirements they deal with the dynamic aspects of operating an application. Operational requirements can be further subdivided into:
  - *User-application interaction* describing the actual interaction between the user and the application. This includes the different movements and steps involved to carry out a specific task within a distinct application context with the goal of achieving a certain objective as efficiently as possible. In this context the ergonomics have to be considered.
  - *Application support and maintenance* covering all requirements related to installing, supporting, updating and maintaining applications in large-scale content-rich organizations.
- *System requirements* derived from the system context an application is in. This includes the definition of specific (system and user) interfaces to control particular system components. Further, all requirements placed on applications that originate from a specific system configuration are part of the system requirements. This also includes the integration with legacy and third-party systems.

Whereas the core set of functional requirements develops further in a relatively constant manner throughout the product life cycle, the operational requirements change from organization to organization and sometimes even differ between departments within the same organization. Thus, application components should be adaptable to facilitate the configuration of application views according to individual user needs.

Representatives of the users should be involved from a very early stage of the design process onwards. This includes the initial application design but is even more important at the introduction and configuration of applications within a project.

In summary, applications should be constructed from flexible and configurable components that can be adapted during the implementation phase within a project. The configuration of applications has to allow tailoring the application view according to usage restrictions and user requirements. Functional and operational requirements have to be considered at the design of application components.

9.1.1.1 Access Rights and User Roles

Access rights play a special role in large-scale distributed systems such as CMS. They restrict the amount and kind of information particular users or user groups are able to read, retrieve, and change. Access rights are not a binary concept (i.e. either one can or cannot access applications and items) but can determine in detail which content object can be accessed by whom, and what kind of access (i.e. read, write, or both) is granted. The kind of interaction with a content object is also determined by access rights. For instance certain user might be allowed to see preview copies of a content object but should not be able to access the hi-res material. Rights might pertain only to individual users or also to a group to which a user belongs. Within an editorial office all users might be allowed to see the raw material produced for a certain program series. However, members of other editorial offices might not be allowed to view this material. Thus, there are individual and group access rights to be considered with each content object. Access Control Lists (ACL) are the traditional means of controlling access to data and resources. In a CMS application environment these can become extensive, and a flexible way to express rights with different granularity within the system has to exist.

The organizational role of a user not only determines access rights but also the configuration of the application environment of a user. The application environment of a journalist or editor is, for instance, different to that of a cataloger. Whereas traditionally applications are installed on a certain machine, in future most applications and their specific configurations should be accessible from almost any computer in the organization. Exceptions are of course applications that require special hardware or devices.

Thus, with each user certain rights are associated in conjunction with the organizational role. The user role and rights determine the information access as well as the application context of a user. A user profile can be used to specify the roles and rights of a user within an application environment. In such a user profile all the relevant information about a user is kept in a computer processable form. The data from such a user profile controls application presentation and information access.

Special treatment is required for users who can adopt multiple roles (for instance administrator and cataloger). In this case it has to be determined if the users should login with one specific role only or if the rights and application context should be an aggregation of the different user profiles.

9.1.2 COMPONENT-BASED APPLICATION DESIGN

The required flexibility and integratability can best be achieved by using a modular approach. Each application module in such a system represents a specific functionality. Modules can be integrated within an application framework to build an application. The granularity and functionality covered by a module depend on the task they support and possible third-party components they are built on. A module can for instance be a VCR control component used for the control of video and audio playback. This module can be used in all situations where video and audio playback have to be controlled. Its position in an application then depends on the application context, and the functional and operational requirements. However, it might not always be possible to define a separate VCR control module. For instance if certain third-party products are used that have an integrated VCR control with the player the layout and presentation cannot be changed. Component-based design not only allows better configurability but also supports the reusability of modules within different applications.

How fine the granular design of a module should be depends on the way different modules can be reused in other application contexts and how flexibly applications should be configured. Further, different modules need to interact to form an application. This interaction can for instance take place using message exchange. Thus, there is also a trade-off between the flexibility and configurability that fine granular application components support and the overall system performance.

In recent years the idea of building systems out of functional blocks or modules has become more and more common. There are a number of technologies that support the so-called componentware approach.

9.1.2.1 CMS Applications and Componentware

The modular application design philosophy is closely related to the idea of component-ware. The original idea of componentware is to build software systems from prefabricated components that are put together to provide the full functionality set (Bergner *et al.*, 1998). Important in this context are the component functionality and behavior, and the interfaces that allow interaction with a component. Components can implement one or more interfaces (export interfaces) offered to other components, and use interfaces offered by others (so-called import interfaces). The aggregation and combination of the different interface types of all components within a system define and constrain the interaction and system functionality. A system (or application) is built by connecting compatible import and export interfaces. The connection type is usually static, i.e. it is determined during the development and integration phase and does not change at runtime.

However, using componentware and a component-based approach is not without problems. For instance, the integration process is not always straightforward and complex 'plumbing' might be required to create an operable system (Sant'Anna *et al.*, 1998). Hence there is a danger of creating sub-optimal applications from components that are not well aggregated within the system context. Large sets of libraries have to be maintained, and the general maintenance of a system assembled from components is a problem. In order to have the flexibility the componentware approach promises while avoiding software monsters it is necessary to strictly control the development process of each component and align it with an overall system build plan. Thus, the flexibility of building adapted applications according to user requirements out of components might require even more rigid control in the development of these components compared to traditional development methods.

There are a number of development tools and infrastructures (such as OCX controls and Java classes) that support the componentware approach. Techniques that support modular application building include CORBA (Orfali and Harkey, 1997), ActiveX (Chappell, 1996), or JavaBeans (Sun Microsystems, 1997). The protocols discussed in Section 4.5

(e.g. SOAP and MOS) can also be used to support component-based application development. A very powerful way to create reusable modules also is using Web technologies or Web services, such as HTML or dynamic HTML. Modules that are provided in this way can be included in applications and Web pages without any kind of client-side installation.

9.2 REUSABLE APPLICATION MODULES

Content management applications have to satisfy the user requirements in the context of a specific workflow. The component-based application approach is most suitable to satisfy the plethora of requirements brought about by specific organizational and environmental contexts. Thus, a particular CMS application may in fact comprise quite a number of different application modules. This section describes some of these modules that have been identified as useful. These modules should be considered as examples of what reusable application modules could look like. They do not represent a set of mandatory or even sufficient components for a CMS. However, a representative of each major type is discussed.

9.2.1 PLAYER

To replay video on audio objects at the desktop applications require media player capabilities to be provided by a player module. The set of functionality such player modules should provide comprises start, stop, and pause of the replay of content. A fast forward/backward navigation (ideally with selectable speed) is also part of the feature set. Further, skip to begin and end of content, and jog and shuttle capabilities (including frame-by-frame stepping back and forth) are required. The player controls include sound volume control and sound balance control. The size of the video playback window should be selectable.

If a player module is used in applications that support working with clips via in- and out-points, or that make use of markers or locators, the module should also provide mark-in, mark-out, remove marks, and add clip. Further, set locator, remove locator, and modify locator names are required in this case.

9.2.2 CLIP LIST

Clip lists are useful modules for building applications that use playlists or edit decision lists. The functionality provided by such a module should include:

- add clip, delete clip;
- provide a configurable set of sequence attributes (such as sequence name, duration, comment, etc.);
- provide a configurable set of clip attributes (such as mark-in, mark-out, record-in, record-out, duration, clip name, comment, content ID, etc.);
- modify sequence and clip attributes;
- rearrange clips in sequence;
- activate/deactivate clips in sequence;
- activate/deactivate (video and audio) tracks in clips.

There may actually be two modules involved here, one that stores and manages the clip list, and one that provides a graphical user interface for it.

9.2.3 TIME LINE

A time-line module is very useful in any kind of rough-cut or editing application. Interesting functionalities for a time line module are, for instance, to be able to display the time line with markers for the start and end frame for each clip. A separate display of video and audio tracks (with the option to activate/deactivate tracks) is also a useful time line feature. Further, a curser that runs along the time line while playing and hence indicates the current position should be part of the time line. It should be possible to use free point-and-click positioning of the cursor and to perform a zoom in/out of the time line. Fast navigation along the time line is another feature that should be provided by the time line module. Inserting clips into other clips at the cursor position should also be possible. Since the time line module needs to be used in the context of other application modules it should also provide functionality that allows interaction with these modules. For instance interaction options for all functionalities that the clip list module provides should be there.

Obviously, a lot of additional functionalities can be imagined for a time line module. Looking at commercially available nonlinear editing solutions gives an impression of what can be done with such a module.

9.2.4 DEVICE CONTROL PANELS

Applications may want to provide functionality for remotely controlling devices, such as videotape recorders. Depending on the device, many interesting functionalities could be provided by such a module. In general, a device control module reflects the functionality of the respective device. For instance for a VTR control panel the functionality should include start, stop, pause playback, and fast forward and backward. More advanced functionality includes jog/shuttle operations. The current timecode should also be displayed within the device control module and it should be possible to wind to a specific timecode. Further, the recording has to be supported by start and stop record. Last but nor least, cue tape and eject tape have to be possible.

Other devices will demand modules that may provide quite different functionalities. This depends entirely on the functionality of the respective device.

9.2.5 METADATA EDITOR

A useful application module for each application that allows entry or modification of metadata is an easily configurable metadata entry form. Functionalities for such a module include:

- providing templates that allow arrangement of data entry fields for all data types in flexible layouts;
- associating default values to fields;
- associating lists of possible values to fields;
- providing means for consistency checks involving arbitrary fields;

- providing fields for automatic computation of values for fields based on rules and/or values of other fields;
- spell checking;
- defining required and optional fields.

Again, this is only a subset of possible functionalities that could be provided by a metadata editor module.

9.2.6 QUERY INTERFACES

Modules that provide query interfaces are also quite interesting building blocks. Depending on the skill set of the users, modules can offer a simple query interface comparable to a Web search engine, and may provide increasing functionality up to a very advanced query tool for experienced users.

Thus, fulltext search interfaces have to be provided by the query interface as the most basic search functionality. Further, the possibility of configuring various attributes for building attribute filters should be supported as well as the possibility of configuring a stratum search interface. In order to provide the query for label concept the query interface module has to allow configuring search interfaces to search for specific labels. Additionally, there should be support for Boolean operators, fuzzy searches, proximity searches, etc., and for selecting values from dropdown lists or thesaurus pick tools. Ideally it should also be possible to save and restore queries.

9.2.7 THESAURUS PICKERS

When thesauri are used to ensure data consistency, a flexible, fast, and easy-to-use tool is required for browsing the thesaurus and selecting the desired terms. Such a tool or module could for example provide the possibility to enter a thesaurus term (together with auto-completion) and to browse the thesaurus in a tree-like structure. Further it should provide details for selected thesaurus terms (such as descriptions) if available. Ideally, such a module should be configurable to provide access to thesauri as well as to simple legal lists.

9.2.8 HIT LISTS

Since a CMS may manage various kinds of content, the requirements of how to present search results in hit lists may differ significantly. It is also quite common for different users or user groups to have different needs with respect to the information displayed with a hit. Hence, a module providing hit lists should be flexibly configurable to fulfill these requirements. Sensible configuration options include:

- activate/deactivate thumbnail display;
- metadata to be displayed for each hit;
- secondary information retrieved from other information systems to be included in the display (online availability of certain instances of the content object, IPR indicator, etc.);
- overall layout of the hit list.

9.2.9 APPLICATION VIEWS INCLUDING KEYFRAMES

Application views using keyframes (or thumbnails) are extensively used within a CSM application set. For instance cataloging clients and retrieval views make use of them to visualize a content object and provide a quick overview of its content. Some general functionality should be available in all views that display keyframes:

- keyframes should be selectable using the mouse. Familiar shortcuts should be supported (e.g. multiple selection);
- it should be possible either to apply any function on selected keyframes, or to select such a function as the default operation for the next selection;
- it should be possible to limit the display of keyframes to those having certain properties;
- it should be possible to play or display the preview copy by clicking on a keyframe (when playback is the current default operation);
- if applicable, it should be possible to display start and end timecodes as well as the duration.

Other more-specific tasks only apply to certain clients. For instance any operation concerned with the manipulation of keyframes is solely part of the cataloging task.

9.3 INPUT APPLICATION CONTROL

Input applications are concerned with the control of the process where content enters the system. Material can enter the system via different channels. This process is referred to as import when metadata and essence are pre-encoded and enter the system in a format that is supported internally. In this case the incoming material is directly associated with a content object ID and can be used without any further processing. Automatic processing (e.g. video analysis or speech recognition) might be performed to generate additional information. However, the material that enters the CMS during the import can be used *as is*.

The case where material has to be recorded or information has to be extracted when it first enters the system is called ingest. In this process the incoming items are converted into formats that can be administered within the system. The ingest process can involve the control of hardware and software devices. In parallel, additional information might be extracted (through automatic analysis) or metadata might be entered manually.

9.3.1 IMPORT

The capabilities of an import application depend on the functionality that is provided by the Import Service. The *Import Client* is a tool that allows scheduling of import of content in file formats from various sources, especially Import Servers. A source may be any external system that provides content as a file for download. Content is picked up from these external sources and imported into the CMS, thereby routing essence to the Essence Management System and metadata into the Data Management System. The Import Service parses the file format and derives essence and metadata from the file as part of this process. The Import Client is the application control component in this context that interacts with other CMS components (such as the Import Service and Import Server) to introduce the material into the system. The kind of interaction and processing required in this context is system dependent. If for instance additional information should be retrieved automatically, a video and/or audio analysis might also be triggered when the material comes in. In these cases the overall process must be controlled by the Workflow Engine. The step-by-step processing is executed by the Task Management Service.

Important features of the Import Client include the possibilities to:

- select the source system;
- select the essence files to be imported;
- select additional formats to be created upon import;
- select additional proxies to be created upon import (e.g. keyframes);
- enter or modify a minimum set of metadata;
- allow the selection of an existing content object imported material should be associated to;
- allow the creation of a new content object;
- manually start and stop an import process;
- schedule single import processes;
- schedule batch import processes;
- add status information and other properties that steer further processing of the imported objects (e.g. appearance in certain folders of the Workspace Management).

In addition, an Import Client should provide means to follow the progress of import processes. Figure 9.1 shows the example of an import interface that allows media files to be imported and a keyframe analysis to be run on the object after import.

For import of content from VTR or other playback devices, essence is recorded onto a suitable streaming server. Playout and recording are scheduled and performed using the media management capabilities of the CMS. Essence and metadata may then be imported from the Streaming Server. In such a case metadata may also be imported asynchronously prior to, during or after a play-in. The Import Client should therefore allow:

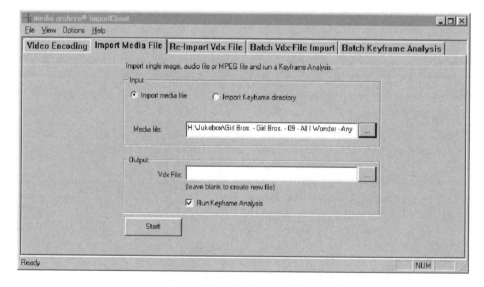

Figure 9.1   An Import Client

- scheduling these exports to studio devices, and for manual import from playback devices;
- control of an external device that performs playback of the essence to be imported.

Besides classic import and export the CMS should support (via the Import Client in conjunction with an Export Client) check-in and check-out for re-versioning of content. Export/import can also be useful to share material between CMS instances that have no network interconnection.

The Import Client is operated by administrative staff such as feed assistants, catalogers or media managers. In a system operating within the boundaries of an editorial office research assistants or junior editors might also use it within their daily work.

9.3.2 INGEST

While import focuses on bringing assets into the system that are already available as files, ingest means acquiring assets from signal. This typically includes encoding and file creation. The *Ingest Client* is a tool to support the recording and ingesting of live signals, such as video or audio feeds. It can also be used to control the recording of material from tape. The client is basically a user interface to the Ingest Service that controls the actual ingest process. An ingest process may involve other services besides the ingest process. In such cases, the overall process is controlled by the Workflow Engine. The step-by-step processing is carried out by the Task Management Service.

Important features of the Ingest Client include the possibilities to:

- provide monitoring capabilities for multiple channels – you can see and/or hear what is currently being provided to the input of an encoder;
- enter basic metadata, such as title, certain ID, etc.;
- create objects within the Data Management, or select existing objects that the recording should be associated with;
- provide crash record capabilities, which means 'start recording now';
- provide scheduled ingest capabilities, which means that you can pre-program multiple recordings from multiple sources over a given period of time (this channel each Monday from 08:00 am to 09:00 am, that channel only tomorrow from 07:45 pm to 08:15 pm, and so on);
- provide event-triggered ingest capabilities, which means that an ingest starts upon reception of a given event, such as a VPS signal or another non-time-based trigger;
- monitor the status of all available channels and recordings.

Other options can easily be imagined. In any case, the Ingest Client is a control interface to access functionality that a CMS, together with other systems, provides, and should not implement too much workflow support by itself.

Figure 9.2 shows an example of an Ingest Client. The Ingest Client has a video window acting as control monitor. Within the Ingest Client the timecode of the recording, which serves as synchronization point for different copies of the essence, is also shown.

As in the case of the Import Client, administrative staff such as feed assistants, catalogers or media managers operate the Ingest Client. In a system operating within the boundaries of an editorial office research assistants or junior editors might also use it

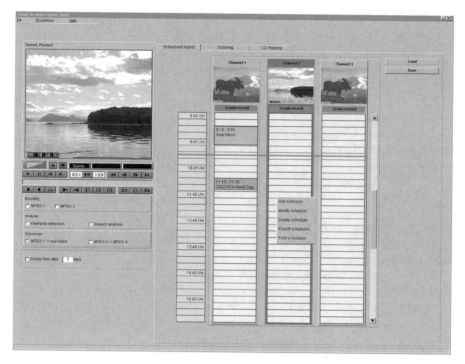

Figure 9.2 An Ingest Client

within their daily work. Since it supports live and real-time processes it has to have a simple, straightforward interface providing only basic controls.

9.4 DOCUMENTATION APPLICATIONS

Documentation applications are tools that support the annotation of content. Thus, these are the essential tools to associate metadata with a content object. These tools also allow updating of existing or automatically generated metadata. They are usually operated by trained staff such as catalogers, archivists, and feed assistants. Two types of documentation applications can be distinguished, namely applications for fast (real-time) annotation of content and applications for in-depth cataloging. With the former speed is of the essence, i.e. it has to be possible to very quickly annotate incoming material to make it searchable within the system. The latter is more concerned with providing accurate and detailed documentation of content objects. In this process content might also be selected for archival. Thus, it has to be possible to mark content for deletion to remove it from the CMS.

9.4.1 LOGGING

The *Logging Client* is a tool that supports fast, online logging of incoming or ingested material such as news feeds. It is usually operated by so-called feed assistants, whose main task is to sift through incoming material and annotate it.

The Logging Client should allow the provision of an initial minimum set of metadata required by editors to retrieve incoming content. Typically, this minimum set of data includes formal data such as title, and time of recording and date of the feed. Further, information that is specific to the single items and topics within the feed such as item number, title, location, approximate length, abstracts, and reference date and time are also part of the minimal metadata set added in the Logging Client. While logging, the Logging Client needs to have instant and full access to any portion of the incoming essence that is already recorded and under the control of the Essence Management System. This means that such a client should not work on a monitor signal (which would be real-time), but on the part of the essence that has already been recorded. This means that the person logging can access the material only after a certain latency (typically some seconds), but s/he can pause, play, and fast-forward through this already recorded portion. This provides much more flexibility in annotation compared to applications that are slaved to real-time.

Figure 9.3 shows an example of a Logging Client interface. To start the logging process the user selects an incoming (or existing) stream for logging from a selection list. Information about streams that have been or are being recorded and are ready for logging is provided by the Workflow Engine. The selection list should allow selection of assets that have not yet been logged, and should provide some information about whether the object is still recording or already fully available. Further, scheduled ingests where the recording has not yet started can also be shown in the list of recordings. The Logging Client can access this information and display it within a selection list from which a user can chose the item s/he wants to annotate.

After selection of a feed the logging processes commences. The video or audio of the object to be logged is played back. The feed assistant or logger adds additional object-related metadata. If possible, entry fields should contain default values automatically derived from equipment or recording devices (e.g. date and time), from the entry logic (e.g. item number, timecode information) or automatically retrieved information (e.g. metadata carried in the media stream). The main task of the feed assistant operating the Logging Client is to enter a verbal description of the events happening in the video. This description should be linked to timecodes. Thus, the Logging Client should allow the set markers to allow segment-based documentation. Annotating conventions within each institution may determine how items should be annotated. There might be controlled word lists that also facilitate the search process. These word lists might also be incorporated in the Logging Client and used in an automatic word completion mode.

A vital factor in working with a Logging Client is time. Ideally the incoming information should be annotated close to real-time. Thus, fast and easy access and straightforward handling of the application are key to success. During manual annotation keyboard shortcuts should assist a fast and simple mapping of descriptive data to time segments. In order to support the work of feed assistants optimally their operational requirements have to be studied in detail. A flexible and freely configurable application interface is crucial since for instance the best way to structure the layout or optimal hot-key combinations depends not only on organizational requirements but also on personal preferences.

The Logging Client should also provide means to physically create items out of programs depending on a segmentation provided by the logger.

Despite the fact that the Logging Client provides a fairly basic functionality, it is one of the applications for which configurability is essential. Ideally, each control, entry field, and monitoring component should be freely and independently configured.

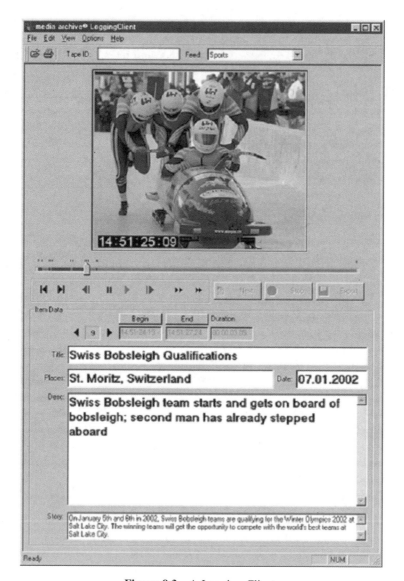

Figure 9.3 A Logging Client

9.4.2 CATALOGING

The *Cataloging Client* is a tool that displays the results of the analysis and indexing services and enables the user to modify the results of the automatic processes and add or modify additional formal or descriptive information. The metadata might be part of the CMS data model or of any information system interfaced by a Data Manager. Further, metadata not part of the CMS data model can also be modified by an application plug-in that has access to the respective information system.

It is a tool mainly used by trained catalogers and archivists. The Cataloging Client is intended to support in-depth documentation of material intended for long-term storage.

This process is typically a *longer than real-time* process, i.e. a detailed description of the content using a sophisticated metadata scheme is usually used in this context.

At the start of the documentation the user has to select the content object that should be annotated in further detail. Therefore the client should have multiple views to support the task of selecting and annotating appropriate content. Three different ways to access a content object for documentation have been identified:

- select a content object to be cataloged from the standard retrieval access;
- select a content object to be cataloged from a to-do list (ideally user dependent);
- enable fast navigation through the content.

Subsequently the user can work on the main task, i.e. associate detailed documentation to the entire or selected parts of a content object. It is important in this context that the documentation can be associated with specific aspects of a content object as well as with the object as a whole. Thus, stratified documentation should be supported as well as multiple levels of temporal and spatial segmentation. The Cataloging Client has to provide different views for the entry of object-related and segment-related metadata. Whereas object-related metadata is very often purely textual, the entry of segment-related metadata is very often supported by audiovisual proxies. Thus, apart from the purely textual documentation the Cataloging Client should also allow modification of all information about a content object that is present in the CMS. This includes:

- adding, deleting or exchanging keyframes or proxies;
- modifying the shot structure as derived from the Analysis Service (for instance by adding, deleting or moving shot boundaries);
- modifying speech recognition results and feeding these changes back to the recognition engine for improving analysis results;
- adding, deleting or modifying all textual information in all strata;
- adding, deleting or modifying the value of any attribute.

The manipulation of keyframes and the views presenting them include:

- editing keyframe properties;
- adding or deleting keyframes by point-and-click operations;
- adding keyframes using the Browsing Client;
- adding, moving or deleting shot or stratum boundaries by drag-and-drop;
- adding a configurable number of strata for cataloging and display purposes.

In addition there should be formal data views dedicated to the display and modification of textual metadata, organized according to template descriptions for different genres.

Visual, audiovisual and textual navigation tools or views should facilitate the navigation through content objects. Examples are, for instance, views that allow displaying the media object itself (e.g. by clicking on keyframes). This includes the playback of a browse version of the video or audio object, or the presentation of a low-resolution graphic.

Further, several levels of abstraction of the media object that provide a good overview of the material at one glance (e.g. via hierarchical keyframe levels) are also a good way of supporting the navigation process.

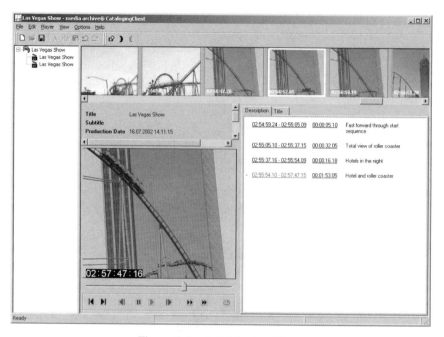

Figure 9.4   A Cataloging Client

For the Cataloging Client it is also important that the user interface can be customized according to the needs of individual users. This can be static configuration of the layout of the application but there is also a requirement for a dynamic change of the view, for instance the kind and number of relevant strata that should be displayed. In general the entire functionality offered by the Cataloging Client should be easily accessible, for example using menu bars, scroll bars, mouse and keyboard shortcuts, and optionally external control pads. A status bar should provide information feedback during operations.

Figure 9.4 shows an example cataloging view that combines display of keyframes that are associated with a selected stratum on top, together with an annotation area for the stratum and a player window that replays the browse proxy. On the left-hand side is screen space reserved for hierarchical display of content object hierarchies (e.g. series – program – item) if available. This allows for fast navigation through the hierarchy via point-and-click.

The requirements that a cataloging client may have to meet go far beyond this. There may be requirements to support proprietary thesauri, legal lists, genre definitions, etc. There may be requirements concerning rules that require that certain metadata take certain values depending on values of other metadata. There may be restrictions with respect to what values a metadata field may take. All this implies that a cataloging client application must be very flexible in the way in which it interacts with the CMS and how it may support business rules.

9.5 RETRIEVAL APPLICATIONS

The retrieval applications are probably the most important user interfaces that a CMS may provide. Via this retrieval interface, users formulate and submit queries to the system, receive result sets as hit lists, and have the possibility to browse through the details that the system can provide for each object that has been retrieved. Often, a retrieval interface also permits launching other applications that allow working with these objects. The retrieval interface should also allow accessing the Workspace Management in order to provide means for collecting items for projects and to access certain system functionality.

Retrieval applications subsume all application views concerned with the search, presentation, and preparation of content. Content is neither enriched nor changed in this context, i.e. no metadata or essence is added, removed or altered. However, the view and details of the content presentation can be customized depending on user requirements.

Various different user groups ranging from unskilled occasional users to trained and specialized content management experts operate retrieval applications. Thus, different input, search and operation modes have to be supported to accommodate all these users. In the context of retrieval three general operations can be distinguished, namely search and presentation, browse and rough-cut, collection and ordering.

9.5.1 SEARCH AND PRESENTATION

The main retrieval tool is concerned with search and the presentation of the search results. This allows the user to sift through and select appropriate hits for further examination. The *Retrieval Client* is the principal tool that supports the functionality necessary for:

- querying the CMS including selected or all databases and information systems interfaced via metadata managers;
- displaying the results as hit lists;
- assessing the quality of the hits in various detailed views.

Querying and search for information can also include other, third-party systems. In this case a request is sent to all systems that are integrated using message exchange. Depending on the selected integration mode (see Section 8.1.1) the result of this process is either displayed in a separate view or as part of the general hit list.

Since the Retrieval Client is used by most of the users and on most of the clients, it is sensible to provide this as a Web-based interface. Hence, a Retrieval Client should basically be a Web browser that renders the interface that is provided by the CMS back-end, namely the Web Retrieval Service. Let us consider the most important requirements for such a client.

The hit list has to contain all necessary information to allow the user a first assessment of the relevance of a hit. The hit list does not only have to contain textual information but can also include a visual representation of the content object. From the general hit list it has to be possible to directly access the detailed views for each content object in the list. The detailed views can be all kinds of proxy representations. If IPR and rights clearance are important in the application context the retrieval system also has to support interaction with the rights department. This interaction can for instance be based on a simple email exchange, or a direct query to the rights management system might be used. The granularity of the information about the rights situation should be on shot level.

The retrieval tool should also support interaction between users and the organization of content. Thus, messaging systems that allow the inclusion of references to content

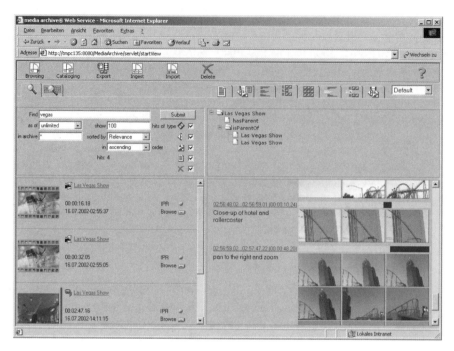

Figure 9.5   A Web-based retrieval interface

objects and user selections need to be supported as well. Figure 9.5 shows an example of a retrieval interface with:

- a simple query interface;
- a hit list including thumbnails, selected metadata, plus information about IPR status and online availability of browse proxies;
- a view that shows relations that the selected object has with other objects;
- a storyboard view of the content object, showing text and keyframes linked to a segmentation provided for a 'description' stratum.

Alternative search interfaces and detail views can be selected via tabs. In addition, a toolbar allows other applications to be launched. When launched, these applications will automatically load the selected content object.

9.5.1.1 Search and Querying

Querying the system is of particular importance. Depending on the user's skills and roles the requirements on search and retrieval vary. Some users prefer to have a very simple query interface, e.g. just one single line for fulltext query. Thus, there should be a simple query, which offers an interface that is comparable to the search interface of an Internet search engine, i.e. one single line for entering fulltext queries. Within the query line familiar operators should support the inclusion or exclusion words and phrases. In addition to the query line there should be options to limit the search to a certain time range and to specify the maximum number of hits to be displayed. It should be possible to store popular queries via drag-and-drop in a suitable folder of the Workspace Management.

For more advanced users, the user interface should allow accessing the labeled queries provided by the Data Management. Such an interface could even allow building a list of labeled queries and combining them via Boolean operators. This kind of query should allow picking a label from a list of available labels, picking an operator from a list of available operators, and entering a search term into a search field. Further, it should be possible to qualify this selection by selecting a NOT operator if required, and to link this query line to the next attribute query line by selecting AND or OR.

Other users may want to make use of sophisticated search facilities, limiting the range of the fulltext query as well as accessing all the attributes of the database model and including them in the query. Even more experienced users may want to see a much more elaborate query interface that allows them to enter values that specific attributes should take, and that provides means to combine these attribute-based queries by Boolean operators. Such a query interface makes use of the underlying data model and hence depends very much on the specific model used by the enterprise. A CMS must provide means to customize the query interfaces to adapt them to the data model in use.

All users may be interested in a query for segments, making use of stratified documentation. Thus, the user interface provided with the Retrieval Client should be flexible and configurable enough to support the entire range of users, e.g. by providing a configurable search interface. The main query modes that should be supported by the retrieval client are:

- simple fulltext queries considering the entire textual information stored in a CMS;
- simple fulltext queries against labels, supporting all the query functionalities that state-of-the-art fulltext search engines offer, such as fuzzy search, proximity search, use of semantic networks, cross-lingual search, and so on;
- attribute-based queries, linking attribute–value pairs together via Boolean operators, and supporting regular expressions in values;
- segment queries, which are fulltext queries against strata.

All these kinds of queries (or multiples of each kind) should be combinable via Boolean operators. For each query it should be possible to select the databases the query should be submitted to from a list of systems interfaced by Data Managers.

A very useful addition to query options as described above are default queries that can be submitted by pressing a button. A good example is a 'What's New' button that asks for all objects that have been added to the CMS within a given time frame. Basically, this could be one of the stored queries provided by the data management, presented by the user interface in a way that it is accessible via a single click.

9.5.1.2 Presentation of Search Results

The result of a query is presented in a hit list where the hits can be ordered according to a relevance ranking. The attributes that are displayed with each hit (identifying content objects in the hit list) should be configurable by the user. The hit list should show relevant metadata elements and maybe a key image representing the object. For other media types than moving images, other kinds of keyframes may be selected as reference to a hit or a segment. Examples are thumbnails of image objects, short sound bites for audio objects, or a short abstract for text objects.

The hit list should be configurable to show also all segments where the query has matched the segment metadata. In the case of a video object this could for instance be a list of all segments where the query matches the image content description. In this case the segments could be displayed showing the image content description and the keyframes for all these segments. This view provides immediate visual feedback on the query.

Further, certain properties (e.g. usage restrictions or the availability of different essence formats) can be graphically depicted in the hit list. In the case of property rights this might be colored bullets (green: free to use, yellow: usable for news, red: copyright protected, grey: rights situation unknown). This rudimentary rights information either has to be part of the CMS or needs to be provided by an integrated rights management system that is queried. Moreover, the user may want to know if the material is currently available online. This can also be depicted using graphical means indicating if essence is online, in transit, near-online, offline or unavailable.

The hit list should also allow dragging objects into Workspace Management folders, and it should allow selecting objects for browsing of metadata and other information that the CMS can provide about the object.

However, the configuration of the hit list and the kind of data it actually displays should be entirely dependent on the requirements of the individual user or the organizational policy. In the latter case a desktop and application standard determines what applications should look like and what information should be provided. If such a standard exists within an organization it has to be possible to configure the retrieval view and hit list content accordingly.

9.5.1.3 Detailed Content Object Representation

The hit list is only used to provide an overview and to allow the selection of a specific content object to examine it in more detail. Depending on the nature of the content object and media type, and its representation in the system, several views are provided. The most basic view is the one presenting the textual metadata. There could be views that show all format data, or information about the formats in which the essence is available, or the carriers that could be loaned from the archive. Other detailed views are for instance a keyframe-only representation of the content object or a view where metadata and audiovisual information is combined as for example in the storyboard view that shows the stratified documentation (see Figure 9.5). By displaying the annotations that are linked to a given segment together with the keyframes that represent this segment, such a view 'tells a story'. For video objects, light-table-like views that show all the keyframes for a selected object are useful (Figure 9.6). This provides a full overview of the image content of an object at a single glance. Further, views that embed players, thus providing preview and pre-listening capabilities, are useful enhancements.

Ideally, there should also be a view dedicated to display copyright data at segment level. This actually depends on the kind of integration between the CMS and the rights management system. If the information is available it should be graphically depicted in conjunction with the audiovisual representation of a content object.

The kind of views available to certain users or user groups should be a configurable system option. It may depend on the kind of material to be presented, the information

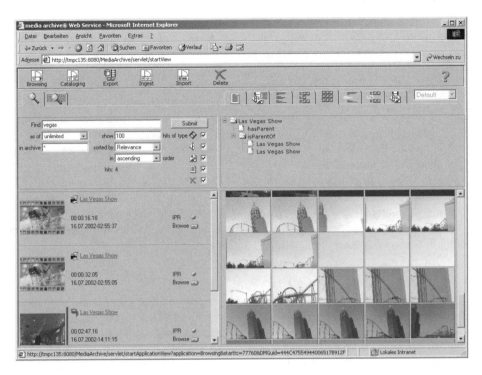

Figure 9.6 A light-table view in a retrieval application

available, access rights, and the role of the user. Many additional useful views can be imagined. It is important that a CMS provides a flexible technology platform that allows implementing different views according to customer requirements.

A content object may be part of a hierarchy (e.g. series – program – item) or may have other relations to other assets. Thus, the user interface used in retrieval should provide a means to navigate through these hierarchies and relations.

9.5.1.4 Advanced Retrieval Client Features

The Retrieval Client is a generic tool that should support all kinds of user/system interaction related to content search, presentation, and selection. There are a number of additional interesting functionalities the Retrieval Client can offer. For instance, it should provide an interface to submit a mediated search request, i.e. the search request should be passed to the archive stating some details about the desired content object. It should also be possible to exchange hit lists between users. This feature is required either to present the results of mediated research to the client or to support user collaboration.

The Retrieval Client should also support the rights clearance process. Therefore it should be possible to submit a list of selected clips or content items to the rights department for clearance. The result are posted back to the submitting users and should than be graphically visualized in the user's workspace.

Further, it should be possible to create links to arbitrary objects in project folders. Additionally it should be possible to share folders representing queries, hits or project with peer users. In order to support collaboration it is also important that the information can be sent in a messaging environment. Note, this should be separate from the normal email since it usually refers to a context given within the CMS. In order to support this, the client needs to have an interface to the messaging service.

However, retrieval and browsing through keyframes representing the content objects is only a first step. The Retrieval Client must also offer the option to access low-resolution copies of the objects found by, for example, clicking on a keyframe and launching an external player. It is also quite helpful when there is some additional intelligence included in this approach, e.g. when an external player is launched by clicking on a keyframe, the player should start replay from the timecode that is represented by the keyframe.

Further, rough-cut on audiovisual proxies as the first step in the pre-production phase should be possible. However, these are advanced functions represented by separate application units. Finally, it should be easy to invoke other applications easily, for example by launching an application by clicking on a button that represents the application, thus loading the object into this application.

9.5.2 BROWSE AND ROUGH-CUT

In order to play back audiovisual proxies a player with similar functionality to standard VCR and software video and audio players is required. This functionality is provided by the *Browsing Client*, which grants access to audiovisual content objects. It displays videos and images and replays audio objects. At the playback of audio and video the client has to provide a means for random positioning of the player and should offer the usual navigation controls with the standard trick modes such as:

- play, stop, pause, fast forward, slow motion, fast backward, slow backward;
- select forward and backward playing speed;
- several levels of skipping through an asset:
  - by a given number of frames;
  - by a given number of seconds or milliseconds;
  - for video objects, skipping by one frame forward or backward is quite a popular functionality.
- jog/shuttle;
- free selection of current position, often implemented via a slider;
- jump to begin/end of an essence object;
- jump to timecode;
- set marks or locators;
- jump to mark;
- display timecode;
- switch to left/right audio only (useful for selecting between original soundtrack and mixed soundtrack in television);
- audio balance control for left/right audio.

Further, provisional equipment allows a variation of the speed with which an audiovisual object is played back. A free selection of forward and backward replay speed is desirable. With regard to the presentation of audio and video, users within a broadcast environment are used to certain operation modes, devices, and functionality. It should be possible to emulate the custom VCR controls or to use the standard control devices. The options to control audio and video players are:

- control via standard mouse operations (i.e. point-and-click);
- keyboard shortcuts for quick and easy control of the video playback;
- a separate control pad that includes functions such as jog shuttle;
- a device for gesture control using mouse or light pens.

The Browsing Client is mainly deployed in a professional media production environment. This implies that it has to provide the sophisticated feature set users are accustomed to in this domain. Thus, standard computer-based video players are not sufficient in this context. However, while a browsing player may provide quite a large number of features, in certain contexts it is more important that it launches fast, works smoothly, provides immediate feedback to commands, and is intuitive and easy to use, so that the user can focus on the asset and not on the application.

9.5.2.1 Rough-Cut

Apart from playing back and navigating in audiovisual content objects it should also be possible to select content items and relevant segments. Further, the combination of these segments in an edit decision list (EDL) for further production is required. The *Rough-Cut Client* is an extension to the Browsing Client that provides access to a rough-cut facility to preselect interesting material and arrange it in a rundown order. This allows forming the outline of a new program within the CMS application context. Further, this rundown order should be translatable into an EDL that can be stored and reloaded. It should be possible to play back originally created EDL. In order to support the post-production process the Rough-Cut Client should allow translating the internal EDL into various EDL formats so that it can be used on a professional nonlinear editing (NLE) systems.

This is an important step in pre-production, which can be done from the editor's desktop using the Rough-Cut Client. An important issue is frame accuracy. To ensure frame-accurate synchronization between preview and contribution quality material, the Rough-Cut Client has to offer an automatic synchronization with the default master copy that is represented in the CMS (i.e. in the Essence Management). Further, manual synchronization by selecting a master copy known to the Essence Management should also be supported. This should be enhanced by the possibility to manually enter a timecode offset and a reference to a master copy.

It is not intended to replace NLE with the Rough-Cut Client. It rather should be a simple tool that does not try to compete with an offline editor, but focuses on selecting clips from the CMS and arranging them into a sequence that is a first draft of the planned new program. Thus, its functionality is limited to a set of basic functions that include:

- the possibility to set mark-in and mark-out points;
- the possibility to annotate edits;
- the display of the current EDL (including media object title, timecodes, annotations);
- the manipulation of the current EDL (including shifting items up and down and activating and de-activating items);
- allowing access of a shared EDL in multiple player windows;

- assembling a joint EDL from multiple media players;
- the option to show different players as source and record monitors;
- the option to show and manipulate the sequence in a visual time-line (may include thumbnails for video tracks and waveform display for audio tracks);
- replaying an EDL (only of enabled items);
- saving and restoring sequences locally and in the Workspace Management for sharing;
- importing/exporting EDL in various formats (e.g. AVID ALE).

Some of these functionalities are only useful in video rough-cutting, while others apply to both video and pure audio processes. For certain usage scenarios it may be useful to even add a little bit more functionality to the client that allows the user to introduce hints for the craft editing step, such as:

- split audio editing, hinting at the desired relation of audio and video tracks in the new program;
- voice over, allowing recording of an audio track at the journalist's desktop;
- mark-up of SMPTE wipe patterns as a hint to the desired edit effects.

It should be possible to control the Rough-Cut Client with the same control set as is used for the Browsing Client. Figure 9.7 shows a Rough-Cut Client. It displays a source monitor window on the top left and a record monitor on the top right. Above each window mark-in, mark-out, and duration of the source clip and the assembled sequence are presented, and below each monitor window there is the respective control button toolbar. The bottom part of the application hosts a visual time line, together with sliders and buttons that enable access to edit, zoom, and positioning functionalities.

The requirements for such a client may differ considerably from organization to organization, and even between workgroups in the same organization. While one group may only be interested in collecting clips and has no need for any functionality that resembles editing, another group may want to prepare the production much more thoroughly during rough-cut. Hence, the application should provide a good level of configurability to match the user interface displayed to the requirements of the respective user group. Most important, again, is ease of use, fast response, and smooth operation of the application.

9.5.3 COLLECTION AND ORDERING

In retrieval it should not only be possible to search, inspect, and select appropriate content objects but there should also be facilities to organize search results. A good way to do this is the provision of a folder hierarchy where the user can drag-and-drop queries and search results into this workspace. Such a folder hierarchy should provide means to synchronize the contents of multiple folders into one folder, to eliminate duplicates, and to highlight elements that appear as new in the target folder. This folder hierarchy might be a separate view within the Retrieval Client or a fully integrated window within the application layout.

Information in this folder is private to its user. However, it should be possible to share specific items, folders or project information they are representing with other users or user groups. Thus, access rights have to be considered in this context.

In larger organizations the delivery of essence typically involves an entity (often the archive) that checks whether the asset requested has intellectual property rights or other

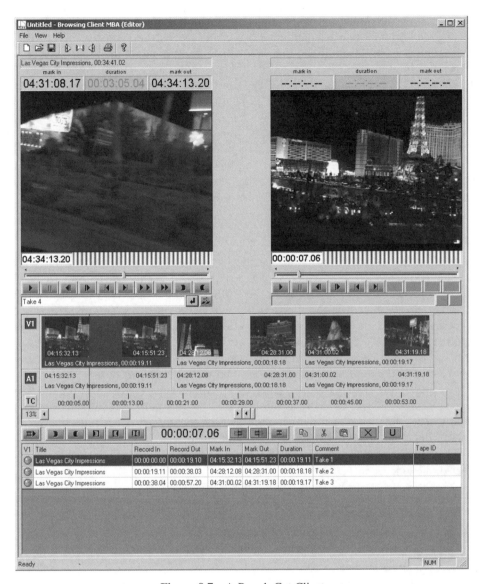

Figure 9.7   A Rough-Cut Client

legal restrictions that do not allow delivery. Further, it checks whether there are other rights that may restrict use of the content object in the planned context and it ensures that the asset is delivered at the desired time in the desired format to the desired location. This may include format conversions, digitization from or playout to tape, etc. This is supported by the *Order Management*, and placing of orders as well as managing the orders should be assisted by user interfaces that make the functionality of the Order Management easily accessible.

It is sensible to integrate the process of ordering into the tree display presented by the Workspace Management. Creating a new order as a sub-folder to an 'orders' section in this tree is a concept that is easy to understand. Adding of assets or EDL to this order via drag-and-drop is also straightforward. An order could be submitted via an option in the context menu of the folder that holds the current order. If the order management holds status information for each order, the folder could change its icon whenever a status change occurs, thus providing direct feedback to the user that has placed the order. Hence, with respect to ordering assets, there is probably no need for a dedicated client.

The people processing an order do have a need for dedicated user interfaces, however. They may need to decide whether an asset is to be delivered as a file or as a tape, they may have to create sub-orders that trigger creation of collection tapes from EDL or digitization of an asset from tape to file, and so on. The functionality of the Order Management has been described in Section 6.5.9. The user interface provided for the Order Management must make this functionality available at the desktop in an efficient way.

9.6 EXPORT

Material in the system has to be exchanged with other systems or is sometimes taken 'off-site' and hence has to be exported. The *Export Client* is a tool that allows scheduling export of content in file formats from the CMS to various targets. The capabilities of an export client depend on the functionality that is provided by the Export Service. A target may be any external system that can accept content as a file for upload. Content is exported from the CMS and uploaded to these external targets, thereby sourcing essence from the Essence Management and metadata from the Data Management. The Export Service creates the file format using essence and metadata as part of this process.

An export process may involve other CMS components besides the Export Server and the Export Service. In such cases the overall process must be configured in (and is controlled by) the Workflow Engine and is processed step by step by the Task Management Service. Important features include the possibilities to:

- select the essence to be exported;
- select the metadata that should go with the essence via templates including the:
  - possibility to add/modify descriptive metadata;
  - possibility to add status information and other properties that steer further processing of the exported objects (e.g. appearance in certain folders of the Workspace Management).
- select the target systems;
- manually start and stop an export process;
- schedule single export processes;
- schedule batch export processes.

For export of content to VTR or similar linear recording devices, essence is staged on a suitable streaming server. Subsequently playout and recording are scheduled and performed using the media management capabilities of the CMS. The Export Client should allow scheduling these exports to studio devices.

Besides classic import and export the CMS should support, via the Import and Export Clients, check-in and check-out for re-versioning of content. Export/import can also be useful to share material between CMS instances that have no network interconnection.

Administrative staff such as feed assistants, catalogers or media managers operate the Export Client. In a system operating within the boundaries of an editorial office research assistants or junior editors might also use it within their daily work.

9.7 SYSTEM ADMINISTRATION

System monitoring and administration is vital for a CMS. This task usually being performed in the background is the administration and daily support and maintenance of the CMS. Since a modular CMS consists of multiple services that may be distributed over a multitude of servers, an application tool is required to enable administrators and media managers to perform this job. These are skilled users who know the system environment as well as the actual infrastructure. The *Administration Client* is the tool they use to perform system monitoring and maintenance tasks. Examples of the functionality it should provide are:

- starting, stopping, and restarting each single service separately on each server, all services on a certain server and the overall system;
- monitoring each service, including active access of services (ping interface), access to internal self test procedures that a service provides, access to dependency trees (a service may depend on other services to be able to perform its task), and access to protocol and log files that services may create and update locally on their host server;
- setting policies and watermarks, e.g. for the Cache Server;
- defining templates, e.g. for ingest or analysis;
- modifying database structures;
- managing user and group rights;
- introducing system components and remote brokers.

The system status should be presented in a graphical user interface, using easy to understand paradigms, such as traffic lights and tree structures.

Figure 9.8 shows an example of a system monitor application. The tree on the left-hand side shows all system components. The status of the component is indicated by a traffic light symbol (green: healthy, red: problems), The tabs on the right-hand side of the application give access to the properties of the selected system component, allowing the user to evaluate dependencies of this component to other system components, and to perform system tests for the selected component.

Basically, the Administration Client should be a platform for a number of client controls, each of which is designed to monitor and configure a certain service or server component of the CMS. Hence, any such component that is added to the CMS needs to provide a well-defined administration interface, and a client control that enables an administrator to perform monitoring, configuration, and maintenance operations using this interface. The component then is responsible for updating its configuration information in the Configuration Service.

9.8 OTHER APPLICATION COMPONENTS

Apart from this essential set of applications there are a number of advanced application views that are required in the context of media production and management. They are

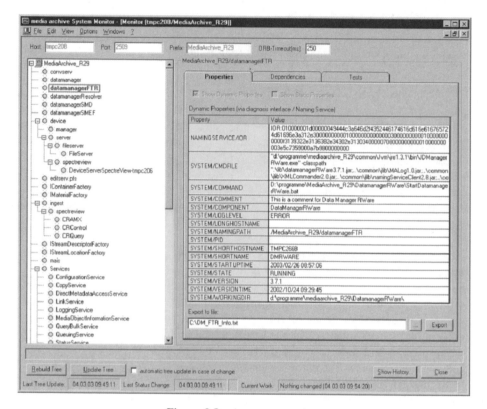

Figure 9.8 A system monitor

sometimes provided by other third-party systems. Thus, the CMS has to offer views that can be integrated in such tools or provide a framework in which they can be presented in the CMS context. In the following, examples of such tools are introduced.

9.8.1 MEDIA PRODUCTION PREPARATION

The creation process of a content object starts with an idea that is sketched out and submitted for funding. This may result in the planning and drafting for a new project with a request for commissioning. Possible applications in this context are the Planning and Drafting Client and the Commissioning Client.

9.8.1.1 Planning and Drafting

Before a new production is commissioned, there are at least two prior steps involved: planning and drafting. For a broadcaster, the planning stage includes for instance a rough plan of the overall yearly broadcast schedule. This comprises the scheduling of the kind of program that should be transmitted giving also day and time references for the weekly schedule (e.g. a 20 minute news show each day at 19:00, a 45 minute political magazine each Monday at 20:15, etc.). Further, more detailed planning of the schedule for each quarter, month, week, and day takes place and should be supported by an application module.

The resulting program plan and schedule includes slots for 'real' programmes that need to be produced to fill these slots. This is where drafting comes in. Drafting means that producers create a proposal for a program that may fill such a slot. This work includes creating a synopsis and providing approximate costs for the production.

Both processes can and should be supported by a CMS. If this is the case metadata can be captured at the very start of the content creation process. From the application point of view, the applications involved handle only metadata but have a need for retrieval capabilities for background research. There is also a need for workflow support, i.e. a draft for a program must be submitted for commissioning, and the result (approved or declined) must be fed back.

Thus, the *Planning and Drafting Client* is a tool to support the processes of planning and drafting as part of the pre-production processes. It:

- may create a new object in the Data Management;
- does not add essence to this object;
- may add first formal metadata, like a working title, an expected duration, a cost estimate, an indication of whether the project is a single piece or a series, an indication of the format, etc.;
- needs to provide support for scripting, comparable to a word processor, in order to create a synopsis or detailed scripts;
- needs to have an interface to the Messaging Service in order to enable notification of the commissioners and to receive results.

This tool should support the initial drafting as a preparation step to commissioning as well as more detailed scripting after commissioning as a planning step towards production.

9.8.1.2 Commissioning

From the application point of view, commissioning is a straightforward task. All drafts that have been submitted for approval are listed, and the commissioner can access each proposal, review it, and decide whether the proposal should be accepted or not.

If a proposal is accepted a budget and a production number are assigned, and the production gets a 'go ahead' label. Here obviously an interface between the Enterprise Resource Planning (ERP) system and the CMS is desirable since managing costs is not a task of a CMS.

Thus, the *Commissioning Client* is a client to support commissioners in deciding whether a certain proposal may be accepted for funding. It needs to have an interface to the Messaging Service in order to receive notifications for new proposals. Subsequently it should allow browsing through the list of proposals. Therefore all the relevant metadata created in the drafting stage has to be made available to the commissioner.

In order to support the commissioning decision it has to be possible to create additional metadata such as funding details, production numbers, etc. Since this process is closely related to the financial data held in the ERP it should have an interface to the organization's accounting system in order to establish an account for the new project and to generate a production number referencing that account.

Further, it needs to support the documentation of the commissioner's decision (accepted or rejected, together with an identification of the commissioner. In order to communicate the decision there should be an interface to the Messaging Service to send back responses to requests for commissioning. Ideally this is done in the context of the CMS and not via normal email so that it can be directly associated with the planned project.

9.8.2 RIGHTS CLEARANCE

IPR issues and rights management are crucial in the production and broadcasting work-flow. As already discussed, there are various points in the workflow where the CMS should be synchronized with the rights management system. The *Clearance Client* is a tool that should be used by the rights department to clear the rights situation for a new production at clip level, based on the rough edit decision list. This client should support the responsible person in the rights department to retrieve enough information from the CMS and the rights database to process the clearance request clip by clip. This includes an indication of the rights situation for each clip. The result of this rights review process is communicated back to the inquiring person by sending a message through the Messaging Service that is an integral part of the CMS.

10

Future Trends

What is called content management these days is a task that has been around for some time. Archives and libraries have traditionally performed this work. However, the content managed in this context used to be (and still is) on physical carriers (such as books, files, film, videotapes) and the growth and preservation task (though substantial) could until recently still be handled by traditional archiving methods. In recent years the need for reorganizing the way content is handled and administered has been caused by two main trends, i.e. the ever faster growing amount of multimedia content, and the increasing time pressure in content production and distribution. In the latter case the reuse of existing content has been becoming more and more important. In addition, the number of different media formats and output channels over which content has to be distributed is also rising. Thus, there is now more content in many more different formats to be administer than ever before. Further, the reuse cycles are getting shorter and content management is becoming a central part of the content production and distribution process rather than an end-of-chain task.

Content management provides the means to cope with all these requirements and changes. This is accompanied by new digital encoding and compression formats, developments in the electronic media space (mainly the World Wide Web), and advancing technologies. The latter are mainly based on multimedia technology developed in the IT and computer sector in the past decade. The emerging CMS are also based on this technology. They make use of the advantages of this technology to facilitate the handling of digital formats and can cope efficiently with vast amounts of data.

However, although the term CMS is used in a general way it has so far not been defined what the features and functionality of a generic CMS are. The term CMS is used for instance for a number of IT-based systems that primarily deal with the management of Web pages. Other content or asset management systems are mainly administering documents, and if they are more advanced they sometimes provide more or less rudimentary support for continuous media as well. The focus of this book are professional CMS used in media production and broadcasting. These kinds of CMS are capable of managing a vast number of content objects (represented by a number of different essence copies in different formats and a set of metadata). Further, they can

Professional Content Management Systems: Handling Digital Media Assets   A. Mauthe, P. Thomas
© 2004 John Wiley & Sons, Ltd   ISBN: 0-470-85542-8

interface to special production and broadcast systems and are also able to integrate legacy systems.

Thus, the expression *content management system* has a number of connotations and meanings depending on the context it is used in. There is currently no tendency away from the specialized systems towards more generic CMS. This makes any predictions concerning market and technology trends difficult. Expectations concerning the market growth in this area have not been met either. In fact, so far no single market for content management products has emerged. Rather, CMS are part of sub-markets of different domain-specific markets such as Web publishing technology, document management, or the media production and broadcasting space. Since the requirements in all these sectors are quite diverse, different technological approaches have been taken. Although the technology is mainly based on common IT products, no middleware or platforms have been developed that could host CMS catering for all the different requirements. Thus, there is neither a universal CMS market as such nor a common content management technology that could provide generic support for the handling of universal content objects.

What makes it even harder to forecast the future trends in this sector is the fact that the areas where content has to be managed are also changing. For some time it has been predicted that IT and broadcast technology would converge, that telecommunications networks would be used for content delivery as well as for data and interactive interpersonal communication, and that the entertainment business would become part of the Web (or Internet). None of this has actually happened in the predicted way up to now but the change is imminent. It is therefore currently not possible to discern a clear trend of how content management will develop in the future.

However, there are interest groups and standardization initiatives addressing content and content management related issues. This chapter discusses some selected examples to show different approaches deemed relevant in this context. In keeping with the theme of the book they are mainly concerned with content management in a professional, enterprise environment. The list of examples is not complete but gives a good indication of the range of relevant initiatives in this context. In particular these are the MPEG-21 standardization, relevant discussions within EBU and SMPTE on content and content management related issues, and the IETF initiative on content delivery networks. The latter has been chosen to demonstrate how a future Internet-based distribution platform that manages and places content within a networked infrastructure would look. The chapter, and therefore the book, is concluded with a discussion on experiences gained so far and an outlook on expected future developments.

10.1 MPEG-21: MULTIMEDIA FRAMEWORK

The Moving Pictures Expert Group has recognized the need to define a framework that describes how the different elements of content management fit together in order to provide 'the big picture' (MPEG, 2002c). This resulted in the specification of the MPEG-21 Multimedia Framework. MPEG-21 is concerned with the entire (fully electronic) workflow of digital multimedia content creation delivery and trading. Its aim is to cover interaction with multimedia content and to provide a framework for the transparent usage of various content types and multimedia resources on multiple devices connected through a wide range of networks.

MPEG-21 has seven key elements:

- *Content Handling and Usage* is an interface specification that covers all workflow steps in the content value chain from content creation, over its manipulation, search and storage, to its delivery and reuse. The vision technology and strategy for such a framework was defined and formally approved in September 2001 (MPEG, 2001).
- *Digital Item Declaration* is a scheme for declaring Digital Items by a set of standardized abstract terms and concepts; i.e. it specifies the makeup, structure and organization of essence and content objects called Digital Items. Digital Items can be, for instance, audiovisual objects but also Web pages containing links and script commands.
- *Digital Item Identification* is a framework for the identification and description of entities regardless of their nature and granularity. It does not specify new identifiers for parts or areas for which description schemes already exist such as ISRC for sound recording.
- *Intellectual Property Management and Protection* (IPMP) deals with IPR management and protection at all involved devices and networks. Important in this context is the interoperability between different systems. It includes standardized ways of retrieving IPMP tools from remote locations, and exchanging messages between IPMP tools and between IPMP tools and the terminal. It also addresses authentication of IPMP tools. Further, it has provisions for integrating Rights Expressions according to the Rights Data Dictionary and the Rights Expression Language.
- *Terminal and Networks* deals with the functional interoperability between heterogeneous networks and devices. The actual network and terminal configuration, and management and implementation issues should be transparent to the users. This should facilitate the provision of network and terminal resources on demand to form user communities where multimedia content can be created and shared. This should be achieved by the use of Descriptor Adaptation Engines and Resource Adaptation Engines that together provide the Digital Item Adaptation.
- *Content Representation* specifies how media resources are represented.
- *Event Reporting* defines the metrics necessary to understand performance and event reports in an MPEG-21 system.

Figure 10.1 depicts the transaction and use of an MPEG-21 compliant system and how this is supported by the MPEG-21 key elements. At its most basic level, MPEG-21 provides a framework in which one user interacts with another where the object of the interaction is a piece of content (called a Digital Item in the case of MPEG-21). These interactions include creating content, providing content, archiving content, rating content, enhancing and delivering content, aggregating content, delivering content, syndicating content, retail selling of content, consuming content, subscribing to content, regulating content, and facilitating and regulating transactions that occur from any of the above.

The MPEG-21 standardization commenced in late 1999, but considering the immense task and large scope of the project it is still at a very early stage. At present much effort is spent on IPR management and protection issues covering all the legal aspects involved in this.

MPEG-21 is an IT-driven initiative and has a strong focus on content management in the context of the Web as an open distribution network with features as provided in the Internet. With its strong focus on rights-related issues it aims at enabling trading and e-commerce interaction in such an open environment. Organizational content management

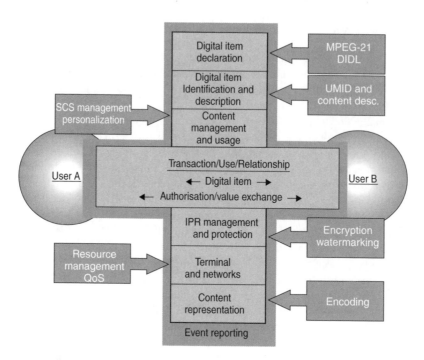

Figure 10.1 User model for transaction supported by the seven key elements

processes are not one of the main objectives of the emerging standard. It is still an open issue if the MPEG-21 framework is suitable and open enough to cover also the requirements of professional CMS such as the ones discussed in this book. Relevant in this context are not primarily internal processes but especially content exchange and trading. It would therefore be important for MPEG-21 to collaborate closely with other standardization initiatives within the broadcast and media production domain (such as for instance SMPTE Metadata Dictionary, P/Meta, and UMID). CMS such as the ones introduced here are going to use for instance the proposed metadata models, dictionary, and globally unique identification scheme standardized by these bodies to exchange or sell content on a business-to-business but possibly also on a business-to-consumer level. MPEG-21 can only succeed if the major content owners and retailers participate in the standardization process and also deploy the standard in future systems. Therefore it needs to be aligned with other standards used in this context at the respective organizations. The standardization process within MPEG-21 still has a long way to go. Whether it will have any impact remains to be seen.

10.2 RELEVANT BROADCAST INITIATIVES

Content management in a broadcast environment has been addressed by the respective interests groups and standardization bodies for some time. The Society of Motion Picture and Television Engineers (SMPTE) and the European Broadcasting Union (EBU) have been developing related standards and codes of practice since the mid-1990s. For instance the joint taskforce for Harmonized Standards for the Exchange of Program

Material as Bitstreams (SMPTE/EBU, 1998) initially defined the term content. Other relevant standards in this area are the already discussed SMPTE Metadata Dictionary (see Section 4.4.3; SMPTE, 2001a) or the P/Meta initiative for the standardization of a common data model within the EBU (EBU, 2001). The initiative that contributed the most to understanding the implication of CMS in broadcast was the EBU Project 'Future Television Archives' (P/FTA).

10.2.1 EBU PROJECT FUTURE TELEVISION ARCHIVES

Closely related to the topic discussed in this book is the EBU project 'Future Television Archives' (P/FTA) (EBU Projects Group, 2001). The project group was set up by the Production Management Committee (PMC) of the EBU. The group was charged with providing an overview of the options available for the technical realization of future television archives, taking into account existing technology, and proposing ways of migrating from the current conservative and static legacy-based content production to new production architectures that could accommodate continuous adaptation to the rapid pace of technological development (EBU Projects Group, 2001).

P/FTA has ensured that its work is in compliance with the final report of the joint EBU/SMPTE Task Force (SMPTE/EBU, 1998), which defined new technical elements needed in a future networked television production, storage, and distribution environment. The new system concept described in the report primarily pursues the following objectives (EBU Projects Group, 2001):

- increased flexibility in production, post-production, and dissemination of content;
- enhanced operational efficiency and overall productivity.

P/FTA has provided a fertile environment for dialog on the multitude of complex issues involved in new system deployment. Participating in the discussions were EBU members, consultants specializing in archival matters, archivist organizations such as FIAT/IFTA, INA, and AMIA, international broadcast organizations, the SMPTE, and a great number of manufacturers of products related to television networking and storage (EBU Projects Group, 2001).

It discussed relevant workflows, media formats, content representation, and the incorporation of related metadata models and information systems into a television archive based on digital formats administered by an enterprise-wide CMS. It also introduced a software architecture and a hardware infrastructure that could be used as reference models for CMS in television and broadcasting (EBU Projects Group, 2001).

The discussions within P/FTA clearly point to the following conclusions (EBU Projects Group, 2001):

- The problems raised by the increasing demands for higher production volumes, faster program throughput, long-term content integrity, and flexible manipulation of content will not be solved by video-centric production technology.
- A new system structure will need to be established where archives become an integral part of a future networked production environment, supported by software-controlled, automated content management systems.

- Investments in the new system structure are justified because program production becomes much more flexible and cost-effective. This in turn helps broadcasters become more competitive as content providers, which will ensure their survival in the longer term.

P/FTA firmly believes that these issues can only be resolved by introducing an IT-based CMS, primarily employing file-based operations and capable of concurrently managing a wide range of media formats, recording formats, and carriers (EBU Projects Group, 2001).

The core part of the P/FTA final report outlines the expected architecture, the services, and functionality of such a CMS. It does not try to re-model current businesses with new technology, but requires establishing new business processes and may open up new business opportunities. Before deploying a CMS, a number of decisions, related to policy, technology, and management of an interim period must be taken. The decisions identified by P/FTA are listed below and are supported by supplements to the final report wherever possible (EBU Projects Group, 2001).

Clearance of Content IPR Conditions

To transform content into an asset, a secure legal platform must be established to allow the exploitation and exchange of broadcast and multimedia content on all foreseeable technical platforms. The legal information required for these transactions should be structured in such a way to permit handling and remuneration of author's rights in an automated way. The methods already established in sound broadcasting should be targeted as a goal.

Content Production and Exploitation

The more frequent and broader internal use of library content in all phases of program production will justify investment in updates of technical tools for preserving and administering content. The success of some broadcasters that exploit film archives clearly indicates that interesting content will not lose its appeal to the public, even when presented decades after production. The advent of new thematic channels is an efficient means and a first step to that end. New distribution channels of the future will allow viewers to pull content on demand from broadcaster-operated libraries. Additional revenue can be derived from metadata and content catalogs used for internal purposes by granting public access to them, pushing selected content to audiences via IP-based file transfer or by allowing download of content or content segments to certified multimedia customers.

A Business Plan Defining the Quality Potential, Target Audience, and Exploitable Lifespan of Each Individual Product

Today's content, produced for consumption through conventional display technology, may soon prove barely acceptable when presented through new display technologies (PDP, DLP). New packaged media, if authored carefully, offer a significantly higher quality potential for display. The breathtaking developments in storage and display technologies for multimedia content have become completely decoupled from the service and quality offerings of broadcasters. To support long-term content exploitation, data volumes and the associated cost for storage and throughput should be adjusted to accommodate the high data rates required to retain the initial quality and post-production potential of the archived content.

Staff Qualification and Shorter Investment Cycles

The change of public expectations, driven by rapidly evolving new consumer technologies, will set new references for quality and functionality. Under these increasingly competitive circumstances, it will become more and more difficult to continue to be the audience's quality of choice. Among the minimum requirements for retaining competition and operational flexibility to allow for a timely response to new consumer demands, Moore's law should be made to work for broadcasters by ensuring that system investment and hardware updates track innovation cycles. This will require the allocation of a predictable annual budget for the purchase of storage media, maintenance, and updates of system hardware and software. The operation of the system will establish new business processes and lead to new job profiles. These new job profiles will in some cases require combining traditional skills with computer knowledge and networked systems experience.

Generation, Distribution and Exploitation of Metadata

Closed-loop generation of metadata related to all relevant aspects of content creation, storage, manipulation, and exploitation must be established. This is a prerequisite for deriving subsequent decisions when tailoring content for specific applications and for archive housekeeping purposes, without an unacceptable amount of manual intervention. Content-related metadata should be derived automatically as far as possible and should be supported by acquisition, production, and storage hardware.

Creating Low-Resolution Proxies of Content

Modern technology offers multi-user access to content in various form such as high and low resolution, keyframe based or textual, opening new opportunities for product diversification and enhanced program volume throughput. Significant time savings can be achieved by adopting a mode of operation which performs straightforward search and editing operations on the low-resolution replica of an original content for program assembly and re-purposing. It will further allow operations via conventional networks within a proven technology, already deployed for audio. Further economic benefits can be derived when selecting an identical quality margin and encoding algorithm for the low-resolution proxy and that needed for Web release.

Introducing Automation for Content Storage

Accommodating content stored on electronic media in robotized libraries will allow close control of the original content by permanently retaining the storage medium under controlled environmental conditions and will obviate the need to remove cassettes by offering authorized users remote access to content. This prevents loss of storage media and reduces the need for multiple security copies. New interfaces allow the routing of content in its native form without loss of quality, generating clones of the stored content for further processing. There is general recognition that the physical and chemical properties of motion picture film and its high quality potential require special treatment in production and preservation.

Adopting an IT-Based Platform for Content Manipulation, Exchange, and Storage

Separating the platforms for content manipulation, storage, and transfer will finally overcome the painful process of selecting a new recording format at regular intervals with the salient time consuming transfer of legacy content, loss of quality, and the commitment to one manufacturer's product. Accommodating content within an agreed and standardized file structure for transfer will lead to an expanded market for competing storage products. Integrating television applications into the large application range of general application IT-based storage platforms will facilitate long-term system planning within a rapidly expanding market and will help to enforce backward compatibility of storage products as technology progresses. It will further provide the choice of the storage technology best matched to a particular application and provide a system management designed to handle diverse storage media on a single platform. New storage system management concepts can adapt to growing needs by offering more efficient ways of using storage volume and by supporting high speed transfer of content in distributed and diversified environments, across different operating platforms. Software-controlled, scheduled 'background migration' operations and automatic monitoring of both media and content integrity will permit control of the floor space of the content library and will provide uninterrupted operations during migration to new storage media.

System Development and Maintenance

To secure previous investments in the system, equipment should be selected that is modular in its design and available from a range of competing manufacturers. Additionally, the technical descriptions of interfaces and protocols used in the equipment must be in the public domain, and they should preferably not be proprietary.

Interesting in this context is the fact that this group also acknowledged the move away from the linear program production chain towards a content-centric model in which the CMS as part of (or operated by) the archive forms the central hub for all content-related processes. In this model the content is placed at the center of the business process supported by the infrastructure for its management and exchange. The business processes exploiting content are pre-production, production, post-production, delivery, and sales. In the pre-production stage content exists as ideas, scripts, production planning documents, etc. This information is exchanged between the participating parties and archived as part of the metadata. During production content is generated and added to the content pool. Post-production refers to the editing and presentation process for the content. In broadcasting this is the actual video and audio editing process, the addition of special effects and graphics, and the preparation for delivery. Although the group has a strong broadcast bias it also acknowledges that the delivery process to the consumer is becoming more and more diverse. Whereas ten years ago it was restricted to broadcasting (terrestrial, cable or satellite), content is now also delivered digitally via various delivery channels such as the Web, digital TV, etc.

Advantages of the integrated content-centric process model using new management and platform technologies are the reduced preservation costs for content, and better management and easier access to the content itself. In the pre-production stage existing content can be more easily considered and new ideas, scripts, etc. can be integrated in the system before the production has even started. During production the new technologies supporting this model allow quicker and easier access to the actual content. Also, all involved parties can observe the progress of the production process and subsequent work steps can be prepared well in advance.

It is hoped that new delivery channels will open new markets for traditional media companies and broadcast organizations. Media sales (which currently have only a minor share of the revenue stream of a broadcaster) are expected to pick up when new technologies make it easier to advertise, access, and deliver content. There is supposed to be a big increase in the volume of business-to-business sales since new channels cannot be served by new productions only. Thus, the role of the archive changes considerably. It is expected to become a more important, central but also profitable department within a broadcaster.

The P/FTA final report proposes an integrated technology approach for this content-centric process and archive model to realize synergy potentials and economies of scale. This is quite similar to the concepts introduced in this book, which are, just as the P/FTA report, mainly based on experience gained with a number of projects in this up to now.

10.3 IETF INITIATIVE FOR CONTENT DISTRIBUTION NETWORKS (CDN) AND CONTENT DISTRIBUTION INTERNETWORKING (CDI)

The IEFT Content Distribution Networking (CDN) and Content Distribution Internetworking (CDI) initiatives are mainly concerned with content distribution and do not have internal content management as their objective. However, the concepts developed in this context are interesting since they require distributed content management. While it may not be possible to use these concepts and the developed protocols unchanged, they give a good insight into a more communication oriented and distributed approach from which elements can become relevant.

The advent of the World Wide Web has caused a change in the use of the Internet. Whereas before it was mainly employed for the transmission of electronic data it is now increasingly used as an information source and for the exchange of content. This has been recognized by the IETF by setting up the Content Distribution Networks (CDN) and Content Distribution Internetworking (CDI) initiatives, which have resulted in the IETF CDI WG. CDN/CDI are intended to serve as platforms for content providers to distribute their content without having to manage an entire infrastructure. The work concerning CDI is still at the requirements stage. At present it has a clear focus on content distribution in a Web context. Issues being addressed by the working group include a model for CDI (Day *et al.*, 2002), architectural questions (Green *et al.*, 2002), distribution requirements (Amini *et al.*, 2002), and CDI Authentication Authorization and Accounting (AAA) requirements (Gilletti *et al.*, 2002). The central aspect in this context is the location, download, and usage tracking of content in CDN/CDI.

Content networks can be regarded as overlay networks operating on top of the actual communication subsystem. Such networks use high-level protocols such as HTTP, RTP, RTSP or emerging communication protocols such as SOAP. Thus, CDN operate over existing IP-based networks and protocols using the infrastructure already in place. A CDN deals with routing and forwarding of requests and responses for content. The basic problem CDN try to solve is to allow a requestor for content (so called content sink) to locate the content of interest and get it delivered. Content items (especially audiovisual content) can be very large in volume and require considerable bandwidth and storage. Hence dissemination mechanisms are required to distribute the load within a CDN. CDN provide mechanisms to deal with these problems. Key concepts are caching proxies, server farms, and routing mechanisms that consider network conditions and load, and are aware of replicas in the CDN.

Sever farms are a group of servers that appear to the outside as one single origin site. Multiple copies of the content might be stored on different servers of a server farm. Requests can then be served by different servers of the server group, thus balancing the load across all available resources. Server farms also provide a degree of resilience since requests are routed away from servers that fail. In a CDN server-farm-like configurations are placed at strategic locations commonly occupied by proxies[1] to host replicas of content. These structures are called surrogate servers. Distribution mechanisms move content from the origin server to the surrogates considering dynamic information about the network state. Requests are routed towards the most appropriate surrogate server considering network conditions and load information.

An accounting infrastructure within a CDN tracks and collects data on request routing and the distribution and delivery of content. Usually the CDN service provider acts on behalf of the content publisher to provide the infrastructure (including accounting infrastructure) for the distribution of their content. Within a CDN content from multiple content publishers might be hosted. The service provider benefits from economies of scale serving different content publishers whereas the content publishers have a larger audience through an increased reach.

Conventionally a CDN is operated by a single service provider (i.e. in a single network context). In order to improve the scalability and increase the reach of a CDN, Content Distribution Internetworking (CDI) was proposed. This allows resource sharing over network boundaries between multiple networks and multiple providers. In such a construct each participating CDN can still hide internal details from others and remain independent in its decision about content location and delivery. Thus, a CDN is a black box to its neighboring CDN. The internetworking between CDN is achieved by three major concepts, namely advertising content, replication of content, and signaling.

The concepts and protocols developed in the context of the IEFT CDN/CDI initiatives are relevant for distributed content management since the placement and handling of content within the CDN is its main objective. Although the protocols are being designed having mainly Web content in mind some of the concepts are relevant to professional content management and point to a more distributed, less centralized organization of a CMS. For instance departmental CMS could be considered as CDN that are interconnected using CDI technology. Storage and other technology, of course, would have to be adapted. There is still a considerable amount of research in linking the different concepts. However, considering the success and proliferation of the use of IP-based technology it could be expected that these new developments may be also swiftly adopted and used in contexts they have not been intended for (such as professional, enterprise-wide content management).

10.4 EXPERIENCES AND OUTLOOK

CMS technology for the enterprise-wide, professional management of content has reached a stage where the first systems are now being productively deployed in day-to-day operations. In the development of these systems a number of experiences and insights have been gained that help develop CMS technology further. We are now at a stage where it is interesting to examine the experiences and major issues that have occurred (i.e. the lessons learned), but also to look at the future and relevant developments that are interesting for CMS. The former deals with aspects that have changed the outlook and perception of features and/or technologies that support content management. Future developments refer to areas that are expected to enrich the features set of a CMS by enhancing its functionality

---

[1] Note, proxies in this context are cache proxies and not related to the representation of content objects as introduced in Section 4.1.

or improving its operation efficiency. It is important to consider both in order to derive the relevant trends and manage the expectations when developing CMS further.

10.4.1 CMS-RELATED ISSUES

The initial expectation in relation to CMS was that it would be the platform for future digital, tapeless production, management, and distribution of multimedia content over existing and upcoming delivery channels. The actual management should be highly automatic using computerized analysis, documentation, handling, and administration tools whenever possible. At the same time the systems should be generic and ubiquitous, allowing the easy exchange of content across organizational boundaries with no or little manual intervention regarding all elements of a content object (i.e. essence and metadata). Further, systems should be scalable, and format- and media-agnostic. This means that it has to be possible to use the same (or at least compatible) technology to manage a relatively small number of audiovisual, low-bandwidth, low-resolution objects as well as the archive of the world's largest broadcasters (containing all kinds of media in various resolutions and possibly stored on multiple different carriers). The content objects in this context may be audio recordings, video, and images, but also Web pages, documents, etc. Furthermore, a CMS should be flexible enough to support and adapt to all kinds of different use cases and workflows. Although a lot of progress has been made, most of these expectations have (at least partially) been disappointed. The following subsections outline what the issues in this context are.

Another issue in the deployment of CMS in a professional environment is that IT-based CMS are perceived as inherently unreliable. In 24×7 operations any downtime can have disastrous consequences. More established systems (such as automation systems or broadcast servers) have built-in mechanisms to cope with hardware and software failures. Though at the core these systems nowadays often also use IT technology they are seen as being much more stable. Thus, it is a matter of system design and investing in the right places to make IT-based CMS as reliable as necessary for a specific deployment. This together with a sensible project implementation plan considering scheduling as well as technical feasibilities should ensure that CMS are successfully introduced into a content-rich organization.

10.4.1.1 Automatic Media Analysis and Processing Tools

Since mid-1990 automatic processing of audiovisual content has seen a rapid development. There are tools that process video to detect transitions and edit effects, extract keyframes and metadata, and allow the classification of video objects. These processes are largely based on the analysis of low-level video and image descriptors such as color histograms, shape and texture recognition, motion vectors, etc. They can relatively reliably discover features that can be derived from these low-level descriptors. This has been even recognized by standardization bodies and has found its way into the MPEG-7 Video and Audio part. However, since most tools have no notion of media semantics there are limits to what kind of information they can retrieve. Thus, the results they produce are often inaccurate or do not match the user's expectations. For instance if users are searching a certain object or person using image similarity recognition they expect images that just show this object or person in different situations. What they get back is a number of images where the mathematical and statistical values of the low-level descriptors are within a given range. Hence, there may be a whole set of images the user cannot relate to his or her initial query. This, of course, disappoints the users' expectations and hence they perceive such tools as 'unusable.'

Automatic indexing also does not always produce 100% accurate results. For instance tennis matches can be relatively easily classified. However, an automatic indexing tool may also classify an advert in which features tennis as a part of the product presentation. Similarly, comedy shows using the news format for fake news presentations would conventionally be classified as news.

Automatic metadata retrieval based on audio and speech analysis also has usually only an accuracy of 95% or less. Since they try to match every discovered phrase to a known expression the transcript can be distorted. Especially with texts that use a specific wording or specialist language the hit rate can become very low.

Thus, current automatic media analysis and processing tools cannot be used without human qualification. It is also important that they are deployed in the right way and to manage the expectations of the users. For image or audio similarity search additional textual information might have to be used to establish a context. Automatic indexing and metadata retrieval will have to be revised by a human expert to ensure the right result. However, these tools can be deployed usefully within CMS and provide added value if their capabilities are used in the right way within an integrated system.

10.4.1.2 CMS as Generic, Open Platform

Many see the CMS as the instrument that 'glues together' the different systems that are involved in media production, broadcasting, and archiving. As such it has to be able to easily integrate all kinds of proprietary hardware and software solutions, legacy systems, different products from alternative vendors, etc. Further, it should also be able to handle all kinds of different media types transparently while still providing advanced support for automatic processing or specific applications. This request can often not be fulfilled since products and devices to be integrated as well as legacy systems require very specific support. Often their interfaces are not well defined or open to integration. Thus, a CMS (however flexible and open it might be) may not be able to integrate all the related systems and act as a platform because of the missing support from the peering systems. In most cases the CMS can also only provide the functionality the special integrated device has. If a browse video encoder, for instance, does not support timecodes, the CMS also can not provide imprinted timecodes in the browse video. Also, each integration of a new system component may require very specific adaptations. However, by developing a better understanding of characteristics of the systems and the requirements of the users and workflow a higher abstraction level can be reached. This allows providing integration hooks and a generic feature set that can be supported. The system architecture presented in this book is already a first step in this direction.

Despite the fact that content objects are handled as digital data (i.e. bits and bytes) in the system, a transparent and format-agnostic media handling is not possible if advanced processing has to be supported as well. Even to choose the appropriate means of delivery the CMS has, for instance, to be able to interpret the media type. Continuous media can be streamed at a certain data rate depending on its format. Proprietary formats also require proprietary tools since their internal structure is not disclosed. The presentation of content may also depend on the capabilities of the reception device. A high-end device might be able to receive full-resolution, full-bandwidth video, whereas a mobile client can only accept a very low-bit rate browse version. Regarding structured content formats (such as Web pages) it is even more difficult to (automatically) select the relevant content parts. Images might be deemed dispensable for low-capability mobile devices. However, they might contain the most valuable information and it would be better to transmit a scaled down version than just the accompanying text. Thus, a CMS can either just provide very basic support for the handling of digital data or more sophisticated operations for which it requires knowledge about the structure, syntax, and sometimes even the semantic of the content.

Different description schemes and metadata models also often hamper the introduction of a CMS. Even within the same organization (sometimes within the same department) a multitude of incompatible information systems and databases can be found. A CMS cannot alleviate this situation as long as no common, basic metadata set has been agreed upon. The CMS can then help to make the different systems more easily accessible and to integrate them into applications. However, the problem of defining the appropriate set of metadata persists and has to be solved by the standardization bodies and interest groups, or at least for an organization within a CMS project.

The requirement to have a scalable system that accommodates small, departmental-level requirements as well as the needs of large archives is also quite challenging. The former might be based on plain IT infrastructure components whereas the latter involves the integrated of sophisticated storage infrastructures, large legacy databases, interface to ERP, automation systems, NLE and newsroom systems, etc. Using the same architecture for both might be an overkill in the case of the small-scale system. In these circumstances it would be better to use a system with the same application and system interfaces but a reduced and adapted set of back-end components. The technology for the different systems has to be carefully chosen to make the system interoperable but still make it possible to accommodate the special requirements of each setup.

10.4.1.3 Workflow Considerations

The main goal of a CMS is to support users during creation, production, search, and consumption of content. Different content-rich organizations, professional groups, and individual users have different requirements towards the system and the interaction with the various system components. Further, there are established workflows that should be captured and reflected by the CMS while still providing scope for change.

Initially on both sides (i.e. with the system developers as well as with the users) there was a lack of understanding of the requirements and capabilities of the system. Further, the expectations were sometimes unduly high and could not be fulfilled by the actual implementation. This led to disappointments on both sides. To the users the CMS seemed to be too restrictive and not flexible enough, whereas the user requirements were deemed to be unreasonable to the system designers. This has been recognized: the component-based approach, for instance, allows much more flexible systems to be built. Workflow engines also provide support for a set of predefined workflows. Further, early involvement of the user in a CMS project helps to capture their requirements near the beginning. This helps to take their concerns into account but also to explain the potentials and limits of the technology. It is important to form a common understanding of what the system can and cannot do and to make it clear that it will facilitate the work of media professionals and support distribution but will not replace everything by automatic processes.

10.4.2 FUTURE DEVELOPMENTS

CMS today can be used in practical applications in content production, administration, transmission, and sales. However, there is scope for further development and improvement of the components and subsystems a CMS is composed of. In particular these are:

- better support for automatic content processing;
- better support to cope with the vast amount of content and content-related metadata;
- more flexible infrastructures accommodating different system types and requirements;
- better support for IPR management and protection.

The first three points deal with improvements and further developments. However, it is not expected that these are straightforward enhancements of the current technology. New concepts are needed to overcome the shortcomings of the current approaches. The latter point refers to the growing importance of rights and rights protection. Although there are a number of rights management systems and protection technologies (e.g. watermarking) being developed, this aspect has so far not received a very high priority in professional content management, though with the emerging new distribution channels and more digital content formats being used in the public domain this is becoming a major issue that also has to be addressed in the context of CMS. Thus, the CMS will eventually become a true Digital Asset Management System.

It is also expected that the understanding between system designer and users for the different capabilities and requirements will grow. Vendors have learned from experience and users have also recognized that the new technology will help rather than hinder, but also that it will not be able to do the miraculous.

10.4.2.1 Developments in Automatic Content Processing

Existing automatic content analysis and processing tools will be improved and developed further on the basis of current technology. However, to meet the requirements and expectations of human users it is necessary to also consider the context and semantics of the media alongside the solely technical low-level descriptors. How this can be done automatically is currently being investigated. Whereas the low-level descriptors are entirely mathematic/technical, the semantic attributes go beyond purely analytical processes into the knowledge domain. Substantial research has been done in this area but it has not yet really been applied together with automatic analysis tools in content management.

Besides the media semantic other input parameters may also be considered in automatic content processing to improve their results. For instance, to facilitate image similarity retrieval not only the example image may be used as input parameter but also the user context (e.g. the user role, department, interests, etc.), which can be derived from the user's profile (i.e. no additional input from the user at the search stage would be required). This could help to rank the hits when similar images are found by linking the additional information about the content and comparing to the relevant factors from the user's profile.

For example, if a political editor uses the image of a person all hits that contain persons from the political domain may be ranked first.

Another example would be to use pre-computed information about a specific domain during the automatic analysis of video. For instance crowd scenes in sport footage rarely contain relevant information. Since they represent, however, a significant change in image content they feature frequently among the produced set of keyframes. Additional classification information could be used to suppress these frames or give them a lower priority.

Thus, there are two kinds of developments that can further improve automatic content processing and analysis processes, namely the use of context variables and additional input parameters to further refine the automatic processing results, and the exploitation of media semantics in addition to the low-level descriptors. Whereas the former is probably more an engineering task, the latter still requires some considerable research. This requires an interdisciplinary approach since media semantics cannot be captured by purely technical means.

10.4.2.2 Information and Metadata Management

With the rapid growth of the amount of content and the development of more and more elaborate but incompatible description schemes, it is becoming increasingly difficult to handle the large amount of information that is present in a CMS. Even a more detailed description in this context does not necessarily mean better search results. On the contrary, it might become more difficult to locate the relevant information. Thesauri and specialized description and classification schemes do not necessarily help since only skilled users are able to apply them. Tests at a large European broadcaster have shown that editors find it harder to find relevant content when it is documented by trained catalogers compared to the documentation provided by one of their junior editors. Thus, a more 'natural' way to represent and locate knowledge is required.

One way to achieve this is to capture the meaning of a term describing content and to reflect this in the way it is represented. Ontologies provide the means to do exactly this. They can help to depict the knowledge relating to a specific subject area and build a network that represents this knowledge space. For example, when looking for specific medical content on bacteria that cause a particular disease (e.g. diarrhoea) it is not necessary to know the exact name of the bacteria. Using a description of the symptoms, the ontology can provide the relevant domain-specific term alongside a detailed explanation and links to relevant content objects. It is important to note that this process goes beyond current search processes and encompasses an entire knowledge domain from which the relevant information is retrieved. This also allows getting more detailed information or following up certain topics when found relevant. In contrast to hyperlinks in Web pages, however, these links are not set at the discretion of a user but represent links derived from knowledge representation.

Since this requires a lot of domain-specific (but also general) knowledge it is a relatively labor-intensive task to create such ontologies. In future this task will be facilitated by automatic tools and processes (Faatz and Steinmetz, 2002). However, some manual work will still be required. This does not necessarily have to be performed by skilled personal only. Since an ontology is a 'living and learning' system it can be improved by its usage. Therefore it is necessary to encourage the use of it within the CMS. This also reflects the natural use and development of language. Therefore, in the long term ontologies might replace special cataloging rules and even thesauri.

Further research is required on how to effectively build ontologies but also how to translate the knowledge from existing databases and information systems into such knowledge networks.

10.4.2.3 Future Infrastructures

CMS will always be large systems that comprise a substantial number of complex modules. Different devices, services, and third-party components are becoming part of such a system. Further, it can already be seen that content-rich organizations such as broadcasters might not have one single large CMS but that there are going to be a number of interacting autonomous systems. The interaction between different CMS can even go beyond organizational boundaries. Thus, the structure and architecture of a CMS will have to develop further. The two important aspects here are how to compose a CMS (i.e. how to organize the interface to the different modules), and the interaction between different largely independent systems. Ideally all of this is transparent to the user, i.e. the user does not necessarily have to be aware where content is located or how it is accessed.

Regarding the organization of the different CMS modules the currently very detailed architecture consisting of the core elements Essence Manager, Device Manager, and Data Manager and a number of services may develop into an even more generic architecture that distinguishes only between devices (including SAN as storage devices) and components that host (or provide) metadata.

The structure of the Device Server (introduced in Section 6.4.2) has all the features that are required for such a generic component. It considers a device to be a unit that stores essence and certain metadata related to an essence object, and that may have storage management capabilities, may provide file level access and file transfer functionalities, and may also have interactive features such as stream playout, recording capabilities etc. Such a device can be regarded as a kind of essence management, specialized to record, store, and provide a certain kind of essence. Within this broad feature description, a 'device' can be almost anything, e.g. a video server, an archive system, a simple file system, an FTP server, a disk recorder, a videotape recorder, a library management system that holds numerous videotapes and hosts multiple videotape recorders, and so on. Applying this concept as a design principle will lead to a distributed essence management, allowing integration of any kind of system that holds essence into the overall CMS concept, sharing metadata, but distributing the essence. The key to this approach is to ensure unique identification of essence objects on each device via a unique material identifier such as SMPTE UMID, provided that the same identifier is applied to all instances and formats of a given essence. In such a case, distributed essence and related metadata are linked together via this identifier.

This is a more abstract view that allows easy integration of all kinds of system devices relevant in the CMS context. However, even if the concept of essence management and device management is taken to a next, higher abstraction level, the basic concepts and design principles presented in this book (such as the Service Group and Broker Manager concept) remain of prime importance. These principles ensure scalability and integration of federated databases and information systems.

It is expected that more inter-system communication will take place than so far, i.e. different CMS will be linked and content can be exchanged between them. This also includes CMS becoming more distributed with each distribution component holding a certain subset of the entire content of a content-rich organization. In order to do this CMS can be organized as autonomous collaborating systems. Each system controls the storage and administration of essence and metadata independently. However, in order to increase content availability it may be replicated at other systems. This can be done proactively or by employing caching-like algorithms. It is important that the collaborating CMS share a set of common interfaces and also a core set of metadata. Related technologies and concepts that should be looked at in this context are for instance developed in the IETF CDN/CDI initiative. Also peer-to-peer principles provide interesting ideas that could be exploited. Further, the GRID initiative (Foster *et al.*, 2002) has also been developing concepts that may be relevant for content management and distribution.

In general the organization of large-scale systems might in future be done using the concept of autonomous systems coordinated and controlled by rules rather than building on strictly hierarchical client/server topologies only. The storage infrastructures may be, for instance, organized as autonomous storage farms based on peer-to-peer principles that provide a service to other components. These might still perceive the storage units as servers; however, they are internally not hierarchically organized or controlled. The idea is to provide even better fault tolerance and resilience employing these concepts.

It is still a research issue to what extent such autonomous structures can be deployed efficiently and what their potentials are. Using these concepts could also erode the boundaries between the production and distribution system, which might be based on data communication technology. However, this will not happen if IPR-related issues have not been resolved.

10.4.2.4 IPR Management and Protection

CMS so far have only provided little (if any) support for IPR management and have made only rudimentary provisions for the protection of IPR (for instance in the form of a watermarking service). The former has very often been prohibited by internal rules within content-rich organizations that prevent disclosure of rights-related content on a large scale to a wider public. The latter was not required since the content was mainly handled internally and transmitted via traditional channels (i.e. terrestrial, cable or satellite) where the problem of rights protection is not as severe as in the new digital media space.

However, it is expected that with the move away from traditional broadcaster, production studio, and publisher operations towards an organization where multimedia content production, publishing, and distribution over various channels is more integrated this will become a more relevant and urgent problem area. Rights management can be relatively easily encompassed by existing CMS since the provisions for a tighter integration are already there.

More problematic is the rights protection issue. The technology developed in this context is by no means mature enough and has not reached a stage where it could be deployed in a wider context. It is currently not even clear what kind of approach is of real technical and practical relevance. Watermarking has so far been the most favored technology, though no scheme has been developed that could sustain an attack in the long-term. Further, the effort to calculate and bring watermarks into an image cannot (up to now) be done in real-time. Other technologies such as fingerprinting might be more appropriate in this context. Since the new distribution channels have not been designed and built, and the imminent change in the IPR rules and regulations has not taken place, it will probably also be very difficult to develop the right technology. There is still scope for basic research in this area and it might still take some time until the technology but also the legislation is mature enough to be deployed in a system that exploits the full potential of digital multimedia content distribution.

References

AAF Association (2000) Advanced Authoring Format (AAF) Specification, version 1.0, Developer Release 4, AAF, http://www.aafassociation.org/.

K. Ahmed, D. Ayers, M. Birbeck, J. Cousins, D. Dodds, J. Lubell, M. Nic, D. Rivers-Moore, A. Watt, R. Worden, A. Wrightson (2001) *Professional XML Meta Data*. Wrox Press, Birmingham, AL.

Aldus Corporation (1992) 'TIFF–Tagged Image File Format, Revision 6.0, Final', Aldus Corporation.

L. Amini, S. Thomas, O. Spatscheck (2002) Distribution Requirements for Content Internetworking, Internet Draft (work in progress), <draft-ietf-cdi-distribution-reqs-00.txt>, February 2002.

BBC (2002) The Standard Media Exchange Framework (SMEF™), http://www.bbc.co.uk/guidelines/smef/, August 2002.

BBC Technology Ltd (2002) SMEF Data Model v1.7: Introduction, Diagrams and Definitions, London.

K. Bergner, A. Rausch, M. Sihling (1998) Componentware–the big picture, 1998 International Workshop on Component-Based Software Engineering (CBSE), Kyoto, Japan.

P. Biron, A. Malhotra (eds) (2001) XML Schema Part 2: Datatypes, W3C, http://www.w3.org/TR72001/REC-xmlschema-0-20010502/datatypes, May 2001.

T. Bray, J. Paoli, C. Sperberg-McQueen, E. Maler (2000) Extensible Markup Language (XML) 1.0 (Second Edition), W3C Recommendation, http://www.w3.org/TR/REC-xml, October 2000.

D. Brickley, R. Guha (eds) (2003) RDF Vocabulary Description Language 1.0: RDF Schema, W3C Working Draft, http://www.w3.org/TR/rdf-schema/, January 2003.

D. Chappell (1996) *Understanding ActiveX and OLE*. Microsoft Press.

J. Clark (ed.) (1999) XML Path Language, Version 1.0, W3C, http://www.w3.org/TR/xpath, November 1999.

J. Cowan (2002) Extensible Markup Language (XML) 1.1, W3C Candidate Recommendation, http://www.w3.org/TR/xml11/, October 2002.

G. Davis, S. Chawla (1999) Wavelet-based Image coding: an overview. In *Applied and Computational Control, Signals, and Circuits* (B. N. Datta, ed.). Kluwer Academic, Boston, MA.

M. Day, B. Cain, G. Tomlinson, P. Rzewski (2002) A Model for Content Internetworking (CDI), Internet Draft (work in progress), <draft-ietf-cdi-model-01.txt>, February 2002.

DCMI (1999) Dublin Core Metadata Initiative (DCMI), Dublin Core Metadata Element Set, Version 1.1: Reference Description, http://purl.org/dc/documents/rec-dces-19990702.htm, (see also Dublin Core Metadata for Resource Discovery, Internet RFC 2413, http://www.ietf.org/rfc/rfc2413.txt).

DCMI (2000) Dublin Core Metadata Initiative (DCMI): Dublin Core Qualifiers, http://dublincore.org/documents/2000/07/11/dcmes-qualifiers/, July 2002.

DCMI (2003) Dublin Core Metadata Initiative (DCMI): About the Dublin Core Metadata Initiative, http://dublincore.org/, February 2003.

B. Devlin (ed.) (2002) Media eXchange Format (MXF) Operational Pattern 1a (Single Item, Single Package), version 10b, SMPTE and Pro-MPEG, Standard Proposal.

EBU Projects Group (2001) Future Television Archives: Report of the EBU Project Group P/FTA, Draft Version 1.00.

W. Effelsberg, R. Steinmetz (1998) *Video Compression Techniques*. dpunkt Verlag, Heidelberg.

European Broadcasting Union (1997a) Specification of the Broadcast Wave Format, EBU Technical Document 3285.

European Broadcasting Union (1997b) Specification of the Broadcast Wave Format, A format audio data files in broadcasting, Supplement 1: MPEG Audio, EBU Technical Document 3285–Supplement 1.

European Broadcasting Union (2001) PMC Project P/META (Metadata exchange standards), http://www.ebu.ch/pmc_meta.html, Geneva, Switzerland, January 2001.

A. Faatz, R. Steinmetz (2002) Ontology enrichment with texts from the WWW, in Proceedings of ECML–Semantic Web Mining 2002.

D. Fallside (ed.) (2001) XML Schema Part 0: Primer, W3C, http://www.w3.org/TR72001/REC-xmlschema-0-20010502/primer/, May 2001.

I. Foster, C. Kessleman, J. Nick, S. Tuecke (2002) Grid services for distributed system integration, *IEEE Computer*, No. 36, June 2002.

D. Gilletti, R. Nair, J. Scharber, J. Guha (2002) Content Internetworking (CDI) Authentication, Authorization, and Accounting Requirements, Internet Draft (work in progress), <draft-ietf-cdi-aaa-reqs-00.txt>, February 2002.

M. Green, B. Cain, G. Tomlinson, S. Thomas, P. Rzewski (2002) Content Internetworking Architectural Overview, Internet Draft (work in progress), <draft-ietf-cdi-architecture-00.txt>, February 2002.

M. Gudgin, M. Hadley, N. Mendelsohn, J. -J. Moreau, H. Nielsen (eds) (2002) Soap Version 1.2, Part 1: Messaging, W3C, http://www.w3.org/TR/soap12-part1, December 2002.

D. Hillman (2001) Using Dublin Core, http://dublincore.org/documents/2001/04/12/usageguide.

A. Hung (1993) PVRG-MPEG CODEC 1.1, Documentation of the PVRG-MPEG-1 Software Codec, Stanford University, California. ftp: havefun.Stanford.edcu:pub/mpeg/MPEGv1.1.tar.Z.

Institute for Telecommunication Sciences (1996) Telecommunications: Glossary of Telecommunications Terms, National Telecommunications and Information Administration (NTIA), Institute for Telecommunication Sciences (ITS), http://glossary.its.bldrdoc.gov/fs-1037/dir-036/_5265.htm.

International Electrotechnical Commission (2001) Helical-scan digital video cassette recording system using 6.35 mm magnetic tape for consumer use (525-60, 625-50, 1125-60 and 1250-50 systems), IEC 61834.

International Electrotechnical Commission (2002) Consumer audio/video equipment–Digital interface, IEC 61883.

International Organization for Standardization (1986) Information Processing–Text and Office Systems–Standard Generalized Markup Language, ISO 8879: 1986.

International Telecommunication Union (1995) Recommendation ITU-R BT 601-5: Studio Encoding Parameters of Digital Television for Standard 4:3 and wide-screen 16:9 aspect ratios, ITU.

ISO/IEC (2000) ISO/IEC JTC1/SC29/WG11, MPEG-7 Principle Concept List (0.9), MPEG2000, N3413, Geneva, Switzerland.

JPEG (1993) ISO/IEC JTC1/SC2/WG10 Joint Photographics Expert Group: Information Technology–Digital Compression and Coding of Continuous-tone Still Images, International Standard ISO/IEC IS 10918.

Microsoft Corporation (1996) DCOM Technical Overview, http://msdn.microsoft.com/library/.

Microsoft Corporation (2003) X/Open Distributed Transaction Processing Standard, http://msdn.microsoft.com/library/.

MOS Consortium (2001) Media Object Server (MOS™) Protocol version 2.6, Document Revision WD-2001-08-09, http://www.mosprotocol.com, August 2001.

B. Manjuhath, P. Salembier, T. Sikora (eds) (2002) *MPEG-7: Multimedia Content Description Interface*, John Wiley & Sons, Chichster.

A. Mauthe, W. Schulz, R. Steinmetz (1992) Inside the Heidelberg Multimedia Operating System Support: Real-Time Processing of Continuous Media in OS/2, Technical Report no. 43.9214, IBM European Networking Center, Heidleberg, Germany.

J. Mitchell, W. Pennebaker, C. Fogg (1996) *MPEG Video Compression Standard*. Chapman & Hall, New-York.

MPEG (1996) ISO/IEC JTC1/SC29/WG11, N MPEG 96: Coding of moving pictures and associated audio for digital storage media at up to about 1,5 Mbit/s. http://mpeg.telecomitalialab.com/standards/mpeg-1/mpeg-1.htm, June 1996.

MPEG (2000) ISO/IEC JTC1/SC29/WG11, N MPEG 00: MPEG-2: Generic coding of moving pictures and associated audio information. http://mpeg.telecomitalialab.com/standards/mpeg-2/mpeg-2.htm, October 2000.

MPEG (2001) ISO/IEC JTC1/SC29/WG11 Coding of Moving Pictures and Audio, TR 21000-1: 2001: MPEG-21 Vision Technologies and Strategies, Part.

MPEG (2002a) ISO/IEC JTC1/SC29/WG11, Coding of Moving Pictures and Audio, N4668: Overview of the MPEG-4 Standard, http://mpeg.telecomitalialab.com/standards/mpeg-4/mpeg-4.htm, March 2002.

MPEG (2002b) ISO/IEC JTC1/SC29/WG11, MPEG-7 Overview, Version 8, http://mpeg.telecomitalialab.com/standards/mpeg-7/mpeg-7.htm, Klagenfurt, Austria, July 2002.

MPEG (2002c) ISO/IEC JTC1/SC29/WG11 Coding of Moving Pictures and Audio, N4801: MPEG-21 Overview version 4, http://mpeg.telecomitalialab.com/standards/mpeg-21/mpeg-21.htm, Fairfax, USA May 2002.

SMPTE/Pro-MEPG Forum (2002) Media eXchange Format (MXF) 7 Parts: File Formats, OP1, DV, GC, Formate, GC D10, Mapping, GC D11 Mapping, GC SDTI-CP Mapping, Version 10, http://www.g-fors.com/mxf.htm, July 2002.

F. Nack, L. Hardman (2002) Towards a syntax for multimedia semantics, CWI Technical Report INS-R0204.

Object Management Group (2002) Common Object Request Broker Architecture (CORBA/IIOP), version 3.0.2, http://www.omg.org/technology/documents/formal/corba_iiop.htm.

R. Orfali, D. Harkey (1997) *Client/Server Programming with JAVA and CORBA*. John Wiley & Sons.

F. Pereira, T. Ebrahimi (eds) (2002) *The MPEG-4 Book*. IMSC Press Multimedia Series, Prentice Hall.

J. Postel, J. Reynolds (1985) File Transfer Protocol (FTP), Internet Engineering Task Force (IETF), Network Working Group, Request for Comments 959.

M. Sant'Anna, J. Sampaio do Prado Leite, A. do Prado (1998) A generative approach to componentware, 1998 International Workshop on Component-Based Software Engineering (CBSE), Kyoto, Japan.

SMPTE/EBU (1998) Task Force for Harmonized Standards for the Exchange of Program Material as Bitstreams–Final Report: Analyses and Results.

Society of Motion Picture and Television Engineers (1997a) Universal Labels for Unique Identification of Digital Data, Proposed SMPTE Standard for Television, SMPTE 298M, White Plains, NY.

Society of Motion Picture and Television Engineers (1997b) 10-Bit 4:2:2 Component and 4fsc Composite Digital Signals–Serial Digital Interface, SMPTE 259M-1997 Television, White Plains, NY.

Society of Motion Picture and Television Engineers (1998a) 6.35 mm type D-7 component format–video compression at 25 Mb/s–525&60 and 625/59, SMPTE Standard for Television Digital Recording, SMPTE 306M.

Society of Motion Picture and Television Engineers (1998b) 6.35-mm Type D-7 Component Format–Tape Cassette, SMPTE Standard for Television Digital Recording, SMPTE 307M.

Society of Motion Picture and Television Engineers (1999) Data structure for DV-based audio and compressed video 25 and 50 Mb/s, SMPTE Standard for Television, SMPTE 314M.

Society of Motion Picture and Televisions Engineers (2000) Serial Data Transport Interface (SDTI), SMPTE 305.2M-2000 Television, White Plains, NY.

Society of Motion Picture and Television Engineers (2001a) SMPTE Metadata Dictionary, RP210.2 (including RP210.1) Merged Version, post trail publication of RP210.2, http://www.smpte-ra.org/mdd/RP210v2-1merged-020507b.xls, White Plains, NY, December 2001.

Society of Motion Picture and Television Engineers (2001b) Data Encoding Protocol using Key-Length-Value, Proposed SMPTE Standard for Television, SMPTE 336M, White Plains, NY.

R. Steinmetz (2000) *Multimedia Technologie: Grundlagen, Komponenten und Systeme*, 3rd edn. Springer Verlag, Heidelberg.

R. Steinmetz, K. Nahrstedt (1995) *Multimedia: Computing, Communications and Applications*. Prentice Hall.

W. Stevens (1994) *TCP/IP Illustrated*, vol. 1: *The Protocols*, Addison-Wesley.

M. Stonebraker (1996) *Object-Relational DBMS–The Next Great Wave*. Morgan Kaufmann.

Sun Microsystems (1997) *Java Beans 1.01*. Sun Microsystems.

H. Thompson, D. Beech, M. Maloney, N. Mendelsohn (eds) (2001) XML Schema Part 1: Structures, W3C, http://www.w3.org/TR72001/REC-xmlschema-0-20010502/structures, May 2001.

D. Tidwell, J. Snell, P. Kulchenko (2001) *Programming Web Services with SOAP*, O'Reilly.

J. Ullman (1988) *Database and Knowledge–Base Systems*, vol. 1. Computer Science Press.

G. Wallace (1991) The JPEG still picture compression standard. *Communications of the ACM* **34**(4).

WAVE (1991) Multimedia Programming Interface and Data Specifications 1.0, Resource Interchange File Format, Waveform Audio File Format (WAVE), IBM Corporation and Microsoft Corporation.

W3C (2003) Architecture Domain: eXtensible Markup Language (XML), http://www.w3.org/XML/, February 2003.

Acronyms

Access Control List (ACL)
Access Units (AU)
Advanced Authoring Format (AAF)
Analog-to-Digital converter (ADC)
Application Programming Interface (API)
Asset Management Systems (AMS)
Audiovisual objects (AVO)
Authentication, Authorisation and Accounting (AAA)
Auxiliary video data (VAUX)
Binary Format (BiM)
Binary Formats for Scenes (BIFS)
Broadcast Wave Format (BWF)
Common Internet File System (CIFS)
Compact Disc Digital Audio (CD-DA)
Component Object Model (COM)
Content Distribution Internetworking (CDI)
Content Distribution Networking (CDN)
Content Management System (CMS)
Database Management System (DBMS)
Description Definition Language (DDL)
Description Scheme (DS)
Digital Asset Management Systems (DAM-S)
Digital Audio Tape (DAT)
Digital Rights Management (DRM)
Digital Video (DV)
Digital-to-Analog (DAC)

Professional Content Management Systems: Handling Digital Media Assets   A. Mauthe, P. Thomas
© 2004 John Wiley & Sons, Ltd   ISBN: 0-470-85542-8

Discrete Cosine Transformation (DCT)
Distributed Component Object Model (DCOM)
Distributed Transaction Processing (DTP)
Document Type Definitions (DTD)
Domain Name Service (DNS)
Document Style Semantics and Specification Language (DSSSL)
Document Type Definition (DTD)
Dublin Core (DC)
Dublin Core Metadata Element Set (DCMES)
Dublin Core Metadata Initiative (DCMI)
Edit Decision List (EDL)
Electronic News Gathering (ENG)
Enterprise Resource Planning (ERP)
European Broadcasting Union (EBU)
eXperimentation Model (XM)
eXtensible Markup Language (XML)
Fast Fourier Transformation (FFT)
File Transfer Protocol (FTP)
Forward Discrete Cosine Transformation (FDCT)
Functional Design Specification (FDS)
Future Television Archives (P/FTA)
Graphic Interchange Format (GIF)
Group of pictures (GOP)
Hertz (Hz)
Hypertext Markup Language (HTML)
Initial Track Information (ITI)
Intellectual Property Management and Protection (IPMP)
Intellectual property rights (IPR)
Interface Design Specification (IDS)
Joint Photographics Experts Group (JPEG)
Key-Length-Value (KLV)
Basic encoding rules (BER)
Kilohertz (kHz)
Library management systems (LMS)
Lightweight Directory Access Protocol (LDAP)
Logical Unit Numbers (LUN)
Low-level descriptors (LLD)
Mass Data Storage Systems (MDS)
Media Asset Management Systems (MAM-S)
Media eXchange Format (MXF)
Media management system (MMS)
Media Object Server Protocol (MOS)
Memory in cassette (MIC)
Moving Picture Expert Group (MPEG)
Multimedia Description Schemes (MDS)
Network Attached Storage (NAS)

Network Disk Control Protocol (NDCP)
Network File System (NFS)
Newsroom Control System (NCS)
Nonlinear editing systems (NLE)
Optical Character Recognition (OCR)
Object descriptors (OD)
Program as Broadcasted (PasB)
Primary Domain Controller (PDC)
Pulse Code Modulation (PCM)
Redundant array of independent disks (RAID)
Relational database management system (RDBMS)
Remote Procedure Call (RPC)
Resource Description Framework (RDF)
Resource Interchange File Format (RIFF)
Storage Area Network (SAN)
Serial Data Transport Interface (SDTI)
Serial Digital Interface (SDI)
Server Attached Storage (SAS)
Standard Generalised Markup Language (SGML)
Signal-to-noise ratio scalability (SNR)
Simple Object Access Protocol (SOAP)
Skims
Society of Motion Pictures and Television Engineers (SMPTE)
Software development kit (SDK)
Standard Interchange Formats (SIF)
Standard Media Exchange Framework (SMEF)
Standard Query Language (SQL)
Storage Area Network (SAN)
Tagged Image File Format (TIFF)
Unique Material Identifier (UMID)
Unique Programme Identifier (UPID)
Video Disk Control Protocol (VDCP)
Visual Texture Coding (VTC)
WAN (Wide Area Network)
Waveform Format (WAVE)
Waveform Audio File Format (WAVE)
World Wide Port Number (WWPN)
World Wide Web (WWW)
XML Linking Language (XLink)
XML Schema Definition (XSD)
XML Path Language (XPath)
XML Query Language (XQuery)

Index

AAF Association, 122, 127, 128, 130
AAF Class Model, 120, 130
AAF File Structure, 128
AC coefficients, 48
Access Control List (ACL), 176, 248
Access rights, 43, 170, 175, 176, 232, 248, 265, 268, 275
Access Units (AU), 99
Accounting & Licensing Service, 176
ActiveX, 171, 228, 229, 243, 249
Adaptive intraframe spatial compression, 59
Administration Client, 271
Advanced Authoring Format (AAF), 123, 127–130, 133, 156, 162, 199
Agency feeds, 7, 19, 242
Airplay, 11
Airtime sales, 225
AMIA, 279
Analog-to-Digital converter (ADC), 63
Analysis Service, 162, 164, 202, 259
Application design principle, 246
Application domains, 6, 10, 98, 246
Application modules, 17, 34, 171, 228, 245, 246, 250, 251, 275

Application Programming Interface (API), 137, 148, 150, 151, 153, 154, 162, 228, 230, 237, 239–243
Application Server, 182, 187, 188, 197, 202, 205, 207, 209
Application support, 182, 247
Archive, 1, 2, 5, 7, 8, 10–12, 15, 17–20, 22, 23, 27–29, 33, 34, 62, 74, 81, 92, 122, 142, 144, 145, 149, 150, 155, 159, 165–169, 177, 183, 184, 192–194, 196, 203, 205, 207, 209, 210, 215, 218, 226, 231, 232, 239, 240, 264, 265, 268, 275, 279–283, 285, 287, 290
Archive Management Server, 142–145, 205
Archive Manager, 142, 179
Archive Transfer Server, 142, 145, 147, 179, 188, 203, 205
Archiving format, 42, 199, 217, 232
Arithmetic coding, 45, 65
ASCII, 67, 70, 72, 107, 131
Aspect ratio, 46
As-run log, 153

Professional Content Management Systems: Handling Digital Media Assets   A. Mauthe, P. Thomas
© 2004 John Wiley & Sons, Ltd   ISBN: 0-470-85542-8